Intense Electron and Ion Beams

S.I. Molokovsky   A.D. Sushkov

Intense Electron and Ion Beams

With 140 Figures

 Springer

Professor Dr. Sc. Sergey Ivanovich Molokovsky
Professor Dr. Sc. Aleksandr Danilovich Sushkov
State Electrotechnical University, Electronics Department
Prof. Popov street 5, 197376 Saint Petersburg, Russia
SIMolokovsky@eltech.ru, ADSushkov@eltech.ru

Library of Congress Control Number: 2004117893

ISBN-10 3-540-24220-1 Springer Berlin Heidelberg NewYork
ISBN-13 978-3-540-24220-1 Springer Berlin Heidelberg NewYork

This work is subject to copyright. All rights are reserved, whether the whole or part of the material is concerned, specifically the rights of translation, reprinting, reuse of illustrations, recitation, broadcasting, reproduction on microfilm or in any other way, and storage in data banks. Duplication of this publication or parts thereof is permitted only under the provisions of the German Copyright Law of September 9, 1965, in its current version, and permission for use must always be obtained from Springer. Violations are liable to prosecution under the German Copyright Law.

Springer is a part of Springer Science+Business Media.

springeronline.com

© Springer-Verlag Berlin Heidelberg 2005
Printed in Germany

The use of general descriptive names, registered names, trademarks, etc. in this publication does not imply, even in the absence of a specific statement, that such names are exempt from the relevant protective laws and regulations and therefore free for general use.

Typesetting and prodcution: PTP-Berlin, Protago-T$_{E}$X-Production GmbH, Berlin
Cover design: *design & production* GmbH, Heidelberg

Printed on acid-free paper   SPIN 10967867   57/3141/YU   5 4 3 2 1 0

Preface

Intense charged-particle beams currently have many different applications. High-current electron beams are used in various microwave tubes: O-type tubes (klystrons, travelling wave tubes, etc.), M-type tubes, gyrotrons, and free-electron lasers. High-current electron and ion beams are applied as tools in the installations of charged-particle beam technology. High-current electron and ion accelerators have been developed for industrial applications and physical experiments.

This book is devoted to intense charged-particle beams used in vacuum tubes, particle beam technology, and installations for physical experiments (free-electron lasers, accelerators, etc.). The following scope of topics is considered in this book:

- Physics and basic theory of intense charged-particle beams
- Methods of electric and magnetic field computation as applied to the development of the systems used for the charged-particle beam formation and focusing
- Methods of analytic and numerical analysis of charged-particle beam motion in self-consistent fields
- Computation and design of the charged-particle guns for initial formation of pencil, hollow and strip beams Computation and design of the systems for focusing (transport) beams by different types of electric and magnetic fields
- Computation and design of the systems for formation of charged-particle probes used in charged-particle beam technology
- Multiple beam charged-particle systems, peculiarities of their design, and computation
- Peculiarities of relativistic beam motion and focusing
- Methods of experimental investigation of intense particle beams

Therefore, this book contains both the physics and theory of intense charged-particle beams and the design of their optical systems for beam formation and focusing. The methods of successive approximation, as well as the method of synthesis, are considered. The corresponding algorithms, formulas, and graphics are presented. The book reflects the authors' experience in the development and experimental investigation of charged-particle optical systems with an emphasis made on the systems for microwave tubes and particle-beam technology.

The book is recommended to people who are studying or work with vacuum electronics and charged-particle beam technology: students, postgraduate students, engineers, research workers. The material presented in the book grew out of lectures given by one of the authors over many years to students and postgraduate students of the electronics faculty of St. Petersburg State Electrotechnical University. Two editions of this book have been published in Russian.

At preparation of the English version of this book a chapter devoted to multiple-beam systems was added and supplementary references were included. Chapters 2–5, 7–11 and the Appendix were written by Professor S.I. Molokovsky, Chap. 12 by Professor A.D. Sushkov, and Chaps. 1 and 6 were prepared in collaboration.

The authors express great appreciation to the head of the Radioelectronic Department Dr. V.B. Yankevich and our colleague Dr. V.A. Ivanov for their help preparing the English version of this book and for their continued interest in it. The authors would also like to express deep gratitude to the staff and managers of Ingredient Ltd. (St. Petersburg) for their support of this work.

S.I. Molokovsky, A.D. Sushkov　　　　　　　　　　St. Petersburg, May 2005

Contents

1 Introduction to Particle-Beam Formation 1
   1.1 Application and Parameters
         of Intense Charged-Particle Beams 1
   1.2 Optical Systems of Electron Devices 4
         1.2.1   Electron Guns 4
         1.2.2   Electron Guns with Beam Current Control 6
   1.3 Focusing System of Electron Devices 9
         1.3.1   Magnetic Focusing with Uniform Field 9
         1.3.2   Magnetic Periodic Focusing 11
         1.3.3   Periodic Electrostatic Focusing 12
         1.3.4   Hollow-Beam Focusing 13
   1.4 Optical Systems of Particle-Beam Technology 14
         1.4.1   Installations of Electron Beam Welding
               and Melting 14
   1.5 Ion Optical Systems 16

2 Methods of Fields Calculation 19
   2.1 General Equations of Electrostatic Fields 19
   2.2 Electrostatic Field Calculation. Dirichlet Problem 20
         2.2.1   Method of the Separation of Variables 20
         2.2.2   Method of Green Function 23
         2.2.3   Method of the Integral Equations 26
         2.2.4   Method of Finite Difference 27
   2.3 Calculation of Electrostatic Field: Cauchy Problem 30
         2.3.1   Method of Analytical Continuation 31
   2.4 Computer Calculation of Electrostatic Fields 36
         2.4.1   Method of Finite Difference 36
         2.4.2   Method of Integral Equations 38
         2.4.3   Method of Green Function 41
   2.5 General Equations of Magnetic Field 43
         2.5.1   Magnetic Vector Potential 44
         2.5.2   Scalar Magnetic Potential 46
   2.6 Numerical Calculation of Magnetic Field 47
         2.6.1   Field of a Solenoid 47
         2.6.2   Field of a Ring-Shaped Magnet 48

VIII   Contents

        2.6.3   Technique of Solution
                of Nonlinear Magnetic Problems 50

3 Fundamentals of Charged-Particle Motion 55
   3.1   Equations of Motion in Newton Form 55
   3.2   Law of Energy Conservation 56
   3.3   Lagrange Equations of Motion 58
   3.4   Hamilton Equations 59
   3.5   Hamilton–Jacobi Equation 60
   3.6   Equations of Motion in Axially Symmetric Fields 62
        3.6.1   Equations of Motion in Newton Form 62
        3.6.2   Lagrange and Hamilton Equations of Motion 66
        3.6.3   Hamilton–Jacobi Equation of Motion 67
        3.6.4   Motion of Electrons in Cylindrical Magnetron 67
   3.7   Motion in Planar Two-Dimensional Fields 73
        3.7.1   Equations of Motion in Newton Form 73
        3.7.2   Lagrange and Hamilton Equations 73
   3.8   Numerical Calculation of Charged-Particle Trajectories 74
        3.8.1   Method of Taylor Series 74
        3.8.2   Runge–Kutta Method of Fourth Order of Accuracy . 75
        3.8.3   Spatial Method of Trajectory Tracing 76
   3.9   Electrostatic Charged-Particle Lenses 77
        3.9.1   Types of Axially Symmetric Electrostatic Lenses ... 77
        3.9.2   Focusing Properties of Electrostatic Lenses 77
   3.10 Magnetic Lenses 86
        3.10.1 Thin Magnetic Lens Approximation 86
        3.10.2 Thick Magnetic Lens 87
        3.10.3 Parameters of Some Magnetic Lenses 88
        3.10.4 Magnetic Lenses with Permanent Magnets 91

4 Motion of Intense Charged-Particle Beams 95
   4.1   Peculiarities of Intense Particle-Beam Motion 95
        4.1.1   Effects of Self-Fields and Velocity Spreading 95
        4.1.2   Mathematical Description
                of Multiple-Velocity Beams 96
   4.2   Simplified Models of Charged-Particle Beams 98
        4.2.1   Monovelocity Models 98
        4.2.2   Hydrodynamic (Laminar) Model of Flow 100
        4.2.3   Quasihydrodynamic Flow Model 102
   4.3   Methods of Solution of Motion Equations 102
        4.3.1   Method of Curvilinear Coordinates 102
        4.3.2   Method of Expansion in Series.
                Paraxial Trajectory Equation 104
   4.4   Approximate Method of Space-Charge Account 105

		4.5	Motion of Charge-Beams in Channels Free from External Fields 107

- 4.5 Motion of Charge-Beams in Channels Free from External Fields 107
- 4.6 Influence of Residual Gases on Electron-Beam Motion 110
  - 4.6.1 Effect of Space-Charge Neutralization by Positive Ions 110
  - 4.6.2 Process of Neutralization of the Electron Beam in a Long Channel 110
  - 4.6.3 Effect of Ion Background on Electron-Beam Motion 112
  - 4.6.4 Ion Focusing 112
  - 4.6.5 Scattering of Electron Beam by Molecules of Residual Gas 114
- 4.7 Estimation of Effect of Thermal Electron Velocities 117
  - 4.7.1 Calculation of Transverse Particle Displacement 117
  - 4.7.2 Calculation of Current-Density Redistribution 119
- 4.8 Methods of Solution of Self-Consistent Problems 121
  - 4.8.1 Method of Successive Approximation 121
  - 4.8.2 Method of "Step by Step" (or Algorithm Modeling of Transient Process) 123

5 Electron Guns 125
- 5.1 The Problem of Electron-Beam Formation 125
- 5.2 Guns for Formation of Strip Beams 126
  - 5.2.1 Parallel Strip Beam 126
  - 5.2.2 Wedge-Shaped Beam 128
  - 5.2.3 Strip-Beam Forming in Crossed Fields 129
- 5.3 Guns for Solid Axially Symmetric Beams 133
  - 5.3.1 Parallel Cylindrical Beam 133
  - 5.3.2 Pierce Gun for Convergent Beam 134
  - 5.3.3 Synthesis of Guns by Ovcharov's Method 138
- 5.4 Guns for Formation of Hollow Axisymmetric Beams 141
  - 5.4.1 Parallel Tubular Beam 141
  - 5.4.2 Convergent Hollow Beam 141
  - 5.4.3 Magnetron Guns for Formation of Hollow Beams ... 142
- 5.5 Electron Guns with Beam-Current Control 145
  - 5.5.1 Guns with Grid Control Electrodes 146
  - 5.5.2 Electron Gun with Diaphragm Control Electrode ... 151
- 5.6 Principle of Computer-Aided Design of Electron Guns 151
  - 5.6.1 Synthesis of Electron Gun 152
  - 5.6.2 Electron-Gun Analysis 152

6 Electron and Ion Sources with Field and Plasma Emitters 157
- 6.1 Some Peculiarities of Field and Plasma Emitters 157
- 6.2 Low-Current Field-Emission Guns 157

	6.3	Multiple-Tip Field-Emission Cathodes 159
	6.4	High-Current Electron Sources with Explosive Emission 160
	6.5	Liquid-Metal Emitters 160
	6.6	Plasma Sources of Charged Particles 161
		6.6.1 Duoplasmatrons 161
		6.6.2 Cold-Cathode Sources 162
		6.6.3 High-Frequency Ion Sources 163
		6.6.4 Microwave Ion Sources 163
	6.7	Extracting of Charge Particles and Beam Forming 165
		6.7.1 Extraction of Ions from Planar Plasma Boundary .. 165
		6.7.2 System of Extraction with Expansion Cup 166
		6.7.3 Probe System of Extraction 167

7 Magnetic Focusing Systems 169

- 7.1 Focusing Systems with Uniform Field 169
  - 7.1.1 Principle of Focusing 169
  - 7.1.2 Balanced (Brillouin) Axially Symmetric Beam 170
  - 7.1.3 Unbalanced Beam in Uniform Field 172
  - 7.1.4 Beam Formed with Partially Shielded Gun 174
  - 7.1.5 Balanced Nonlaminar Beam 176
  - 7.1.6 Magnetic Systems for Uniform Field 180
- 7.2 Systems of Reversed- and Periodic-Field Focusing 181
  - 7.2.1 Principle of Reversed-Field Focusing 181
  - 7.2.2 Estimation of Beam Disturbances 183
  - 7.2.3 Principle of Periodic-Field Focusing 184
  - 7.2.4 Analysis of Periodic Focusing 185
  - 7.2.5 Stability of Periodic Focusing 187
- 7.3 Focusing of Hollow Beams in Uniform Field 188
  - 7.3.1 Equilibrium Hollow Beam 189
  - 7.3.2 Injection of Hollow Beams in Uniform Field 191
  - 7.3.3 Boundary Trajectory Pulsation 192
- 7.4 Focusing of Strip Beams 193
  - 7.4.1 Uniform-Field Focusing 193
  - 7.4.2 Periodic Magnetic Focusing 195
- 7.5 Computer-Aided Design of Magnetic Systems 196

8 Electrostatic Focusing Systems 199

- 8.1 Focusing System for Solid Axially Symmetrical Beams 199
  - 8.1.1 Equilibrium Cylindrical Beam 199
  - 8.1.2 Focusing System with Periodic Potential Distribution 201
  - 8.1.3 Periodic System of Apertured Disks 203
  - 8.1.4 System of Unipotential Electrostatic Lenses 203
- 8.2 Systems for Focusing of Strip Beams 210
  - 8.2.1 Equilibrium Strip Beam 210

	8.2.2 Periodic Electrostatic System 211
8.3	Focusing Systems for Hollow Beams 211
	8.3.1 Periodic Electrostatic Focusing 211
	8.3.2 Centrifugal Electrostatic Focusing 213

9 Optical Systems of Technological Installations 215
- 9.1 Electron Probe of Welding Installation 215
  - 9.1.1 Effect of Thermal Velocities 216
  - 9.1.2 Aberrations of Optical System 216
  - 9.1.3 Effects of Self-Fields and Scattering of Electrons ... 219
- 9.2 Optical System for Electron-Beam Lithography 219
  - 9.2.1 System with Thermal Cathode 219
  - 9.2.2 System with Cold Field Emitter 221
- 9.3 Ion Probes 221

10 Intense Relativistic Charged-Particle Beams 223
- 10.1 Relativistic Equations of Motion 223
  - 10.1.1 Law of Energy Conservation 224
  - 10.1.2 Equations of Motion in Axially Symmetry Fields ... 224
- 10.2 Intense Relativistic Beams in Vacuum Channels 225
  - 10.2.1 Electron Beam Spread under Self-Fields Action 225
  - 10.2.2 Beam Current Limitation 227
  - 10.2.3 Brillouin Relativistic Electron Beam 229
- 10.3 Neutralized Relativistic Beams 232
  - 10.3.1 Beams in Space Free from External Fields 232
  - 10.3.2 Neutralized Magnetically Confined Beams 232
- 10.4 Beam-Motion Computation 232
  - 10.4.1 Axially Symmetric Lens Systems 233
  - 10.4.2 Quadruple Lens Systems 234

11 Multiple-Beam Electron-Optical Systems 237
- 11.1 Peculiarities and Application of Multiple Beams 237
- 11.2 Multiple-Beam Guns and Magnetic Systems 239
  - 11.2.1 Electron Guns 239
  - 11.2.2 Magnetic Focusing Systems 241
- 11.3 Peculiarities of Multiple Systems 243
  - 11.3.1 Analysis of Stray-Fields Action 243
  - 11.3.2 Transverse Fields 244
- 11.4 Estimation of Effects of Transverse Fields 248
- 11.5 Analysis of Beam Interaction 250

12 Methods of Experimental Investigation of Beams 255
- 12.1 Classification of Methods 255
- 12.2 The Pin-Hole Chamber Method 258
- 12.3 Application of Modified Pin-Hole Chamber 261

Appendix 265

References 269

Index 279

1 Introduction to Particle-Beam Formation

1.1 Application and Parameters of Intense Charged-Particle Beams

In the period 1920–1940 charged-particle beams, mainly electron beams, were used in such devices as cathode ray tubes, X-ray tubes, electron microscopes, and charged-particle accelerators. Low-current beams with value of currents within the range microamperes–milliamperes are typical for these devices. The space-charge density in such beams is extremely small and its action on charged-particle motion is negligible. Charged-particle dynamics and optics of such beams are the subject of classical optics [1–6].

Later, beginning in the 1950s the charged-particle beams found many other applications. High-current electron beams are used in various microwave tubes: O-type tubes (klystrons, traveling-wave tubes, etc.), M-type tubes and in the installations of electron-beam technology for electron-beam welding, melting and refining of metals. High-current accelerators were developed for industrial applications.

Charged-particle beams with high values of currents of order of amperes and tens of amperes are used in these beam applications. The space-charge density of such beams is rather high and the motion of charge particles is considerably affected by the electric self-fields created by the space charge.

The necessity to take into account the self-beam fields led to the development of the new direction of charged-particle optics known as charged-particle optics of intense beams. The first monograph on the subject was written by Pierce in 1954 [7]. Since then, many monographs on intense charged-particle beams, devoted to their physics, theory, computation, simulation, design and application, have been published [8–20].

An important parameter of the beams, which characterized the beam intensity is beam perveance that is determined by the expression $P = I/U_a^{3/2}$, where I is the beam current expressed in amperes, U_a is accelerating voltage expressed in volts.

If the beam current is expressed in microamperes, the above parameter sometimes is called as microperveance.

It is known from experiments and computations, that in electron beams the space-charge effects become appreciable at perveance values $P \geq 0.1$ μA/V$^{3/2}$.

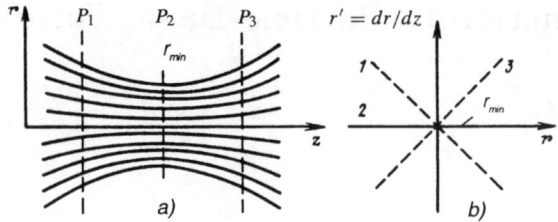

Fig. 1.1. Trajectories (**a**) and phase curves (**b**) of a laminar charged-particle beam

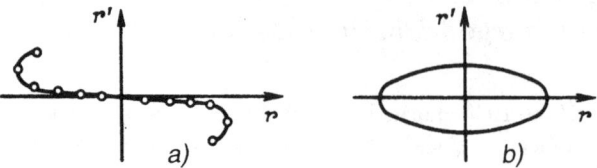

Fig. 1.2. Phase figures: **a** nonlinear phase curve, **b** phase ellipse

The electron beams used in microwave tubes have perveance values up to 20 µA/V$^{3/2}$, i.e., several orders more than values of beam perveance of cathode ray tubes.

Perveance values of the installations of charge particle beam technology lay in the range 0.01–20 µA/V$^{3/2}$.

Beside the perveance, the following set of parameters is used in the theory and the practice of intense charged-particle beams: beam-compression, phase curve, phase ellipse, emittance and brightness. They are applied for the description of the structure, degree of order and quality of charge particle beams.

The concept of the phase curve and its application for the beam-structure analysis is illustrated by drawings presented in Figs. 1.1 and 1.2. The axially symmetric beam shown in Fig. 1.1a is a laminar one whose trajectories are not crossed. Any point of a separate trajectory is characterized by the three parameters: coordinates r and z and the trajectory slope $r' = dr/dz$.

Considering a transverse plane $z = \text{const}$ (for instance, plane P$_1$ in Fig. 1.1) one can find a set of points, where beam trajectories cross the plane, and fix the radial coordinates r and trajectory slopes r' at these points. Then, plotting the values r and r' for this set of points at coordinate plane $r - r''$, known as transverse phase space or phase plane, one obtains the phase curve for the given transverse plane. The phase curves plotted for several transverse planes (for instance, P$_2$, P$_3$, etc.) create the phase-curve family or phase diagram, which is used for the description of the beam structure, the beam degree of order and beam behavior.

If the phase curves of a beam $r' = f(r)$ are single-valued functions of r, they describe a laminar beam, whose trajectories are not crossing.

1.1 Application and Parameters of Intense Charged-Particle Beams

For the beam shown in Fig 1.1a the phase curves are straight lines (Fig. 1.1b). They describe motion of the beam in the following way. The linearity of the phase curve at plane P_1 indicated that the beam is uniformly convergent in this plane. Furthermore, it is uniformly compressed up to the minimum cross section at plane P_2, where its initial convergence is completely balanced by the action of space-charge forces. The phase curve at this plane is the line located at r-axis of the phase plane r, r'. At plane P_3 it is line 3, showing a uniformly divergent beam.

Aberrations of charged-particle optical system, in particular, gun aberration, lead to crossing of beam trajectories and give nonlinear phase curves like that shown in Fig. 2a. This curve obtained for a transverse plane in the vicinity of minimum beam cross section indicates crossing of peripheral beam trajectories.

As the thermal velocity spread is taken into account particles with a range of transverse velocities exist at any point of beam cross section and instead of a phase curve a figure of finite area is formed. In many cases this figure can be approximated by an ellipse (Fig. 2.1b). There exist two parameters that are used to describe beam quality: emittance ε and brightness B.

The emittance of a beam is determined as phase ellipse area A divided by π:

$$\varepsilon = A/\pi \ .$$

The brightness of a beam is determined as

$$B = I/\pi r^2 \Omega \ ,$$

where I – beam current, πr^2 – beam cross section, Ω – solid angle, which determines beam divergence.

More details concerning the above-mentioned parameters and their application can be found in the Appendix.

In general, a charged-particle optical system includes two main parts. One of them provides initial formation of the charged-particle beam with setup parameters: beam current, perveance and shape of beam cross section. It is usually named as "charged-particle gun".

Another part is traditionally referred to as a "focusing system". But its real functions can be different depending on the optical system application. Focusing systems of most installations of particle-beam technology are really intended for beam focusing, that is, obtaining a very small beam cross section (point focusing). While focusing a system of linear beam microwave tubes (klystrons, traveling-wave tubes, etc.) should provide transport of electron beams through extended channels with minimum beam-current loss. In this case their name does not correspond to their function. This is why they also can be named as transport systems, as is commonly done for the accelerator technique.

1.2 Optical Systems of Electron Devices

1.2.1 Electron Guns

In electron devices intended for electromagnetic-wave generation electron beams are used as an "instrument" for transformation of electric energy of a power supply into energy of electromagnetic waves. In microwave tubes with linear beams (O-type tubes) the energy of a power supply initially is transformed into kinetic energy of an electron beam during beam acceleration in the gun region. A part of this energy is transformed into electromagnet field energy as a result of interactions of the electron beam with a resonator or slow wave system. Electron beams of different cross section are used in the tubes: solid and hollow axially symmetric beams, strip beams and beams with variable cross section (in quadruple focusing systems).

To obtain a high level of output power, electron beams with high current density are required. In many cases the value of beam current density considerably exceeds the permissible current density of modern cathodes. The system of electron-beam formation (electron gun) should provide beam compression, which is characterized by a coefficient of compression equal to the ratio of beam current density j to cathode current density j_k, $C_j = j/j_k$. It is attained in guns with convergent beams, which are usually developed using Pierce's principle (see Chap. 5).

At values of beam perveance $P \leq 0.5$ μA/V$^{3/2}$ beam compression can be of the order of 10^2 or greater, as in the gun developed in the State Electrotechnical University of St. Petersburg [8] shown in Fig. 1.3. Its design includes a concave spherical cathode (2) with heater (1), cathode focusing electrode (4) and anode electrode (3).

The Pierce principle allows easy design of electron guns with convergent beams with values of perveance up to 1 μA/V$^{3/2}$. Designing guns for higher

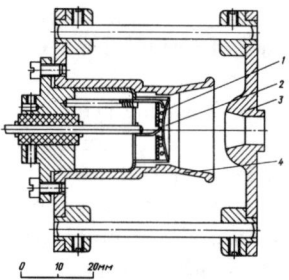
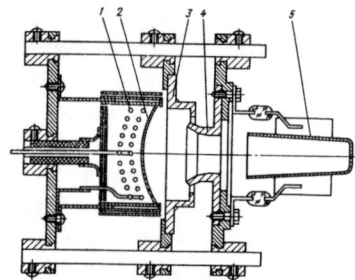

Fig. 1.3. Electron gun of Pierce type: $U_a = 50$ kV, $P = 0.5$ μAV$^{3/2}$, $C_j = 10^2$ 1 – heater, 2 – cathode, 3 – anode, 4 – cathode focusing electrode

Fig. 1.4. Modified gun of a high perveance: $U_a = 20$ kV, $P = 3$ μA/V$^{3/2}$, $C_j = 6$. 1 – heater, 2 – cathode, 3 – cathode focusing electrode, 4 – anode, 5 – collector

perveance of order of $(2-3)\mu A/V^{3/2}$ required considerable modification of the cathode focusing electrode geometry (Fig. 1.4).

Formation of strip beams with initial beam compression can be arranged with electrode configuration shown in the drawings of Figs. 1.3 and 1.4 if we consider them as drawings of cross sections of electrode systems with planar symmetry, which have some extent in the direction normal to the plane of the drawing.

Additionally, it is possible to imagine that these electrodes can be used for formation of hollow, initially compressed beams. For this it is necessary to rotate these figures around some assumed axis that is parallel to their axis of symmetry.

A modification of an electrostatic electron gun for formation of a hollow beam with a value of perveance equal to $15\mu A/V^{3/2}$ is presented in Fig. 1.5 [21].

A special class of electron guns intended for formation of hollow high-perveance beams is magnetron guns (Fig. 1.6) [22]. A hollow beam is formed in the region between two coaxial conical electrodes, the inner one being the cathode electrode, immersed in a magnetic field, the magnetic lines of which are approximately parallel to the surface of the cathode electrode. The motion of electrons in the vicinity of cathode surface is similar to that in a coaxial magnetron at operating conditions when the value of the magnetic induction exceeds the critical one, $B > B_{cr}$. However, inclination of electrode surfaces relative to the axis of symmetry results in electron-beam motion in the direction of this axis and formation of a hollow beam that is emitted from the interelectrode ring gap.

A variant of the magnetron gun is widely used in a new class of microwave tubes, known as gyrotrons (see Fig. 1.12).

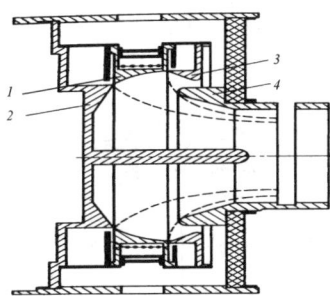
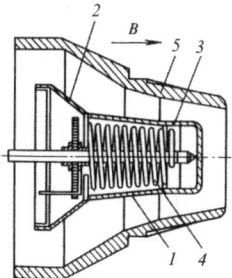

Fig. 1.5. Electron gun for hollow beam: $U_a = 100$ kV, $P = 15$ $\mu A/V^{3/2}$
1 – cathode, 2 – heater, 3 – cathode focusing electrodes, 4 – anode

Fig. 1.6. Magnetron electron gun: $U_a = 50$ kV, $P = 10$ $\mu A/V^{3/2}$
1 – cathode, 2,3 – cathode focusing electrodes, 4 – heater, 5 – anode

1.2.2 Electron Guns with Beam Current Control

Modulation of the power carried by electron beams is necessary for some electron-beam applications such as UHF tubes, electron-beam technology and so on. This can be done by different methods. The most obvious is variation of the anode voltage of a gun. It is known as the method of "anode" modulation. However, it requires a modulator unit designed for high values of voltage and power. The second disadvantage of the method is that the power modulation is accompanied by the variation of electron velocities. This is why the method of "grid" modulation is preferable in practice when the beam power modulation is achieved by the variation of the potential applied to an electrode other than the anode. This can be the cathode focusing electrode or an additional modulating electrode specially introduced in a gun design. In this case the beam-power modulation is produced by the variation of the beam current, the beam velocity being kept constant. Under certain conditions the voltage and the power required for modulation can be reduced to small portions of the corresponding beam parameters.

There are several types of controlled electron guns, which differ in the design of the control electrode (Fig. 1.7a–f) [23]. A focusing electrode insulated from the cathode (a), a focusing electrode together with a pin passing through the central hole in the cathode surface (b), a wire or laminar grid (c), a thick or thin diaphragm (d,e), and first anode in a magnetron type gun (f) can serve as the control electrode.

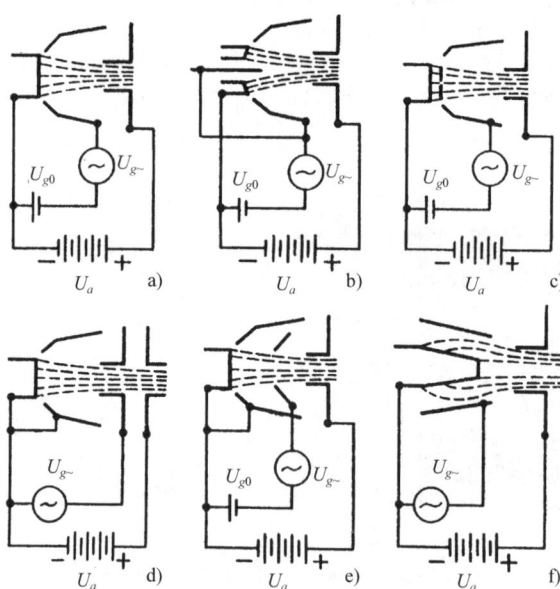

Fig. 1.7. Types of electron guns with beam current control: U_{c0} – bias voltage, $U_{g\approx}$ – control voltage, U_a – anode voltage

The effectiveness of beam-current control is characterized by the following parameters: relative value of cutoff voltage of the control (modulating) electrode, which provides the cutoff beam current, U_{gc}/U_a or triode amplification factor $\mu = U_a/|U_{gc}|$, control-electrode voltage U_{gm} at which the beam current reaches its rated value I_m; coefficient of current control $K = U_a/(U_{gm} + |U_{gc}|)$.

Comparative analysis shows the low effectiveness of beam-current control (K_c is not more then several units) of all systems with the exception of guns with a grid control electrode [23].

A drawback of grid-controlled guns operating with a positive value of U_{gm} ($U_{gm} > 0$) is beam-current interception by the grid. This can be eliminated in tetrode guns with a so-called shadow grid (2) (Fig. 1.8) [24–26]. This grid, having zero potential, screens the control grid (3) working with positive potential, due to this the control grid does not intercept the beam current. Detailed analysis of the operation of grid-controlled guns is given in Chap. 5.

It should be noted that triode guns allow forming electron beams with variable perveance. Let us take as an example an electron gun with a thick diaphragm as the control electrode. As the penetration factor of this electrode is low, formation of electron beam in the interelectrode space "cathode–control electrode" does not depend on the value of anode voltage U_a. The value of beam current is determined as $I = P_1 U_g^{3/2}$, where P_1 – perveance of "primary gun" created by the cathode and control electrode, U_g – control-electrode voltage, which is supposed to be positive, $U_g > 0$.

The resulting beam perveance will be equal to $P = I/U_a^{3/2}$ or $P = P_1(U_g/U_a)^{3/2}$. Therefore, electron-beam perveance is a function of the ratio of U_g/U_a. Increasing the ratio leads to increasing beam perveance. Triode

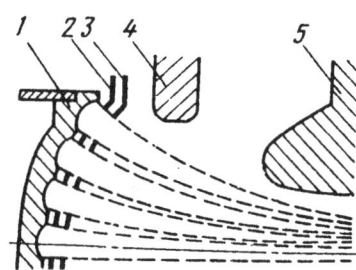

Fig. 1.8. Schematic of a gun with shadow grid: 1 – cathode, 2 – shadow grid, 3 – control grid, 4 – focusing electrode, 5 – anode

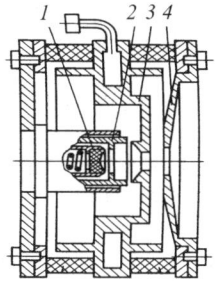

Fig. 1.9. Gun with "longitudinal beam compression": 1 – cathode, 2 – focusing electrode, 3 – control electrode (first anode), 4 – anode

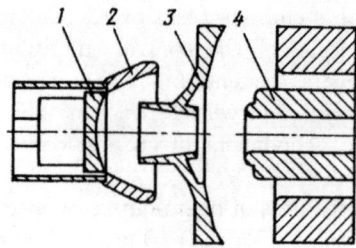

Fig. 1.10. Triode gun with two anodes: 1 – cathode, 2 – focusing electrode, 3 – first anode (control electrode), 4 – second anode (anode)

electron guns operating with a ratio $U_g/U_a > 1$ are named, some times, as guns with "longitudinal compression".

Design of an experimental triode gun operating with longitudinal compression is shown in Fig. 1.9. The perveance of its primary gun P_1 is equal to 1.5 µA/V$^{3/2}$. With potentials of control and anode electrodes being equal to $U_g = 7 \times 10^3$ V and $U_a = 1.5 \times 10^3$, respectively, the electron gun forms an electron beam with current $I = 1$ A and perveance $P = I/U_a^{3/2} = 15$ µA/V$^{3/2}$ [27].

A triode gun forming a power electron beam of high brightness was developed for a free-electron laser. Its main parameters were: beam current $I = 10^3$ A, potential of control electrode (first anode) – several hundreds kV, anode potential $U_a = 1$ MV, beam brightness –10^6 A/cm^{2}sr (Fig. 1.10.) [28].

Systems of Beam Formation for Microwave Grid-Controlled Tubes

Electron-optical design principles have been recently used in grid-controlled tubes (triodes, tetrodes) in order to improve their operating parameters and, in particular, to minimize beam-current loss. A tube design includes a number of separate cells arranged at a cylindrical surface. The cell is optimally designed to form an electron beam with the required operating parameters.

Structures of typical cells used in grid control tubes are shown in Table 1.1.

In these structures: C – cathode, F – focusing electrode, G_1 – control grid, G_2 – screen grid.

In all cells with the exception of cell $C\,3$, a directly heated strip cathode is used. In some cells the cathode is placed inside the special focusing electrode F. Grid electrodes G_1, G_2 perform their usual functions. All electrodes are designed in such a way to provide proper electron-beam formation and control with a minimum loss of beam current.

The electrode structure placed in the $C\,3$ unit of the table is a grid-controlled tube designed on the basis of a magnetron-type gun.

Table 1.1. Types of cells used in grid-controlled tubes

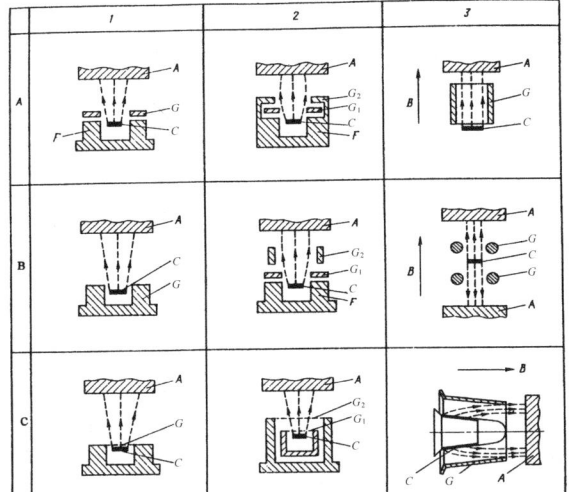

1.3 Focusing System of Electron Devices

The real function of these systems, as has been mentioned above, is transport of an electron beam through extended channels with minimum beam-current loss.

There exist a variety of focusing system at present. They can be classified in the following way:

- On type of fields used: electrostatic, magnetic or combined fields;
- On space-field distribution: uniform axially symmetric, axially periodic, longitudinal periodic and reversed field;
- On shapes of beam cross section: solid axially symmetric, hollow axially symmetric, strip and beam of variable cross section.

The most important of them, from the practical point of view, will be considered, in brief, below.

1.3.1 Magnetic Focusing with Uniform Field

A magnetic system of uniform field used in a power klystron is shown in Fig. 1.11. The magnetic field is created by a set of solenoids (5) enclosed in a magnetic shield (4). The magnetic field inside the shield is practically uniform. The electron gun of the klystron is completely screened from the magnetic field and the electrostatic field created by gun electrodes (1, 2, and 3) produces the initial electron-beam formation.

Beam electrons passing through the channel in the shield interact with local nonuniform field and obtain angular velocity. Interaction of this com-

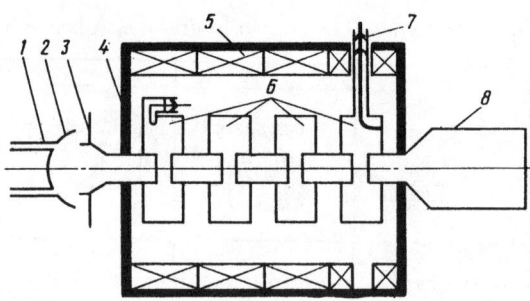

Fig. 1.11. Schematic drawing of a klystron: 1 – cathode, 2 – focusing electrode, 3 – anode, 4 – magnetic shield, 5 – solenoid, 6 – cavities, 7 output, 8 – collector

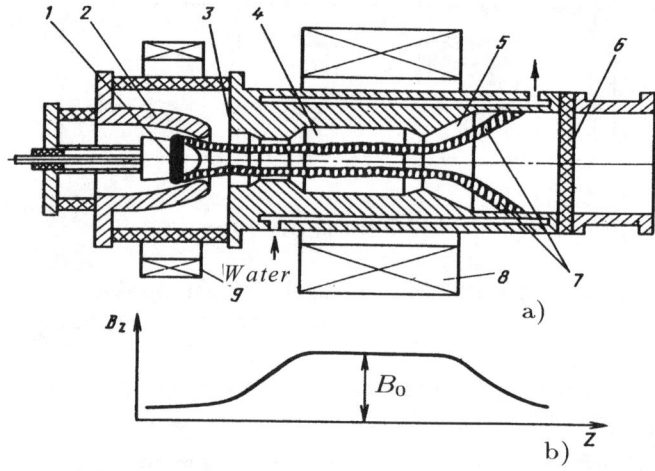

Fig. 1.12. Schematic drawing of a gyrotron **a**: 1 – emitting belt of gyrotron cathode, 2 – first anode, 3 – second anode, 4 – cavity, 5 – output waveguide and collector, 6 – ceramic window, 7 – electron beam, 8 – main solenoid, 9 – auxiliary solenoid; **b**: curve of magnetic-field distribution, B_0 – induction of uniform field

ponent of velocity with the longitudinal component of the uniform magnetic field creates a radial inward force that just balances the two outward forces: one of them produced by the space charge, another is centrifugal force. In principle, a balanced beam of a constant radius can be obtained over the entire tube length. This type of focusing is known as Brillouin focusing [7].

A schematic drawing of a gyrotron-type tube with its magnetic focusing system is presented in Fig. 1.12. Two solenoids (8) and (9) produce the magnetic field. The curve of magnetic-field distribution $B_z = f(z)$ shows that the magnetic field in the central part of the tube is uniform $B_z = B_0 \approx$ const and smoothly decreases in directions to the magnetron electron gun and to the tube output window (6). Such a configuration of magnetic field provides formation and focusing of the hollow electron beam with electrons rotating around the force lines along which they are moving, with rotational velocity being 1.5–2 times more than the longitudinal one, as is required for effective gyrotron operation [29].

1.3.2 Magnetic Periodic Focusing

An example of a magnetic periodic system used in a traveling-wave tube is shown in Fig. 1.13. It consists of ring-shaped magnets (4) magnetized in the longitudinal direction being opposite for neighboring magnets and magnetic pole pieces of a soft magnet material (5).

Such a system provides a high concentration magnetic field inside the region of beam propagation with minimum stray magnetic fluxes. This allows for a considerable reduction (by several times) of the weight of the periodic magnetic system as compared with a uniform-field system.

Axial magnetic-field distribution is usually described by sine or cosine curves, for instance $B_z = B_m \cos 2\pi z/L$, where B_m is field amplitude, L – period of the system. A magnetic focusing force being proportional to B_z^2 will be directed toward the beam axis irrespective of the sign of B_z. Beam focusing in a periodic system can be described in the following way. Let us suppose that in the region of strong magnetic field ($B_z \approx B_m$) the magnetic focusing force exceeds the space-charge repulsion force. Then electrons will be deflected here toward the axis and the beam will converge. In the region of weak magnetic field the space-charge repulsion force will be predominant and electrons will be deflected away from the axis. As electrons enter the strong magnetic field region they again undergo the focusing action and so on. Therefore, the periodic magnetic focusing field can balance the diverging effect of the space charge on the average for field period L. Evidently, this beam pulsation is unavoidable for this type of focusing, although their amplitude can be made sufficiently small. Periodic magnetic focusing systems are

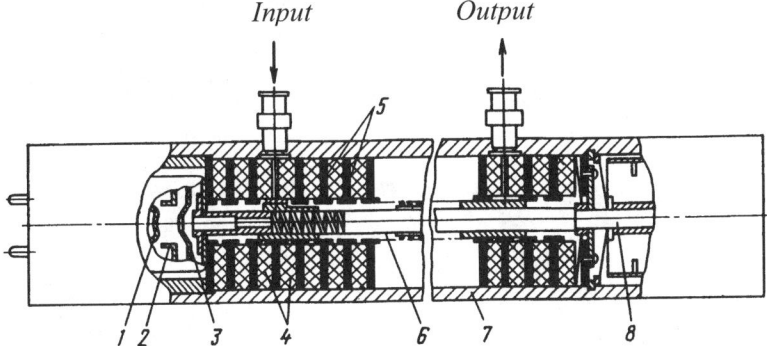

Fig. 1.13. Traveling wave tube with periodic magnetic focusing system: 1 – cathode, 2 – focusing electrode, 3 – anode, 4 – permanent magnets, 5 – magnetic pole pieces, 6 – slow-wave system, 7 – vacuum envelope, 8 – collector

traditionally used for traveling-wave tubes. Recently, G. Caryotakis has developed a high-power relativistic klystrons with a permanent magnet periodic system at Stanford SLAC Laboratory.

1.3.3 Periodic Electrostatic Focusing

A periodic electrostatic focusing system used in klystron-type tubes is presented in Fig. 1.14. It includes three unipotential (Einzel) lenses that are created by incorporating focusing electrodes (7) in the gaps of drift tubes, that are usually connected to the cathode of the tube. These lenses are convergent ones.

Focusing of the electron beam (actually transport of the beam) is performed in the following way. In the regions between lenses electrons move under the action of the space-charge field and are deflected away from the tube axis. Entering the lens-field region they undergo the focusing force action and are deflected toward the axis. The lens-focusing forces balance, an average, the space-charge repulsion and provide beam transport through klystron drift tubes.

Electrostatic periodical focusing is successfully realized in a traveling-wave tube with a double helix slow-wave structure, with the focusing effect

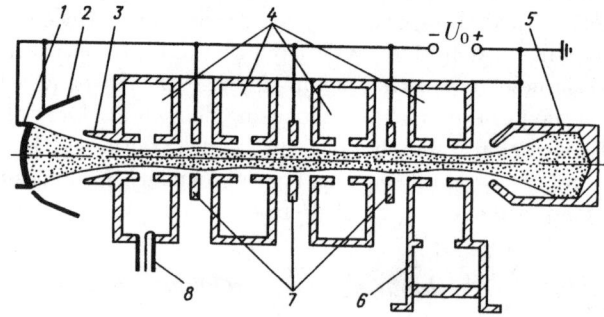

Fig. 1.14. Periodic electrostatic focusing in a klystron: 1 – cathode, 2 – cathode focusing electrode, 3 – anode, 4 – cavities, 5 – collector, 6 – output, 7 – lens focusing electrode, 8 – input

Fig. 1.15. Double-helix slow-wave structure: potential difference $\Delta U = U_2 - U_1$ creates periodic electrostatic field, which provides periodic electrostatic beam focusing

being obtained due to the difference of potentials $\Delta U = U_2 - U_1$ applied to the helices (Fig. 1.15). the value of this difference is to be matched with the beam perveance to provide minimum beam-boundary pulsation.

1.3.4 Hollow-Beam Focusing

Magnetic and electrostatic fields of different space distributions can be used for hollow-beam focusing, however, there exists a peculiarity of balanced focusing, which is caused by the specific properties of the space-charge field distribution in hollow beams. Applying the Gauss theorem to a cylindrical volume confined by the inner beam boundary (the beam length is supposed to be much greater than the mean beam radius) one finds that the radial component of the field is equal to zero at the inner beam boundary. Therefore, for obtaining balanced motion of inner charged-particles a radial focusing force should vanish at this boundary or it should be balanced by some other force, for instance, a centrifugal force or external electrostatic force. A scheme of hollow-beam focusing (or more exact to say, beam transport) is shown in Fig. 1.16 [30,31]. Electrons are accelerated from the ring-shaped cathode (1) by the anode electrode (3) that simultaneously is a part of the magnetic circuit. Passing through the ring gap in the magnetic circuit they cross the radial magnetic field and obtain some angular momentum. Farther on they are moving in the interelectrode space of coaxial diode (5), being under the action of the three radial forces: the centrifugal force arising due to electron rotation, the force of the electrostatic field, directed toward the axis of electrode system and the force of the space-charge field, the latter vanishing for inner electrons. It is evident that by proper choice of the gap magnetic field intensity and potential difference of coaxial electrodes it is possible to provide an exact balanced motion of the inner electrons. For balance motion of all

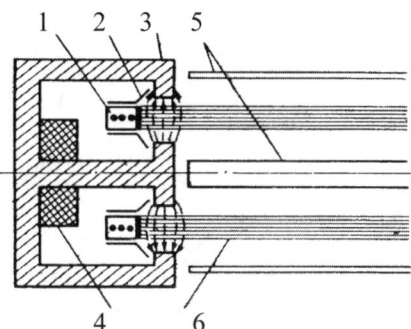

Fig. 1.16. A scheme of hollow-beam focusing: 1 – ring-shaped cathode, 2 – focusing electrode, 3 – anode (simultaneously it is a part of a magnetic circuit), 4 – magnetic coil, 5 – electrodes of the cylindrical diode, 6 – hollow beam

electrons of the beam some additional condition should be fulfilled [31]. A different scheme of hollow-beam focusing will be considered in Chap. 7.

1.4 Optical Systems of Particle-Beam Technology

1.4.1 Installations of Electron Beam Welding and Melting

The electron-optical system of a typical installation of electron-beam welding is presented in Fig. 1.17a. It includes the electron gun, consisting of cathode (1), focusing (forming) electrode (2) and anode (3), magnetic focusing coil (5) and deflecting system (6).

The optical system should provide initial beam formation and subsequent focusing in a small spot to obtain a value of power density of order of 10 kW/mm^2 or more (see Table 1.2) [32]. Electron-beam welding systems are considered to be of two kinds: low voltage with accelerating potential in the range 15–30 kV and high voltage utilizing accelerating potentials in the range 70–150 kV. High-voltage installations allow a smaller beam spot size with higher spot power density to be obtained. For installations with an accelerating voltage in the 120 kV to 150 kV ranges an additional multiple-electrode accelerating system (8) is used (Fig. 1.17b). The function of the deflecting system is to change (or correct) the beam spot position on the welded materials.

Peculiarities of electron-beam welding formation and focusing installations are connected with a rather low operating vacuum (of order $10^{-2}Pa$). This results in a high rate of ionization of residual gases and evaporated metals. Positive ions completely neutralize the electron-beam space charge and produce intensive ion bombardment of the cathode. The latter effect restricts the choice of cathode materials.

The optical system of an electron-beam melting furnace is shown in Fig. 1.18. It contains the same main elements as electron-beam welding systems, that is electron gun and magnetic focusing system. It produces an electron beam of 100 kW power at an accelerating voltage $U_a = 20$ kW. The beam is injected into a vacuum chamber, where the metal to be melted is placed [33,34].

Electron Guns with Plasma Sources

In these guns electrons are extracted from the plasma of a gas discharge. Different types of gas discharge can be used for plasma generation such as cold-cathode glow discharge, arc-type discharge, high-frequency discharge etc.

A schematic drawing of a typical plasma electron source using cold-cathode glow discharge is shown in Fig. 1.19 [35]. A high-voltage anode plasma glow discharge is sustained at gas pressure in the range 1–100 Pa

1.4 Optical Systems of Particle-Beam Technology

Table 1.2. Parameters of electron-beam welding installations

Accelerating voltage, kV	Beam current, mA	Beam diameter at focus, mm	Power density at focus, kW/mm^2	Type of cathode
20	150	0.6–0.8	10	Lanthanum hexaboride
25	500	1.2–1.3	5	Lanthanum hexaboride
40	50	0.25–0.3	30	Metallic
60	35	0.6–0.7	5–7	Lanthanum hexaboride
80	0.3; 0.6	0.03	32; 64	Metallic
100	15	0.1	150	Metallic
150	20	0.1	300	Metallic

Metallic cathode – cathodes of metals Ta or W

and accelerating voltage 1–10 kV. A cold cathode (1) is the emitter of electrons, pure metals, alloys, or metal ceramics being used as cathode materials.

Electron emission results from bombardment of a cathode surface by positive ions emitted from an anode plasma. Electrons are extracted from the gas-discharge region through the hole in the anode electrode.

The configuration of the electron beam forming by the gun significantly depends on the shape of the plasma boundary in the cathode–anode region.

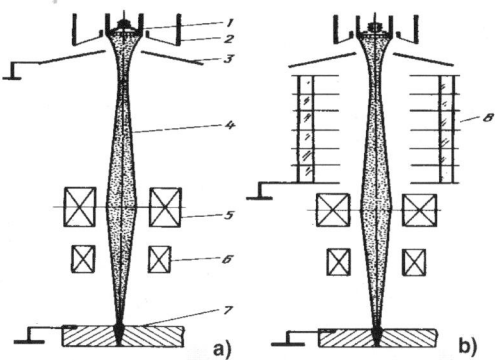

Fig. 1.17. Schemes of electron-beam welding installations: 1 – cathode, 2 – cathode focusing electrode, 3 – anode, 4 – electron beam, 5 – magnetic focusing solenoid, 6 – deflecting system, 7 – metal pieces to be welded, 8 – accelerating section

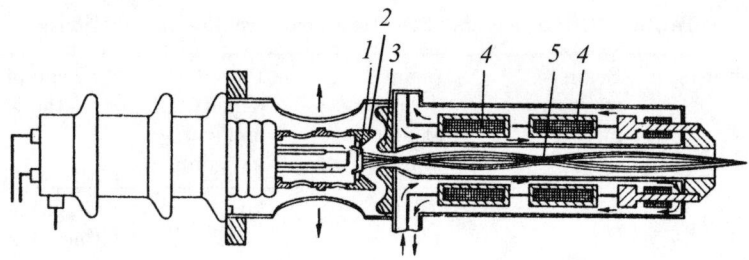

Fig. 1.18. Optical system of electron-beam melting furnace: 1 – cathode, 2 – cathode focusing electrode, 3 – anode, 4 – magnetic coils, 5 – beam

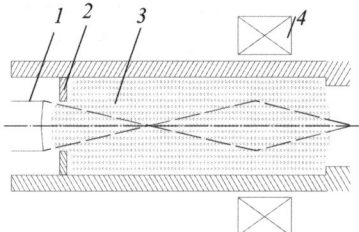

Fig. 1.19. Scheme of plasma electron source: 1 – cold cathode, 2 – anode, 3 – plasma, 4 – focusing coil

Plasma electron guns with cold-cathode glow discharge allow a convergent electron beam with a focus spot area about of one square millimeter with spot power density of order 10 kW/mm^2 to be obtained. They are used for welding, brazing, evaporating and other technological applications, with these processes being curried out in vacuum or in a control-gas atmosphere.

1.5 Ion Optical Systems

The systems are used in charged-particle beam technology, accelerating techniques and other applications.

A natural source of ions is considered to be a gas-discharge plasma. An arc type discharge with a hot cathode is usually utilized as the source of ions. They will be considered in Chap. 6.

Ion guns with special solid emitters also have been developed [36]. Such a gun, intended for formation of a cesium ion beam, is shown in Fig. 1.20. It includes the following main elements: emitter (4) made of porous tungsten, heater (5), forming (focusing) electrode (3), accelerating electrode (2). The emitter is heated to temperature about 1600 K. Cesium vapor proceeds into the emitter cavity from a cesium evaporator on a fitting pipe (6). Atoms of cesium diffuse in the pores of the emitter and are ionized as a result of thermal ionization. Ions diffuse to the surface of the emitter and then accelerated by the field of the accelerating electrode are creating the ion beam. The process

1.5 Ion Optical Systems   17

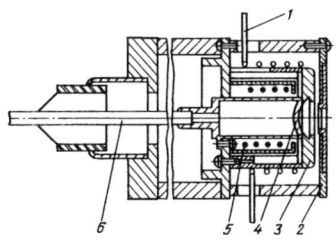
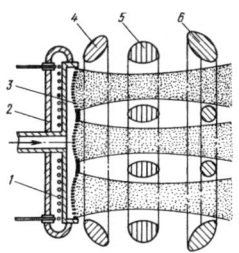
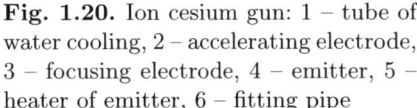

Fig. 1.20. Ion cesium gun: 1 – tube of water cooling, 2 – accelerating electrode, 3 – focusing electrode, 4 – emitter, 5 – heater of emitter, 6 – fitting pipe

Fig. 1.21. Multiple-beam ion cesium gun: 1 – heater of emitter, 2 – thermal screen, 3 – emitter of porous tungsten, 4 – cathode beam-forming electrode, 5 – first accelerating electrode, 6 – second accelerating electrode

of formation of the ion beam in this gun is similar to that of the Pierce-type electron gun.

However, the value of perveance of the gun under consideration is three orders smaller as compared with that of typical electron guns, $P_i = I_i/U_a^{3/2} \approx 1.5 \times 10^{-3}$ µA/V$^{3/2}$. This result is explained by the fact that ions have greater mass and move in an accelerating field much more slowly than electrons. The ratio of their velocities is determined by $v_i/v = (m/m_i)^{1/2}$.

It is easy to show that the ratio of perveance values of ion beam P_i and electron beam P being formed by guns of the same geometry is expressed by a similar equation $P_i/P = (m/m_i)^{1/2}$, for a cesium ion beam it is about of 0.5×10^{-3}. This makes it difficult to design Pierce-type axially symmetric ion guns for high values of beam perveance. The increase of ion-beam perveance is possible by using multibeam ion guns as has been done in an ionic rocket engine, a schematic drawing of which is shown in Fig. 1.21 [36]. The total perveance of ion beams formed by these systems lay in the range $(1-3) \times 10^{-2}$µA/V$^{3/2}$. This corresponds to a value of electron beam perveance equal to 20–60 µA/V$^{3/2}$.

It is important to note that recently multiple-beam systems have found application in microwave O-type tubes (klystrons, TWT, etc.). They will be considered in Chap. 11.

2 Methods of Fields Calculation

2.1 General Equations of Electrostatic Fields

Electrostatic fields in vacuum obey Poisson's equation

$$\nabla^2 U = -\frac{\rho}{\varepsilon_0}, \tag{2.1}$$

where U – electrostatic potential, ρ – space-charge density, ε_0 – dielectric constant, ∇^2 – the Laplace operator, which depends on a coordinate system:

In the rectangular coordinates x, y, z

$$\nabla^2 = \partial^2/\partial x^2 + \partial^2/\partial y^2 + \partial^2/\partial z^2 \ .$$

In the cylindrical coordinates r, θ, z

$$\nabla^2 = \frac{1}{r}\left[\frac{\partial}{\partial r}\left(r\frac{\partial}{\partial r}\right) + \frac{\partial}{\partial \theta}\left(\frac{1}{r}\frac{\partial}{\partial \theta}\right) + r\frac{\partial}{\partial z}\left(\frac{\partial}{\partial z}\right)\right] \ .$$

In the generalized coordinates q_1, q_2, q_3

$$\nabla^2 = \frac{1}{h_1 h_2 h_3}\left[\frac{\partial}{\partial q_1}\left(\frac{h_2 h_3}{h_1}\frac{\partial}{\partial q_1}\right) + \frac{\partial}{\partial q_2}\left(\frac{h_3 h_1}{h_2}\frac{\partial}{\partial q_2}\right) + \frac{\partial}{\partial q_3}\left(\frac{h_1 h_2}{h_3}\frac{\partial}{\partial q_3}\right)\right] ,$$

where h_1, h_2, h_3 – scale coefficient of the coordinate system.

In regions that are free of space charge the electric potential is described by the Laplace equation

$$\nabla^2 U = 0 \ . \tag{2.2}$$

Electric field strength is expressed as

$$E = -\nabla U \ ,$$

where ∇ – Nabla operator which is expressed as follows:

In the rectangular coordinates

$$\nabla = \vec{i}_x \frac{\partial}{\partial x} + \vec{i}_y \frac{\partial}{\partial y} + \vec{i}_z \frac{\partial}{\partial z} \ ,$$

where $\vec{i}_x, \vec{i}_y, \vec{i}_z$ – unit vectors.

In the cylindrical coordinates r, θ, z

$$\nabla = \vec{i}_r \frac{\partial}{\partial r} + \vec{i}_\theta \frac{1}{r}\frac{\partial}{\partial \theta} + \vec{i}_z \frac{\partial}{\partial z} \ ,$$

where $\vec{i}_r, \vec{i}_\theta, \vec{i}_z$ – unit vectors.

In the curvilinear coordinates q_1, q_2, q_3

$$\nabla = \vec{i}_1 \frac{1}{h_1} \frac{\partial}{\partial q_1} + \vec{i}_2 \frac{1}{h_2} \frac{\partial}{\partial q_2} + \vec{i}_3 \frac{1}{h_3} \frac{\partial}{\partial q_3},$$

where $\vec{i}_1, \vec{i}_2, \vec{i}_3$ – unit vectors.

Two types of problems are usually required to be solved in the process of calculation, analysis and design of charged-particle optical systems: Dirichlet problem and Cauchy problem, which are considered later.

2.2 Electrostatic Field Calculation. Dirichlet Problem

In this case, the potential distribution is sought inside the region confined by the electrode system, which geometry and potentials are supposed to be known, as the well as the space-charge distribution. Different analytical and numerical methods are used for the solution of the problem. Some of them are considered below.

2.2.1 Method of the Separation of Variables

Let us consider the application of the method for solution of the Laplace equation for axially symmetric electrode systems [37, 38].

Using the cylindrical coordinates r, θ, z and taking into account the axial symmetry we can write the following equation for the potential distribution:

$$\nabla^2 U = \frac{1}{r} \frac{\partial}{\partial r} \left(r \frac{\partial U}{\partial r} \right) + \frac{\partial^2 U}{\partial z^2} = 0.$$

The solution is sought as the product of two functions of r and z:

$$U(r, z) = R(r) Z(z).$$

Substitution of this expression in the Laplace equation splits it into two ordinary differential equations:

$$\frac{d^2 Z}{\partial z^2} + k^2 Z = 0 \quad \frac{1}{r} \frac{d}{dr} \left(r \frac{dR}{dr} \right) - k^2 R = 0,$$

which are related by a separation constant k and have the following solutions:

$$\left. \begin{array}{l} Z = A_k \sin kz + B_k \cos kz, \\ R = C_k I_0(kr) + D_k K_0(kr); \end{array} \right\} k \text{ – real value}$$

$$\left. \begin{array}{l} Z = A_k \sinh \chi z + B_k \cosh \chi z, \\ R = C_k J_0(\chi r) + D_k N_0(\chi r); \end{array} \right\} k = i\chi \text{ – imaginary value}$$

In these equations: A_k, B_k, C_k, D_k are arbitrary constants, $J_0(\chi r)$ and $N_0(\chi r)$ are the Bessel functions of the first and second kind and zero order,

2.2 Electrostatic Field Calculation. Dirichlet Problem

$I_0(kr)$ and $K_0(kr)$ are the modified Bessel functions of the first and second kind and zero order.

The Laplace equation is a linear one, so its general solution can be obtained by summing of the product of the above functions and can be written as follows:

$$U(r,z) = \sum_k (A_k \sin kz + B_k \cos kz)[C_k I_0(kr) + D_k K_0(kr)], \tag{2.3}$$

$$U(r,z) = \int_k [A(k) \sin kz + B(k) \cos kz][C(k) I_0(kr) + D(k) K_0(kr)] dk ; \tag{2.4}$$

$$U(r,z) = \sum_\chi (A_\chi \sinh \chi z + B_\chi \cosh \chi z)[C_\chi J_0(\chi r) + D_\chi N_0(\chi r)],$$

$$U(r,z) = \int_\chi^\chi [A(\chi) \sinh \chi z + B(\chi) \cosh \chi z \times$$
$$\times [C(\chi) J_0(\chi r) + D(\chi) N_0(\chi r)] d\chi . \tag{2.5}$$

The arbitrary constants, which are involved in these equations, are found to meet boundary conditions and some additional physical conditions following from a problem. For example, if the region where the solution is sought includes the axis of symmetry ($r = 0$), then the arbitrary constants D_k and D_χ should be taken equal to zero.

Example 2.1. Let us apply the obtained results for calculation of the field of an electrostatic lens formed by two semi-infinite coaxial cylinders separated by an infinitesimal gap (Fig. 2.1). The cylinders are at potentials $U_1 = 0$ and U_2. For a given lens the electrode configuration potential at the middle plane of the lens ($z = 0$) is constant and equal to $U_0 = (U_1 + U_2)/2 = U_2/2$. Inside the cylinders the potential has a finite value.

At the cylindrical boundary $r = R$ the potential is changed by a jump at $z = 0$ from $U_1 = 0$ to U_2, that is, $U = U_1$ at $z < 0$ and $U = U_2$ at $z > 0$. This potential distribution can be expressed in terms of Heaviside's unit function changed by a jump from 0 to 1 at $z = 0$, $U(R,z) = U_2 \cdot H(z)$.

Heaviside's function is analytically expressed through a Fourier integral:

$$H(z) = \frac{1}{2} + \frac{1}{\pi} \int_0^\infty \frac{\sin kz}{k} dk .$$

The potential at the cylindrical boundary $r = R$ can be presented as:

$$U(R,z) = U_2 \cdot H(z) = \frac{U_2}{2} + \frac{U_2}{\pi} \int_0^\infty \frac{\sin kz}{k} dk .$$

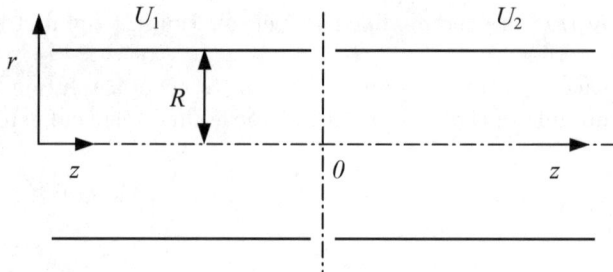

Fig. 2.1. Immersion lens formed by two coaxial cylinders at potentials $U_1 = 0$ and $+U_2$

The solution of the Laplace equation, which meets the above condition can be taken in the form (2.4) with $B(k)$ and $D(k)$ being equal zero, $B(k) = D(k) = 0$:

$$U(r,z) = \frac{U_2}{2} + \int_k \tilde{A}(k) \sin kz I_0(kr) \, dk,$$

where $\tilde{A}(k) = A(k)C(k)$ is an arbitrary constant.

It is found from the requirement that the potential at a cylindrical boundary $r = R$ should satisfy the boundary condition:

$$\frac{U_2}{2} + \int_0^\infty \tilde{A}(k) I_0(kR) \sin kz \, dk = \frac{U_2}{2} + \frac{U_2}{\pi} \int_0^\infty \frac{\sin kz}{k} \, dk.$$

This gives

$$\tilde{A}(k) = \frac{U_2}{\pi} \frac{1}{k I_0(kR)},$$

So, the ultimate expression will be:

$$U(r,z) = \frac{U_2}{2} + \frac{U_2}{\pi} \int_0^\infty \frac{\sin kz}{k} \frac{I_0(kr)}{I_0(kR)} \, dk,$$

and at the symmetry axis ($r = 0$)

$$U(z) = \frac{U_2}{2} + \frac{U_2}{\pi} \int \frac{\sin kz}{k} \frac{1}{I_0(kR)} \, dk.$$

To obtain the potential distribution a numerical computation of the integral is required.

Another approach to this problem, which gives the solution in the form of a series, is considered in Example 2.3.

2.2.2 Method of Green Function

With reference to the problem of electrostatic field calculation the Green function represents the potential created by unity point charge $q = 1$, located in a point $M_0(x_0, y_0, z_0)$, inside the region enclosed by the surface of electrodes, which potential is supposed to be equal zero [37]:

$$G(M, M_0) = 1/(4\pi\varepsilon_0 R) + \int_S \frac{\sigma dS}{R_i} .$$

The first term in the right part of this equation expresses the potential of the field created by the charge in free space, the second one is the potential created by surface charge induced at electrode surfaces S.

In this equation R – distance from the point $M_0(x_0, y_0, z_0)$ of charge q location to current point $M(x, y, z)$, where potential is determined (later, for short, "point of observation") $R = \sqrt{(x-x_0)^2 + (y-y_0)^2 + (z-z_0)^2}$; R_i – distance from point of observation M and a current point at the electrode surface M_i, $R_i = \sqrt{(x-x_i)^2 + (y-y_i)^2 + (z-z_i)^2}$; $\sigma(x_i, y_i, z_i)$ – density of surface charge induced at electrode surfaces.

The Green function, in accordance with its definition, satisfies the Laplace equation in all space, confined by electrodes, except point M, and becomes zero at the electrode surfaces. The Green function is symmetric concerning its arguments, i.e. $G(M, M_0) = G(M_0, M)$. This physically means that charge q placed in point M creates at point M_0 the same potential as charge q being placed at point M_0 produced at point M.

The knowledge of the Green function allows reduction of the solution of the Dirichlet problem to computation of the corresponding integrals [39–41]:

$$U(M_0) = -\varepsilon_0 \int_S U_S(M) \frac{\partial G(M, M_0)}{\partial n} dS + \int_V \rho(M) G(M, M_0) dV ,$$

where $U_S(M)$ – a given potential distribution over the surface S, $\rho(M)$ distribution of the space-charge density.

Integration is made on coordinates x, y, z of point M, with U_S, $\partial G/\partial n$, and ρ are considered to be a known function of these coordinates.

In view of the symmetry function $G(M, M_0)$ to its arguments, it is possible to write the alternative formula for potential, expressing it as a function of point M

$$U(M) = -\varepsilon_0 \int_S U_S(M_0) \frac{\partial G(M, M_0)}{\partial n} dS + \int_V \rho(M_0) G(M, M_0) dV . \tag{2.6}$$

In this case integration is performed on coordinates x_0, y_0, z_0 of point M_0 [41].

Two examples of the application of the Green function are given below.

Example 2.2. Let us calculate the distribution of potential that is created in a metallic tube of radius "a" by cylindrical column of space charge of radius "b", with density of space charge being function only longitudinal coordinate z:

$$\rho = \rho(z_0) = \rho_0 + \rho_m \cos \frac{2\pi}{L} z_0 ,$$

where ρ_0 – constant component of space-charge density, ρ_m and L are amplitude and period of space-charge variation.

The Green function for a cylindrical channel of radius a, with account of axial symmetry of the space-charge distribution, has the following analytical expression:

$$G = \frac{1}{2\pi\varepsilon_0 a^2} \sum_{i=1}^{\infty} e^{-\frac{\lambda_i}{a}|z-z_0|} \frac{J_0(\lambda_i r/a) J_0(\lambda_i r_0/a)}{\lambda_i/a [J_1(\lambda_i)]^2} ,$$

where J_0 and J_1 are Bessel functions of the first kind of zero and first order, λ_i are the roots of function J_0.

It is seen that function G is symmetric to its arguments r, z and r_0, z_0.

The potential created by the space-charge column is found as

$$U = \int_V \rho(r_0 z_0) G(r, z, r_0, z_0) 2\pi r_0 dr_0 dz_0 ,$$

where integration is performed over the volume V occupied by the space charge.

Substituting the Green function in this integral and changing the order of integration and summation one obtains:

$$U = \frac{1}{\varepsilon_0 a} \sum_{i=1}^{\infty} \frac{J_0(\lambda_i r/a)}{\lambda_i [J_1(\lambda_i)]^2} \int_V \left(\rho_0 + \rho_m \cos \frac{2\pi}{L} z_0 \right) r_0 dr_0 dz_0 .$$

The integration performed on r in the limits from 0 to b and on z in the limits from $-\infty$ to ∞ gives:

$$U = \frac{2ab}{\varepsilon_0} \sum_{i=1}^{\infty} \frac{J_0(\lambda_i r/a) J_1(\lambda_i b/a)}{\lambda_i^3 [J_1(\lambda_i)]^2} \left[\rho_0 + \rho_m \frac{1}{1+(2\pi a/L\lambda_i)^2} \cos \frac{2\pi}{L} z \right] .$$

Example 2.3. Let us apply the Green-function method for calculation of the field of an electrostatic immersion lens, which is created by two coaxial cylinders, separated by infinitaly small gap (Fig. 2.1). The potential of the first cylinder is assumed to be equal to zero $U_1 = 0$, the potential of the second one equal to U_2.

For the potential computation we use (2.6). As in the case under consideration $\rho(M_0) = 0$, it is reduced to:

$$U(M) = -\varepsilon_0 \int_S U_S(M_0) \frac{\partial G(M, M_0)}{\partial n} dS .$$

The Green function of the given problem is the same one as in Example 2.1. Let us find its derivative, that is, the derivative along the normal to the surface S, it is equal to:

$$\frac{\partial G}{\partial n} = \frac{\partial G}{\partial r} = -\frac{1}{2\pi\varepsilon_0 a^2} \sum_{i=1}^{\infty} e^{-\frac{\lambda_i}{a}|z-z_0|} \frac{J_0(\lambda_i r/a) J_1(\lambda_i r_0/a) \lambda_i/a}{\lambda_i/a \left[J_1(\lambda_i)\right]^2} ,$$

and

$$\left.\frac{\partial G}{\partial r}\right|_{r_0=a} = -\frac{1}{2\pi\varepsilon_0 a^2} \sum_{i=1}^{\infty} e^{-\frac{\lambda_i}{a}|z-z_0|} \frac{J_0(\lambda_i r/a)}{J_1(\lambda_i)} .$$

Substitution of this equation into the integral of the above formula for $U(M)$ gives:

$$U(r,z) = \frac{1}{a} \sum_{i=1}^{\infty} \frac{J_0(\lambda_i r/a)}{J_1(\lambda_i)} \int_{-\infty}^{\infty} U_S e^{-\frac{\lambda_i}{a}|z-z_0|} dz_0 .$$

Since $U_S = 0$ at $z < 0$ and $U_S = U_2$ at $z ¿ 0$ (at $z = 0$ $U_S = U_2/2$), one obtains:

$$\int_{-\infty}^{\infty} U_S e^{-\frac{\lambda_i}{a}|z-z_0|} dz_0 = U_2 \int_{0}^{\infty} e^{-\frac{\lambda_i}{a}|z-z_0|} dz_0 .$$

Calculation of the last integral gives the following equations for the potential distribution in two coaxial cylinder lenses:

$$U(r,z) = U_2 \sum_{i=1}^{\infty} \frac{J_0(\lambda_i r/a)}{\lambda_i J_1(\lambda_i)} e^{-\frac{\lambda_i}{a}|z|} \quad \text{for} \quad z \leq 0 ,$$

$$U(r,z) = U_2 \left[1 - \sum_{i=1}^{\infty} \frac{J_0(\lambda_i r/a)}{\lambda_i J_1(\lambda_i)} e^{-\frac{\lambda_i}{a}z}\right] \quad \text{for} \quad z \geq 0 .$$

This solution is more convenient for practical use than that obtained by the classical method of separation of variables (see Example 2.1)

The method of Green function is especially effective for calculation of fields created by space charge varying in time. In this case computation of potential is produced by the equation:

$$U(M,t) = \int_{V} \rho(M_0,t) G(M,M_0) dV ,$$

where $\rho(M_0,t)$ – space-charge density varying in space and time, $G(M,M_0)$ – Green function, which depends on the coordinates of points M and M_0 and the geometry of a region, where the potential is computed, but does not depend on time.

2.2.3 Method of the Integral Equations

Applying the Green theorem it is possible to show that a solution of the Dirichlet problem for the Poisson equation can be presented in the form:

$$U(x, y, z) = \frac{1}{4\pi\varepsilon_0} \int_V \frac{\rho dV}{R_0} + \frac{1}{4\pi\varepsilon_0} \int_S \frac{\sigma dS}{R_i}. \tag{2.7}$$

The first integral of this equation is taken over the volume V, confined by surface S, and represents the potential of space charge $\rho(x_0, y_0, z_0)$, with R_0 being the distance from the current point of observation with coordinates x, y, z to the point of space-charge location x_0, y_0, z_0. The second one expresses the potential of surface charges distributed on surface S with surface density σ. With reference to the calculation of fields of charged particle optical systems surface S is understood as the surface of electrodes forming the systems, with R_i being the distance from the point of observation to a point belonging to the electrode surface and having coordinates x_i, y_i, z_i,

$$R_0 = \sqrt{(x-x_0)^2 + (y-y_0)^2 + (z-z_0)^2}, \quad R_i = \sqrt{(x-x_i)^2 + (y-y_i)^2 + (z-z_i)^2}.$$

For the case of two-dimensional fields ($U(y, z)$, $\partial U/\partial x = 0$) the integral form of the solution is written as:

$$U(y, z) = \frac{1}{2\pi\varepsilon_0} \int_V \rho \ln \frac{1}{R_0} dV + \frac{1}{2\pi\varepsilon_0} \int_S \sigma \ln \frac{1}{R_i} dS,$$

where $R_0 = \sqrt{(y-y_0)^2 + (z-z_0)^2}$, $R_i = \sqrt{(y-y_i)^2 + (z-z_i)^2}$.

Integration is performed over the volume V and surface S having unit extent in the direction of the x-axis. As result of this $dV = dy_0 dz_0 \cdot 1$ and $dS = dC \cdot 1$, dC being the elementary segment of electrode contour C.

Then the above equation for potential $U(y, z)$ can be transformed to:

$$U(y, z) = \frac{1}{2\pi\varepsilon_0} \int_F \rho \ln \frac{1}{R_0} dF + \frac{1}{2\pi\varepsilon_0} \int_C \sigma \ln \frac{1}{R_i} dC, \tag{2.8}$$

where the first integral is taken over surface F, occupied by space charge, the second one along the electrode contour C.

For an axially symmetric field, using the cylindrical coordinate system r, θ, z and integrating on the θ-coordinate one finds from the general equation (2.7) the following formula for potential $U(r, z)$:

$$U(r, z) = \frac{1}{2\pi^2\varepsilon_0} \int_V \frac{\rho K(t_0) dV}{\sqrt{(r+r_0)^2 + (z-z_0)^2}}$$

$$+ \frac{1}{2\pi^2\varepsilon_0} \int_S \frac{\rho K(t_i) dS}{\sqrt{(r+r_i)^2 + (z-z)^2}}.$$

Here, r_0 and z_0, r_i and z_i – coordinates for location of space and surface charges;

$$t_0 = \frac{2\sqrt{rr_0}}{\sqrt{(r+r_0)^2 + (z-z_0)^2}}, \quad t_i = \frac{2\sqrt{rr_i}}{\sqrt{(r+r_i)^2 + (z-z_i)^2}} \ ;$$

$K(t)$ – total elliptic integral of the first kind:

$$K(t) = \int_0^{\pi/2} \frac{d\beta}{\sqrt{1 - t^2 \sin^2 \beta}} \ ; \quad dV = 2\pi r_0 dr_0 dz_0 \ ; \quad dS = 2\pi r_i dC \ ,$$

dC being a segment of the electrode contour.

The above equation for potential $U(r, z)$ can be rewritten in the following equivalent form:

$$U(r,z) = \frac{1}{\pi\varepsilon_0} \int_F \frac{\rho r_0 K(t_0) \, dF}{\sqrt{(r+r_0)^2 + (z-z_0)^2}}$$

$$+ \frac{1}{\pi\varepsilon_0} \int_C \frac{\sigma r_i K(t_i) \, dC}{\sqrt{(r+r_i)^2 + (z-z_i)^2}} \ . \quad (2.9)$$

Here, integration is made over surface F, belonging to the meridian plane $\theta = const$, occupied by space charge and the contour of electrodes C.

The distribution the space-charge density $\rho(r, z)$ is supposed to be a known function of coordinates r and z at this stage of field calculation. On the contrary, the distribution of the surface charge density σ is not known and must be found before the potential calculation.

This can be done in the following way. Let us place the "point of observation" on the electrode contour, the potential of which is supposed to be known, then the left part of (2.9) will be a known function of coordinates and this equation can be considered as an integral equation concerning the unknown value of σ. The solution of this equation can be made by numerical methods with use of computers (see Sect. 2.4).

2.2.4 Method of Finite Difference

The essence of the method consists of replacement of a partial differential equation by a finite difference equation that is obtained by replacement of partial derivatives with their approximate finite difference expressions. Let us consider this procedure on an example of a two-dimensional Poisson equation:

$$\frac{\partial^2 U}{\partial y^2} + \frac{\partial^2 U}{\partial z^2} = -\frac{\rho}{\varepsilon_0} \ . \quad (2.10)$$

The second derivatives including in this equation can be expressed at a point 0 through the values of the first derivatives at the neighboring points a, b, c, d (Fig. 2.2a):

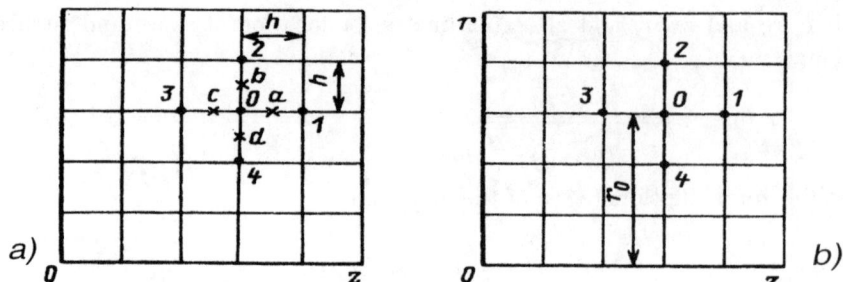

Fig. 2.2. Meshes for derivation of finite-difference equations: **a** two-dimensional field; **b** axially symmetric field

$$\frac{\partial^2 U}{\partial z^2} \approx \frac{1}{h}\left[\left(\frac{\partial U}{\partial z}\right)_a - \left(\frac{\partial U}{\partial z}\right)_c\right], \quad \frac{\partial^2 U}{\partial y^2} \approx \frac{1}{h}\left[\left(\frac{\partial U}{\partial u}\right)_b - \left(\frac{\partial U}{\partial y}\right)_d\right]. \tag{2.11}$$

The first derivatives of these equations in turn can be expressed through the finite differences of corresponding potentials:

$$\left.\begin{array}{l}\left(\dfrac{\partial U}{\partial z}\right)_a \approx \dfrac{1}{h}(U_1 - U_0), \left(\dfrac{\partial U}{\partial z}\right)_c \approx \dfrac{1}{h}(U_0 - U_3) \\[2mm] \left(\dfrac{\partial U}{\partial y}\right)_b \approx \dfrac{1}{h}(U_2 - U_0), \left(\dfrac{\partial U}{\partial y}\right)_d \approx \dfrac{1}{h}(U_0 - U_4)\end{array}\right\}, \tag{2.12}$$

where U_1, U_2, U_3, U_4 – values of potential at points 1, 2, 3, 4 (Fig. 2.2a).

Substitution of (2.12) in (2.11) gives:

$$rot_n \vec{J} = \int_l \vec{J}d\vec{l}/\Delta S, \quad \frac{\partial^2 U}{dy^2} \approx \frac{1}{h^2}[(U_2 - U_0) - (U_0 - U_4)],$$

$$\frac{\partial^2 U}{\partial y^2} + \frac{\partial^2 U}{\partial z^2} \approx \frac{1}{h^2}(U_1 + U_2 + U_3 + U_4 - 4U_0).$$

Then, the finite difference analog of the Poisson equation is expressed as:

$$U_1 + U_2 + U_3 + U_4 - 4U_0 = -h^2\rho/\varepsilon_0.$$

For the two-dimensional Laplace equation it reduces to:

$$U_1 + U_2 + U_3 + U_4 - 4U_0 = 0.$$

The finite-difference analog of the Poisson equation in cylindrical coordinates is written as:

$$U_1 + U_2 + U_3 + U_4 - 4U_0 + \frac{h}{2r_0}(U_2 - U_4) = -\frac{h^2\rho}{\varepsilon_0}, \tag{2.13}$$

where r_0 is the distance from the z axis to the point 0 (Fig. 2.1b).

For the points lying on the axis of symmetry this equation is replaced by:

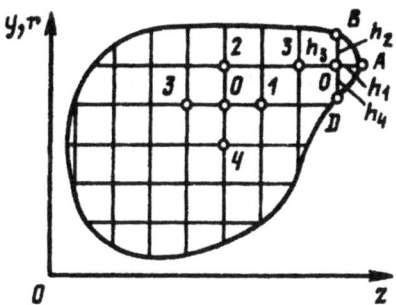

Fig. 2.3. Irregular interior nodes in the region of the complicated boundary

$$U_1 + U_3 + 4U_2 - 6U_0 = -h^2\rho/\varepsilon_0 \ .$$

The finite-difference equation connects values of potential in separate discrete points. For potential computation a regular mesh is introduced in the region enclosed by the electrode system; such a mesh with square cells with pitch $\underline{h}$ is shown in Fig. 2.2.

There are several types of mesh nodes:

- boundary nodes located at the electrode boundary,
- regular interior nodes located inside the region, enclosed by an electrode system, at a distance more than the mesh pitch h from the electrode boundary,
- irregular interior nodes located at a distance less than the mesh pitch h from the boundary.

For every interior node the finite-difference equation is composed, which connects its potential with potentials of neighboring mesh nodes. For regular nodes the regular finite-difference equations are used. They are derived in the way considered above.

There exists a peculiarity of creation of finite-difference equations for irregular nodes. There are several approaches to create these equations, one of which is considered below. Let us take an irregular node shown at Fig. 2.3. Its distances from the boundary are h_1, h_2 and h_4, which are less than the mesh pitch h. In this case, as the neighboring points of the node in the equation the following points are taken: $3, B, A, D$.

The last three of them belong to the boundary, their potentials are supposed to be known and equal to U_B, U_A, U_D. Since they are located at different distances from nodal point 0 the finite-difference equation has the form:

$$\frac{2}{h_1 + h_3}\left(\frac{U_1}{h_1} + \frac{U_3}{h_3}\right) + \frac{2}{h_2 + h_4}\left(\alpha\frac{U_2}{h_2} + \beta\frac{U_4}{h_4}\right)$$
$$- \left(\frac{2}{h_1 h_2} + \frac{2}{h_2 h_4} + \gamma\right)U_0 = -\frac{\rho}{\varepsilon_0} \ , \qquad (2.14)$$

where $\alpha = \beta = 0$, $\gamma = 1$ for a two-dimensional field; $\alpha = 1 + h_4/2r_0$, $\beta = 1 - h_2/2r_0$ $\gamma = 1/h_2 r_0 - 1/h_4 r_0$ for an axially symmetric field.

The finite-difference equation written for interior nodes creates a system of linear algebraic equations, the number of which is equal to the number of unknown values, that is, potentials of interior nodes. Thus, determination of the potential distribution is reduced to solution of this system. Boundary conditions participate in the solution through the values of potential of boundary nodes.

Replacement of the partial differential equation by the finite-difference one produces an error ε, the value of which is of order o fh^2, $\varepsilon \sim h^2$. It is reduced as the mesh pitch h is decreased and the number of mesh nodes is increased. It, in turn, leads to an algebraic equation system of high order. Its solution by the direct method of elimination appears to be impossible and the method of successive approximation, or the iteration method, is used for determination of the potential distribution. This method has several versions and is widely used in computer programs. A technique of the method is considered in Sect. 2.4.

2.3 Calculation of Electrostatic Field: Cauchy Problem

The necessity of solution of such a problem arises in the process of designing and synthesis of charge particle optical systems.

Frequently, it is required to find the electrostatic potential in a region with initial conditions, that is potential and its normal derivative, being given on a surface or line. In such a statement, determination of the electrostatic potential is reduced to solution of the Cauchy problem for the Laplace equation.

A typical example of such a problem is determination of the space potential distribution with potential and its normal derivative being given at the plane or axis of symmetry. In electron optics it is known as the problem of potential series expansions.

In mathematics the Cauchy problem for the Laplace equation is related to so-called incorrectly set up problems, since its solution is unstable with respect to small variations of initial conditions or using approximate, for instance, numerical methods of solution.

When the initial conditions are given in analytical form and the solution is obtained by analytical methods the incorrectness is not an obstacle for the obtaining the solution of this problem.

In the general case spatial methods known as methods of regularization are used for its solution [48].

2.3.1 Method of Analytical Continuation

Two-Dimensional Field

Solutions of the Cauchy problem for the two-dimensional Laplace equation $\partial^2 U/\partial y^2 + \partial^2 U/\partial z^2 = 0$ will be considered below for three different kinds of initial conditions:

I. *Distribution of potential is given along a straight line, with its normal derivative being equal to zero.*

Let us suppose that this line coincides with the real axis of a complex plane $\zeta = z + iy$, then the initial conditions are written as $U|_{y=0} = f(z)$, $\partial U/\partial y|_{y=0} = 0$. In this case the outside potential distribution is found as an analytical continuation of the function $f(z)$ on the complex plane

$$U(y,z) = Re f(z+iy) = \frac{1}{2}[f(z+iy) + f(z-iy)] . \qquad (2.15)$$

This expression, on known properties of functions of complex variables, obeys the Laplace equation and satisfies the initial conditions.

Function $f(z+jy)$ can be expanded in a Taylor series:

$$f(z+iy) = f(z) + iyf'(z) + \frac{(iy)^2}{2!}f''(z) + \ldots + \frac{(iy)^n}{n!}f^{(n)}(z) , \qquad (2.16)$$

where primes mean differentiation with respect to the variable z.

Then, for potential $U(z,y)$ one obtains:

$$U(z,y) = f(z) - \frac{y^2}{2!}f''(z) + \ldots + \frac{(-1)^n y^{2n}}{n!}f^{(2n)}(z) . \qquad (2.17)$$

This row is the convenient form of potential presentation. It permits the potential outside of the axis to be found if its distribution along the axis is known.

II. *Distribution of potential and its normal derivative is given along a straight line.*

As in the previous case this line is supposed to coincide with the real axis of the plane of the complex variable $\xi = z + iy$. In this case the Cauchy initial conditions are written as $U|_{y=0} = f_1(z)$ and $\partial U/\partial y|_{y=0} = f_2(z)$. It is necessary to note that the first of them can be given in equation form $\partial U/\partial y|_{y=0} = f_3(z)$.

Let us introduce the complex potential $W(\xi) = U(z,y) + iV(z,y)$. Its derivative can be written as follows:

$$\varphi(\zeta) = \frac{dW}{d\zeta} = \frac{\partial U}{\partial z} - i\frac{\partial U}{\partial y} .$$

It is seen that at the z-axis $(y = 0)$ the value $\varphi(\zeta)$ $\varphi(\zeta)$ is uniquely determined by the initial conditions written as $\varphi(\zeta)|_{y=0} = f_3(z) - if_2(z)$. Then in an arbitrary point of the complex plane we can find the derivative of the complex potential $dW/d\zeta$ by analytical continuation of functions f_3 and f_2:

$$\frac{dW}{d\zeta} = f_3(z+iy) - if_2(z+iy) = f_3(\zeta) - if_2(\zeta) . \qquad (2.18)$$

The complex potential W is found as:

$$W = \int [f_3(\zeta) - if_2(\zeta)] \, d\zeta + C . \qquad (2.19)$$

Then, the distribution of electrostatic potential corresponding to the given initial conditions is expressed as follows:

$$U = ReW = Re f_1(\zeta) + Im \int f_2(\zeta) \, d\zeta + C . \qquad (2.20)$$

III. *Distribution of potential and its normal derivative is given along a curve.*

We will suppose that this line lies in the plane of the complex variable $\zeta = z + iy$ and is presented in parametric form $\zeta = F_1(u) + iF_2(u)$, where u is a parameter.

The solution of this problem is reduced to solution of the previous one if we find the conformal transformation that maps the curve in question to the real axis of new complex plane $\psi = u + iv$. This can be done in the following way. Let us consider the function $\zeta = F_1(\psi) + iF_2(\psi)$. It is easy to see that this function maps the real axis of the plane ψ to the curve at plane ζ. Indeed, substitution in this equation $\psi = u$ gives $\zeta = F_1(u) + iF_2(u)$, that is, the parametric equation of the considered curve. Then, the inverse function allows the required transformation to be produced and the Cauchy initial conditions at the real axis of the complex plane ψ to be found.

Axially Symmetric Field

The distribution of potential and its normal derivative is supposed to be set up on the z-axis of the rectangular coordinate system x, y, z:

$$U = f(z) \quad \text{and} \quad \partial U/\partial x = \partial U/\partial y = 0. \qquad (2.21)$$

It is required to find the solution of the Cauchy problem for the Laplace equation, which results in an axially symmetric potential distribution. In rectangular coordinates the Laplace equation has the form:

$$\frac{\partial^2 U}{\partial x^2} + \frac{\partial^2 U}{\partial y^2} + \frac{\partial^2 U}{\partial z^2} = 0 . \qquad (2.22)$$

It is possible to be convinced by direct substitution that this equation has a partial solution $U(x, y, z) = f(ix \cos\gamma + iy \sin\gamma + z)$, with γ being an arbitrary constant. At the z-axis ($x = y = 0$) it is reduced to $U = f(z)$ and, therefore, satisfies the first initial condition.

2.3 Calculation of Electrostatic Field: Cauchy Problem

Introducing cylindrical coordinates r, θ, z we can write $x = r\cos\theta$, $y = r\cos\theta$. Substitution of these expressions in the above solution gives:

$$U = f[z + ir\cos(\gamma - \theta)] . \tag{2.23}$$

This equation is considered to be an analytical continuation of function $f(z)$.

However, the potential given by (2.23) depends on angle θ. An axially symmetric solution can be obtained as a superposition of such solutions [1],

$$U = \frac{1}{2\pi} \int_{\theta}^{\theta+2\pi} f[z + ir\cos(\gamma - \theta)]d\gamma .$$

Indeed, on integration the argument γ of the trigonometric function $\cos(\gamma - \theta)$ is changed in the range $0 - 2\pi$, being independent of the value of θ. This means that the value of the integral is the same for arbitrary values of θ.

Introducing the angle $\alpha = \gamma - \theta$ one obtains:

$$U = \frac{1}{2\pi} \int_0^{2\pi} f(z + ir\cos\alpha)d\alpha . \tag{2.24}$$

Evidently, this solution satisfies the first initial Cauchy condition $U = f(z)$ at $r = 0$. The derivative $\partial U/\partial r$ is found to be:

$$\frac{\partial U}{\partial r} = \frac{i}{2\pi} \int_0^{2\pi} f'(z + ir\cos\alpha) \cos\alpha \, d\alpha \Big|_{r=0} = 0 .$$

At $r = 0 \, \partial U/\partial r = 0$, therefore, (2.24) also satisfies the second Cauchy condition.

Substitution in (2.24) Taylor's series expansion of function $f(z + ir\cos\alpha)$ and integration gives:

$$U = f(z) - \frac{r^2}{2^2} f''(z)| + ... + \frac{(-1)^n}{(n!)^n} \left(\frac{r}{2}\right)^{2n} f^{(2n)}(z) . \tag{2.25}$$

This series is known as the series expansion of the axially symmetric potential and is widely used in charged particle optics.

Example 2.4. It is required to find the geometry and potential of electrodes that produce a given periodical potential distribution at the axis of symmetry $U = U_0 + U_m \cos 2\pi z/L$. Since at the axis of symmetry $\partial U/\partial r = 0$, then the solution of this problem is reduced to solution of the Cauchy problem for the Laplace equation, with the two last equations being taken as the initial conditions at the axis of symmetry. Equation (2.24) can be used for this calculation:

$$U = U_0 + \frac{U_m}{2\pi} \int_0^{2\pi} \cos(Z + iR\cos\alpha) \, d\alpha ,$$

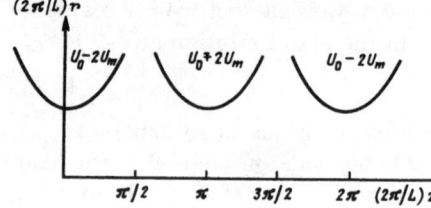

Fig. 2.4. System of electrodes producing periodical potential distribution

where $Z = \frac{2\pi}{L}z$ and $R = \frac{2\pi}{L}r$ are normalized variables.
This gives:

$$U = U_0 + \frac{U_m}{2\pi}\left[\int_0^{2\pi} \cos Z \cos(iR\cos\alpha)\,d\alpha - \int_0^{2\pi} \sin Z \sin(iR\cos\alpha)\,d\alpha\right].$$

The second integral in this equation is equal to zero.
Thus,

$$U = U_0 + U_m \cos Z \frac{1}{2\pi}\int_0^{2\pi} \operatorname{ch}(R\cos\alpha)\,d\alpha = U_0 + U_m \cos Z \frac{2}{\pi}\int_0^{2\pi} \operatorname{ch}(R\cos\alpha)\,d\alpha.$$

Since

$$\frac{2}{\pi}\int_0^{\pi/2} \operatorname{ch}(R\cos\alpha)\,d\alpha = I_0(R),$$

then the ultimate result will be:

$$U = U_0 + U_m I_0\left(\frac{2\pi}{L}r\right)\cos\frac{2\pi}{L}z. \tag{2.26}$$

At the z-axis $(r=0)$ this is reduced to:

$$U = U_0 + U_m \cos\frac{2\pi}{L}z.$$

The obtained result allows calculation of the potential distribution outside the axis and the form of electrodes producing the required axis potential distribution to be found (Fig. 2.4).

Solution of Cauchy Problem for Poisson Equation

In the cylindrical coordinate system r, θ, z for the case of axial symmetry the Poisson equation is written as:

$$\frac{1}{r}\frac{\partial}{\partial r}\left(r\frac{\partial U}{\partial r}\right) + \frac{\partial^2 U}{\partial z^2} = -\frac{\rho(r,z)}{\varepsilon_0}.$$

The problem is set up in the following way. The potential distribution and value of its normal derivative at axis of symmetry $U = U(z)$ and $\partial U/\partial r = 0$

2.3 Calculation of Electrostatic Field: Cauchy Problem

as well as the space-charge density distributions $\rho(r, z)$ are supposed to be known.

A solution of this problem can be found by power series expansion of the potential $U(r, z)$ and space-charge density $\rho(r, z)$:

$$U(r, z) = f_0(z) + f_2(z) r^2 + f_4(z) r^4 + \ldots + f_{2n}(z) r^{2n}$$
$$\rho(r, z) = \rho_0(z) + \rho_2(z) r^2 + \rho_4(z) r^4 + \ldots + \rho_{2n}(z) r^{2n}.$$

Substitution of these rows into the Poisson equation and equating the terms in its left and right parts having the same power of r leads to the solution in the form of a power series [77]:

$$U(r, z) = f_0(z) - \frac{r^2}{4} \left[f_0''(z) + \frac{\rho_0(z)}{\varepsilon_0} \right]$$
$$+ \frac{r^4}{16} \left[\frac{1}{4} \left(f_0^{IV}(z) + \frac{\rho_0''(z)}{\varepsilon_0} \right) - \frac{\rho_2(z)}{\varepsilon_0} \right] + \ldots, \quad (2.27)$$

where $f_0(z)$ – function of the axial potential distribution, $f_0''(z)$ and $f_0^{IV}(z)$ – the second and the fourth derivatives of $f_0(z)$ on the z-coordinate.

In practice, for computation and design of charge particle optical systems it is expedient to present the solution of the problem under consideration as a superposition of two functions [78]:

$$U(r, z) = U_1(r, z) + U_2(r, z).$$

One of them $U_1(r, z)$ is found from solution of the Cauchy problem for the homogeneous equation:

$$\frac{1}{r} \frac{\partial}{\partial r} \left(r \frac{\partial U_1}{\partial r} \right) + \frac{\partial^2 U_1}{\partial z^2} = 0 \quad (2.28)$$

, with the following initial condition set up at the axis of symmetry: $U_1(z) = f_0(z)$, where $f_0(z)$ is the axial potential distribution, and $\partial U_1/\partial r = 0$.

Another one $U_2(r, z)$ is determined from solution of the Cauchy problem for the nonhomogeneous equation:

$$\frac{1}{r} \frac{\partial}{\partial r} \left(r \frac{\partial U_2}{\partial r} \right) + \frac{\partial^2 U_2}{\partial z^2} = -\frac{\rho(r, z)}{\varepsilon_0},$$

with initial conditions $U_2(z) = \text{const} = 0$ and $\partial U/\partial r = 0$ given at the axis of symmetry.

Solution of the Cauchy problem for these two types of equations has been considered above. In many cases, when the space-charge density variation along z-axis is slow enough, it allows neglection of the term $\partial^2 U/dz^2$ in the nonhomogeneous equation. Then, its solution is reduced to computation of the following integrals:

$$U_2(r,z) = \frac{1}{2\pi\varepsilon_0} \int_0^r \frac{q(r,z)}{r} dr$$

$$q(r,z) = -2\pi \int_0^r \rho(r,z) r dr \ . \tag{2.29}$$

If the space-charge cloud has a sharp boundary $R(z)$, so that at $r > R$, the following equations are obtained for potential $U_2(r,z)$ computation:

$$U_2(r,z) = \frac{1}{2\pi\varepsilon_0} \int_0^r \frac{q(r,z)}{r} dr, r \leq R$$

$$U_2(r,z) = U_2(R,z) + \frac{q(R,z)}{2\pi\varepsilon_0} \ln rR, r \geq R \ .$$

In these formulae: $U_2(R,z)$ – value of U_2 at the space-charge cloud boundary ($r = R$), which is found with help of (2.29), $q(R,z)$ is the charge per unit length of the space-charge cloud.

2.4 Computer Calculation of Electrostatic Fields

2.4.1 Method of Finite Difference

For solution of the system of finite-difference equation the method of iterations is used. There are several types of this method.

Liebmann's Method

Let function U satisfy the Poisson equation $\nabla^2 U = f$. The generalized finite-different equation, which is equivalent to this equation can be written as

$$U(i,k) = C_a U(i+1,k) + C_b U(i,k+1)$$
$$+ C_d U(i-1,k) + C_e U(i,k-1) + C_f f(i,k) \ .$$

Here, $U(i,k)$ is the potential of the central mesh node, being marked by indices i and k (Fig. 2.5) $U(i+1,k)$, $U(i,k+1)$, etc. are the potentials of the adjoining nodes; C_a, C_b, C_d, C_e, C_f are coefficients depending on the type of finite-difference equation, the form of mesh cells, node position, etc.

To start the iteration procedure the initial values of potential of the inner nodes are to be assigned. They can be arbitrary enough, for instance taken equal, to zero, but the closer they are to the true values the faster the iteration procedure will converge. Furthermore, the potential values of the first iteration $U^1(i,k)$ are found using the equation:

$$U^1(i,k) = C_a U^0(i+1,k) + C_b U^0(i,k+1)$$
$$+ C_d U^0(i-1,k) + C_e U^0(i,k-1) + C_f f(i,k) \ .$$

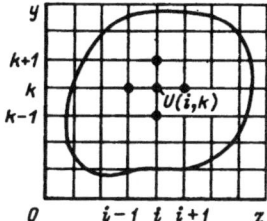

Fig. 2.5. Numbering of mesh nodes for iterative procedure

The potentials of the next iterations are computed by the following general equation:

$$U^n(i,k) = C_a U^{n-1}(i+1,k) + C_b U^{n-1}(i,k+1) \\ + C_d U^n(i-1,k) + C_e U^n(i,k-1) + C_f f(i,k) \; . \tag{2.30}$$

The computations are continued until results of the next iteration $U^n(i,k)$ are close enough to the result of the previous one $U^{n-1}(i,k)$ to provide the required accuracy:

$$\left| U^n(i,k) - U^{n-1}(i,k) \right| \leq \varepsilon \; ,$$

where ε is a small value, which is determined by the required accuracy of the potential computation.

Seidel's Procedure

This procedure differs from the previous one by an algorithm of node-potential computation.

Instead of (2.30) the following is used:

$$U^n(i,k) = C_a U^{n-1}(i+1,k) + C_b U^{n-1}(i,k+1) \\ + C_d U^n(i-1,k) + C_e U^n(i,k-1) + C_f f(i,k) \; .$$

It is seen that the equation for the new node potential $U^n(i,k)$ contains the old node potentials $U^{n-1}(i,k)$, as well as the potentials found in the current iteration U^n. To realize this procedure the definite order of node-potential computation is required. For instance, for the case shown at Fig. 2.5 this algorithm can be arranged to compute the potential of the nodes located on the vertical i-line in the direction from the bottom to the top, and then go to the next $i+1$ vertical line and so on. Seidel's iterations provide faster convergence of the iteration process.

Method of Overrelaxation

This considerably increases the speed of iteration convergence and uses the following equation for node-potential recalculation:

$$U^n(i,k) = \omega[C_a U^{n-1}(i+1,k) + C_b U^{n-1}(i,k+1) + C_d U^n(i-1,k)$$
$$+ C_e U^n(i,k-1) + C_f f(i,k)] + (1-\omega)U^{n-1}(i,k) .$$
(2.31)

where ω is the overrelaxation factor, the rate of convergence being critically dependent on ω. Usually, its value lies in the range 1–2. In many cases the values of ω about of 1.7–1.8 may be recommended.

Accuracy of Finite-Difference Method

There are two types of errors that limit the accuracy of calculation. The first of them is the result of replacement of the partial differential equation by its finite-difference equivalent. The second one arises due to the approximate solution, obtained by iterative methods.

The error of the first kind depends on the mesh pitch h, and for the square mesh is of order of $M_4 h^2 \rho^2$, where M_4 is the maximum value of the fourth partial derivative of potential U, ρ is the radius of a circle that encloses the region where the solution is found [42]. Therefore, the error is reduced as the number of mesh nodes is increased and the mesh pitch is reduced.

The error of the second kind is reduced on increasing the number of iterations. The number of iterations, which is required to reduce the error to an acceptable value depends on the mesh pitch and the type of iteration procedure. For Seidel's iteration it is of order of $1/h^2$ and for the overrelaxation process is of order of $1/h$.

Hence, for increasing accuracy the mesh pitch h is to be reduced, but it results in increasing the number of iterations and computation time.

2.4.2 Method of Integral Equations

This method sometimes is called the surface-charge method, as it exactly expresses the physical nature of the method. The main difficulty of the method consists of determination of the surface-charge density σ distribution. This concerns the necessity to solve the integral equation of the following type:

$$\int_C \sigma \psi dC = U_C - U_\rho ,$$

where U_c – potentials of electrodes, which are supposed to be known, U_ρ space-charge potential, which is calculated for points located at electrode contour C, ψ – the kernel of integral equation.

In the general case, numerical methods are to be used for its solution. For that electrode contour C is divided for n elementary segments h_i (Fig. 2.6) and the integral of the left part of (2.31) is replaced by a finite sum:

$$\int_C \sigma \psi dC = \sum_{i=1}^n \sigma_i h_i \psi_i ,$$
(2.32)

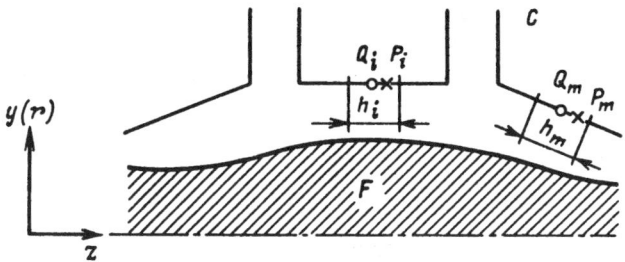

Fig. 2.6. For computation of electrostatic fields by integral equation method

where σ_i and ψ_i are average values of σ and ψ at segment h_i.

In the case of a two-dimensional field the product $\sigma_i h_i$ can be considered as a linear charge of value $q_i = \sigma_i h_i$, where h_i – length of a couture segment, σ_i – average surface-charge density.

In the case of an axially symmetric field it is a ring charge of value $q_i = 2\pi r_i h_i \sigma_i$, where r_i is the r-coordinate of the midpoint Q_i of segment h_i.

Hence these discrete charges located at point Q_i having coordinate z_i, y_i (r_i) replace the continuously distributed surface charge.

If we place at electrode contour n observation points P_m with coordinates z'_m, y'_m (r'_m), then it is possible to write the following system of algebraic equations, being equivalent to the integral equation (2.31):

$$\sum_{i=1}^{n} q_i \varphi_{mi} = U_m \ .$$

In these equations $U_m = U_{Cm} - U_{\rho m}$ – potential at points of observation, which consists of two parts: U_{Cm} – potential at the electrode contour at points P_m and $U_{\rho m}$ – potential produced at these points by space charge; φ_{mi} – Green functions of free space written for point locations of discrete charges Q_i and the observation points P_m: two-dimensional field

$$\varphi_{mi} = \frac{1}{2\pi\varepsilon_0} \ln \frac{1}{\sqrt{(y'_m - y_i)^2 + (z'_m - z_i)^2}} \ ,$$

axially symmetric field

$$\varphi_{mi} = \frac{1}{2\pi^2 \varepsilon_0} \ln \frac{K(t_i)}{\sqrt{(r'_m + r_i)^2 + (z'_m - z_i)^2}} \ ,$$

$K(t_i)$ – total elliptic integral of the first kind of argument t_i

$$t_i^2 = \frac{4 r'_m r_i}{(r'_m + r_i)^2 (z'_m - z_i)^2} \ .$$

Values of space-charge potentials $U_{\rho m}$ are calculated with the help of the following equations:

two-dimensional field

$$U_{\rho m} = \frac{1}{2\pi\varepsilon_0} \int_F \rho \ln \frac{1}{\sqrt{(y'_m - y_0)^2 + (z'_m - z_0)^2}} dF,$$

axially symmetric field

$$U_{\rho m} = \frac{1}{\pi\varepsilon_0} \int_F \frac{\rho r_0 K(t_0) \, dF}{\sqrt{(r'_m + r_o)^2 + (z'_m + z_0)^2}}.$$

In these equations integration is performed over the surface F occupied by a space-charge distribution that is supposed to be known, $K(t_0)$ is the total elliptic integral of the first kind of argument t_{m0}

$$t_0^2 = \frac{4r'_m r_0}{(r'_m + r_0)^2 + (z'_m - z_0)^2},$$

where r_0, z_0 – coordinates of current points of space-charge location.

The obtained system of linear algebraic equations is equivalent to the following matrix equation:

$$\begin{Vmatrix} \varphi_{11} \cdot \varphi_{1i} \cdot \varphi_{1j} \cdot \varphi_{1m} \cdot \varphi_{1n} \\ \cdots\cdots\cdots\cdots\cdots\cdots\cdots \\ \varphi_{i1} \cdot \varphi_{ii} \cdot \varphi_{ij} \cdot \varphi_{im} \cdot \varphi_{in} \\ \cdots\cdots\cdots\cdots\cdots\cdots\cdots \\ \varphi_{j1} \cdot \varphi_{ji} \cdot \varphi_{jj} \cdot \varphi_{jm} \cdot \varphi_{jn} \\ \cdots\cdots\cdots\cdots\cdots\cdots\cdots \\ \varphi_{m1} \cdot \varphi_{mi} \cdot \varphi_{mj} \cdot \varphi_{mm} \cdot \varphi_{mn} \\ \cdots\cdots\cdots\cdots\cdots\cdots\cdots \\ \varphi_{n1} \cdot \varphi_{ni} \cdot \varphi_{nj} \cdot \varphi_{nm} \cdot \varphi_{nn} \end{Vmatrix} \begin{Vmatrix} q_1 \\ \cdot \\ q_i \\ \cdot \\ q_j \\ \cdot \\ q_m \\ \cdot \\ q_n \end{Vmatrix} = \begin{Vmatrix} U_1 \\ \cdot \\ U_i \\ \cdot \\ U_j \\ \cdot \\ U_m \\ \cdot \\ U_n \end{Vmatrix}. \qquad (2.33)$$

Equation (2.33) can be written in the following short form:

$$\|\varphi_{mi}\| \|q_i\| = \|U_m\|. \qquad (2.34a)$$

It can be solved in respect of discrete charges q_i by inversion of matrix $\|\varphi_{mi}\|$, which can be done by standard computer codes of matrix inversion:

$$\|q_i\| = \|\varphi_{mi}\|^{-1} \|U_m\|. \qquad (2.34b)$$

With diagonal terms $\varphi_{ii}, \varphi_{jj}, \varphi_{mm}$ being predominant the matrix is well conditioned. This provides stability of its inversion. At the same time it should be noted that the value of discrete charge q_i essentially depends on the mutual position of points Q_i and P_i (at a segment h_i). It can be shown that there exists an optimal distance between these points $\delta \approx 0.17 h_i$ [10].

As values of q_i are found, the potential as well as components of electric field $\vec{E}$ are easily determined.

2.4.3 Method of Green Function

This method can be used for solution of different electrostatic problems but it is the most efficient for computation of the space charge-field. As has been mentioned above, its computation is reduced to calculation of the integral

$$U_\rho(M) = \int_V \rho(M_0) G(M, M_0) \, dV \, .$$

As space charge is a function of time $\rho(M_0, t)$ one obtains in the quasi static approximation:

$$U_\rho(M, t) = \int_V \rho(M_0, t) G(M, M_0) \, dV \, ,$$

with Green function $G(M, M_0)$ being independent of time.

This means that $G(M, M_0)$ being found for a region of interest can be used for computation of the space-charge field produced by space charge changing with time.

The analytical equation for $G(M, M_0)$ is known only for a limited number of simple regions like circle, cylinder, coaxial cylinder, etc.

The method of Green function has been successfully used for computation of electron motion in a magnetron [49] and a traveling-wave tube [50]. In both cases the real complicated boundary of the region was replaced by a cylindrical boundary, with an analytical Green function being used.

For space-charge field computation in a region with complicated boundaries the method of discrete Green function was developed [51]. A mesh is introduced in the region of field computation (Fig. 2.7). The potential in mesh nodes can be computed by the following equations:

two-dimensional field

$$U_\rho = \sum_k \rho_k G_{sk} \Delta F_k \, ,$$

axially symmetric field

$$U_{\rho s} = 2\pi \sum_k \rho_k G_{sk} r_k \Delta F_k \, ,$$

where $U_{\rho s}$ – potential at mesh node s, ρ_k – average space-charge density in vicinity of mesh node k at elementary area dF_k, G_{sk} – discrete Green function calculated for s and k nodes, summation is performed on all k nodes.

Values of the discrete Green function G_{sk} for all mesh nodes are to be computed in advance, with the symmetry Green function $G_{sk} = G_{ks}$ being taken into account.

An effective method of numerical computation of this function was suggested in [10, 51]. It is based on the method of integral equations and realized in the following way. Let us place at node k a unit charge $q = 1$, then the

2 Methods of Fields Calculation

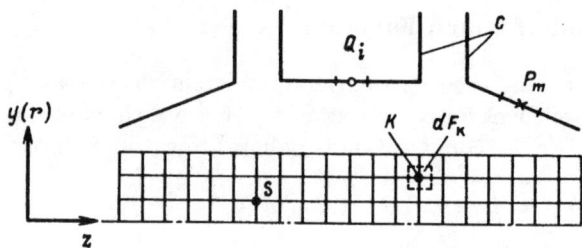

Fig. 2.7. Grid for calculation of discrete Green function

Green function for nodes s and k – G_{sk} is found as a potential at nodes s created by charged q and by surface charge induced at the boundary electrodes:

$$U_{sk} = U'_{sk} + U''_{sk}$$

$$U'_{sk} = q\varphi_{sk}, \qquad U''_{sk} = \sum_{i=1}^{n} q_i \varphi_{si}.$$

In these equations φ_{sk} is the Green function of free space written for points s and k, φ_{si} is that written for points s and i, the last point is the central point of segment h_i of the electrode boundary, where a discrete induced charge q_i is located. The value of q_i is determined by the integral equation method with the help of (2.34b). Potentials U_m in the right part of this equation are found from equality $U_m = -U_{\rho m}$, where $U_{\rho m}$ is the potentials created in observation point U_m by the unit charge q, located in point k, $U_{\rho m} = q\varphi_{mk}$ (note that $\varphi_{mk} = \varphi_{km}$).

The ultimate equation for discreet function G_{sk} can be written in the following matrix form [10]:

$$G_{sk} = \varphi_{sk} - \|\varphi_{si}\| \, \|\varphi_{mi}\|^{-1} \, \|\varphi_{mk}\|, \qquad (2.35)$$

where $\|\varphi_{mi}\|^{-1}$ – matrix inverse of $\|\varphi_{mi}\|$; $\|\varphi_{si}\|$ – matrix row of matrix elements written as the Green's function of free space for node s and points of location of induced charges, $\|\varphi_{mk}\|$ – matrix column with matrix elements written as a Green function for points of observation P_m located at the electrode boundary and mesh node k. It should be noted that in the above equation the first index of subscripts denotes the point of observation, while the second one denotes the point of the discrete charge location.

2.5 General Equations of Magnetic Field

The following system of equations is used for static magnetic field calculation:

Differential form   integral form

$$\operatorname{rot} \vec{H} = \vec{\delta} \qquad \oint_l \vec{H} d\vec{l} = I \qquad (2.36)$$

$$\operatorname{div} \vec{B} = 0 \qquad \oint_S \vec{B} d\vec{S} = 0, \qquad (2.37)$$

where $\vec{B}$ and $\vec{H}$ are vectors of magnetic induction and magnetic field strength, respectively and I and $\vec{\delta}$ are electric current and electric current density, respectively.

These equations are added by so-called material equations, which determine the local connection between $\vec{B}$ and $\vec{H}$.

For vacuum:

$$\vec{B} = \mu_0 \vec{H},$$

where μ_0 – permeability of vacuum.

For magnetic materials:

$$\vec{B} = \mu_0 \left(\vec{H} + \vec{J} \right), \qquad (2.38)$$

where $\vec{J}$ – vector of intensity of magnetization of material.

In the particular case of soft magnetic materials magnetization $\vec{J}$ and magnetic field strength are connected by the ratio:

$$\vec{J} = \chi \vec{H}, \qquad (2.39)$$

where χ – magnetic susceptibility.

Substitution of (2.39) in (2.38) gives the following form of the material equation:

$$\vec{B} = \mu_0 (1 + \chi) \vec{H},$$

or

$$\vec{B} = \mu \vec{H}, \qquad (2.40)$$

where

$$\mu = \mu_0 (1 + \chi).$$

In the general case, values of μ, χ and J are some functions of $\vec{H}$, as a result of this the connection between $\vec{B}$ and $\vec{H}$ is nonlinear and, sometimes, a hysteretic one.

Experimental B–H curves are usually used for its description.

2.5.1 Magnetic Vector Potential

The initial system of (2.36)–(2.38) can be reduced to one differential equation if we introduce vector magnetic potential $\vec{A}$, which is connected to the vector of magnetic induction by the relation:

$$\vec{B} = \operatorname{rot} \vec{A}, \tag{2.41}$$

and satisfies the equation:

$$\operatorname{div} \vec{A} = 0.$$

Taking into account (2.38) we can write:

$$\operatorname{rot} \vec{B} = \mu_0 \left(\vec{\delta} + \operatorname{rot} \vec{J} \right).$$

Substitution of (2.41) into the last equation gives:

$$\operatorname{rot} \operatorname{rot} \vec{A} = \mu_0 \left(\vec{\delta} + \operatorname{rot} \vec{J} \right). \tag{2.42}$$

Using the vector equality $\operatorname{rot} \operatorname{rot} \vec{A} = \operatorname{grad} \operatorname{div} \vec{A} - \nabla^2 \vec{A}$ and the condition $\operatorname{div} \vec{A} = 0$ one finds:

$$\nabla^2 \vec{A} = -\mu_0 \left(\vec{\delta} + \operatorname{rot} \vec{J} \right), \tag{2.43}$$

where ∇^2 – Laplace operator.

Values $\vec{\delta}$ and $\operatorname{rot} \vec{J}$ in the right part of (2.43) are considered to be sources of magnetic field. For further analysis it is convenient to consider $\operatorname{rot} \vec{J}$ as equivalent current density $\vec{j} = \operatorname{rot} \vec{J}$. Then, (2.43) can be written as:

$$\nabla^2 \vec{A} = -\mu_0 \left(\vec{\delta} + \vec{j} \right), \tag{2.44}$$

$$\vec{j} = \operatorname{rot} \vec{J}. \tag{2.45}$$

If vectors $\vec{B}$ and $\vec{H}$ are connected by (2.40), the following equations can be obtained for vector magnetic potential:

$$\operatorname{rot} \left(\nu \operatorname{rot} \vec{A} \right) = \vec{\delta},$$

$$\operatorname{rot} \operatorname{rot} \vec{A} = \mu \vec{\delta} + \left[\nu \operatorname{grad} \mu \times \operatorname{rot} \vec{A} \right], \tag{2.46}$$

or

$$\nabla^2 \vec{A} = -\mu \vec{\delta} - \left[\nu \operatorname{grad} \mu \times \operatorname{rot} \vec{A} \right], \tag{2.47}$$

where $\nu = 1/\mu$.

The fundamental difference between (2.44) and (2.46) consists of the method of account of material magnetization. In (2.44) by use of equivalent (virtual) current density $\vec{j} = \operatorname{rot} \vec{J}$, in (2.47) by coefficient $\nu = 1/\mu$.

2.5 General Equations of Magnetic Field

Solution of the equation for vector potential (2.44), by analogy with Poisson's equation, is written as:

$$\vec{A} = \frac{\mu_0}{4\pi} \int_V \frac{\vec{\delta}+\vec{j}}{R} dV , \qquad (2.48)$$

where integration is performed over the volume occupied by the sources of magnetic field $\vec{\delta}$ and $\vec{j}$.

Equations for Planar Fields

In the rectangular coordinate system x, y, z, assuming that the field does not depend on the x coordinate, and vectors $\vec{A}, \vec{\delta}$ and $\vec{j}$ have only x-components, one obtains from (2.44) and (2.46):

$$\frac{\partial^2 A}{\partial y^2} + \frac{\partial^2 A}{\partial z^2} = -\mu_0 (\delta + j) ,$$

$$\frac{\partial}{\partial y}\left(v\frac{\partial A}{\partial y}\right) + \frac{\partial}{\partial z}\left(v\frac{\partial A}{\partial z}\right) = -\delta ,$$

where $j = \text{rot}_x \vec{J}$ – x-component of vector rot $\vec{J}$.

Axially Symmetric Fields

In the cylindrical coordinate system r, θ, z, assuming that vectors $\vec{A}, \vec{\delta}, \vec{j}$ have only one θ-component and the field is independent of the θ-coordinate, it is possible to obtain the following equations for the vector magnetic potential:

$$\frac{\partial^2 A}{\partial r^2} + \frac{\partial}{\partial r}\left(\frac{A}{r}\right) + \frac{\partial^2 A}{\partial z^2} = -\mu_0 (\delta + j) \qquad (2.49)$$

$$\frac{\partial}{\partial r}\left[\nu\left(\frac{\partial A}{\partial r} + \frac{A}{r}\right)\right] + \frac{\partial}{\partial z}\left(\nu\frac{\partial A}{\partial z}\right) = -\delta , \qquad (2.50)$$

where $j = \text{rot}_\theta \vec{J}$ – θ-component of vector rot $\vec{J}$.

The vector magnetic potential in a point with coordinates r, z can be expressed through the magnetic flux ψ enclosed by the circle of radius r:

$$A(r, z) = \psi/2\pi r ,$$

where $\psi = \int\limits_0^r B_z 2\pi r dr$.

Substitution of the above equation for $A(r, z)$ in (2.49) and (2.50) gives:

$$\frac{\partial^2 \psi}{\partial r^2} - \frac{1}{r}\frac{\partial \psi}{\partial r} + \frac{\partial^2 \psi}{\partial z^2} = -2\pi r \mu_0 (\delta + j) , \qquad (2.51)$$

$$r\frac{\partial}{\partial r}\left(\frac{v}{r}\frac{\partial \psi}{\partial r}\right) + \frac{\partial}{\partial z}\left(v\frac{\partial \psi}{\partial z}\right) = -2\pi r \delta . \qquad (2.52)$$

Components of magnetic induction are expressed through ψ by the equations:

$$B_r = -\frac{1}{2\pi r}\frac{\partial \psi}{\partial z}, \qquad B_z = \frac{1}{2\pi r}\frac{\partial \psi}{\partial r}. \tag{2.53}$$

2.5.2 Scalar Magnetic Potential

The system of initial magnetic field equation (2.36)–(2.38) is reduced to one partial differential equation for scalar magnetic potential U, if magnetic fields are produced by permanent magnets only, with current density $\vec{\delta}$ being equal to zero.

Equation (2.36), now written as rot $\vec{H} = 0$, will be fulfilled if we express $\vec{H}$ through the gradient of magnetic potential U:

$$\vec{H} = -\operatorname{grad} U. \tag{2.54}$$

Substitution of (2.38) into (2.37) with account of (2.54) gives div grad $\vec{U}$ = div $\vec{J}$, or $\nabla^2 U = \operatorname{div} \vec{J}$. The last equation by analogy with the electrostatic field, can be written in the following form:

$$\nabla^2 U = \rho, \tag{2.55}$$

where $\rho = \operatorname{div} \vec{J}$ is considered to be the density of equivalent "magnetic charge".

For soft magnetic materials as $\vec{B} = \mu \vec{H}$ for calculation of ρ the following equations can be obtained:

$$\rho = \operatorname{div} \vec{J} = \operatorname{div}\left(\frac{\vec{B}}{\mu_0} - \vec{H}\right) = \operatorname{div}\left(\frac{\vec{B}}{\mu_0} - \frac{\vec{B}}{\mu}\right) = -\operatorname{div}\frac{\vec{B}}{\mu}.$$

Using the equation of vector analysis this equation can be transformed to give:

$$\rho = \vec{B}\frac{\operatorname{grad}\mu}{\mu^2}. \tag{2.56}$$

In regions where $\mu = \operatorname{const}$, $\operatorname{grad}\mu = 0$ and $\rho = 0$, the magnetic potential obeys the Laplace equation:

$$\nabla^2 U = 0. \tag{2.57}$$

Series Expansion of Axially Symmetric Magnetic Field

The distribution of magnetic induction at the axis of symmetry $B_{z0} = \varphi(z)$ is supposed to be known. Since $B_z = -\partial U/\partial z$, this means that the axial magnetic potential distribution $U = f(z)$ is also known. With the condition of the axis of symmetry $\partial U/\partial r = 0$ being taken into account, determination of the potential outside of the axis is reduced to solution of the Cauchy problem for (2.57).

Its solution has been considered in Sect. 2.3.1 and can be presented by the series:

$$U(r,z) = f(z) - \frac{r^2}{2^2}f''(z) + \ldots + \frac{(-1)^n}{(n!)^2}\left(\frac{r}{2}\right)^{2n} f^{(2n)}(z). \quad (2.58)$$

Then, for the components of magnetic induction one obtains:

$$B_r = -\frac{\partial U}{\partial r} = \frac{r}{2}f''(z) - \ldots - \frac{(-1)^n}{(n-1)!n!}\left(\frac{r}{2}\right)^{2n-1} f^{(2n)}(z)$$

$$B_z = -\frac{\partial U}{\partial z} = -f'(z) + \frac{r^2}{2^2}f'''(z) - \ldots - \frac{(-1)^n}{(n!)^2}\left(\frac{r}{2}\right)^{2n} f^{(2n+1)}(z),$$

or as $f'(z) = -\varphi(z) = -B_{z0}$ and, in general, $f^{(2n)}(z) = -\varphi^{(2n-1)}(z) = -B_{z0}^{(2n-1)}$ the above equations are written as:

$$B_z = B_{z0} - \frac{r^2}{2^2}B''_{z0} + \ldots + \frac{(-1)^n}{(n!)^2}\left(\frac{r}{2}\right)^{2n} B_{z0}^{(2n)} \quad (2.59)$$

$$B_r = -\frac{r}{2}B'_{z0} + \frac{r^3}{2\cdot 2^3}B'''_{z0} - \ldots - \frac{(-1)^n}{(n-1)!n!}\left(\frac{r}{2}\right)^{2n-1} B_{z0}^{(2n-1)}, \quad (2.60)$$

The magnetic flux $\psi = 2\pi \int\limits_0^r B_z r dr$ can be also presented by a series.

2.6 Numerical Calculation of Magnetic Field

Calculation of magnetic fields, produced by sources placed in space with homogeneous magnetic permeability $\mu = \mu_0 = $ const, can be done with the help of (2.48). This is reduced to computation of volume integrals. Let us consider the calculation of axially symmetric fields produced by solenoids and ring-shaped permanent magnets.

2.6.1 Field of a Solenoid

The geometry of a solenoid is shown in Fig. 2.8. A square mesh divides it into a set of elementary ring volumes, which results in an elementary current $\Delta I_k = \delta_k \Delta S_k$, where ΔS_k – cross section of the elementary volume, δ_k – average current density. Since ΔS_k is taken to be small enough, the field created by this elementary volume can be computed as the field of a single ring conductor, placed in the central point of ΔS_k with coordinates r_k, z_k. At a point of observation with coordinates r and z it is possible to write the following expression for a partial magnetic flux $\Delta \psi$, which is enclosed by the ring of radius r:

$$\Delta \psi_k = \mu_0 \Delta I_k \sqrt{rr_k}\frac{2}{t}\left[\left(1 - \frac{t^2}{2}\right)K(t) - E(t)\right], \quad (2.61)$$

where $t = 2\sqrt{rr_k}/\sqrt{(r+r_k)^2(z-z_k)^2}$, $K(t), E(t)$ – complete elliptic integrals of the first and second kind.

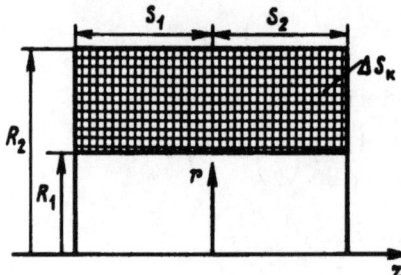

Fig. 2.8. Geometry of a solenoid

The total magnetic flux and components of magnetic induction at an observation point are found by summation partial magnetic fluxes:

$$\psi = \mu_0 \sum_k \Delta I_k \sqrt{rr_k} \frac{2}{t} \left[\left(1 - \frac{t^2}{2}\right) K(t) - E(t) \right] \tag{2.62}$$

$$B_r = -\frac{1}{2\pi r} \frac{\partial \psi}{\partial z} = -\frac{\mu_0}{2\pi} \sum_k \Delta I_k \frac{z - z_k}{r \left[(r + r_k)^2 + (z - z_k)^2\right]^{1/2}} f_1(r, r_k) \tag{2.63}$$

$$B_z = \frac{1}{2\pi r} \frac{\partial \psi}{\partial r} = \frac{\mu_0}{2\pi} \sum_k \Delta I_k \frac{1}{r \left[(r + r_k)^2 + (z - z_k)^2\right]^{1/2}} f_2(r, r_k) . \tag{2.64}$$

Functions $f_1(r, r_k)$ and $f_2(r, r_k)$ included in (2.63) and (2.64) are expressed as follows:

$$f_1(r, r_k) = K(t) - \frac{r_k^2 + r^2 + (z - z_k)^2}{(r - r_k)^2 + (z - z_k)^2} E(t) ,$$

$$f_2(r, r_k) = K(t) + \frac{r_k^2 - r^2 - (z - z_k)^2}{(r - r_k)^2 + (z - z_k)^2} E(t) .$$

Then for the points at the axis of symmetry one obtains:

$$B_z = \frac{\mu_0}{2} \sum_k \frac{\Delta I_k r_k^2}{\left[r_k^2 + (z - z_k)^2\right]^{3/2}} . \tag{2.65}$$

2.6.2 Field of a Ring-Shaped Magnet

A ring-shaped magnet (Fig. 2.9) is supposed to be magnetized in the axial direction. The vector of magnetization $\vec{J}$ has only one z-component J, which is assumed to be uniform over all the magnet volume $J = \text{const}$. The value

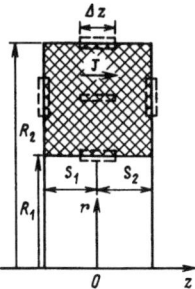

Fig. 2.9. Ring-shaped permanent magnet

of J depends on the magnetic material and the position of the operating point at the demagnetization curve. A method for its determination will be considered later.

As has been noted in Sect. 2.5 the magnetization can be accounted for by introducing equivalent (virtual) currents with density $\vec{j} = \operatorname{rot} \vec{J}$. The projection of this equation on some direction $\vec{n}$ is $j_n = \operatorname{rot}_n \vec{J}$. As is known from vector analysis $\operatorname{rot}_n J$ can be expressed as:

$$\operatorname{rot}_n \vec{J} = \oint_l \vec{J} d\vec{l} / \Delta S \;,$$

where integration is performed on a contour l, enclosing the elementary surface ΔS, with $\vec{n}$ being normal to this surface.

The last equation gives the following algorithm of calculation of virtual current-density components. Elementary oriented surfaces are considered in the magnet volume. Calculation of the circulation integral on the contours enclosing these surfaces allows the values of normal components of vector magnetization J_n to be found. Dashed lines in Fig. 2.9 show several such elementary surfaces. They lie in the meridian plane of the ring magnet. Computation of circulation integrals over the contours enclosing these surfaces gives θ-components of the virtual current j_θ. They are expressed as follows: $j_\theta = \operatorname{rot}_\theta \vec{J} = 0$ – at the interior points of the magnet and points at the end magnet surfaces, $j_\theta = \operatorname{rot}_\theta \vec{J} = \pm J \Delta z / \Delta S$ – at points inside and outside the cylindrical magnet boundaries.

Considering elementary surfaces oriented normal to r- and z-directions one finds that there are no components of $\operatorname{rot} \vec{J}$ of those directions. Therefore, in the case under consideration virtual currents have only a j_θ-component and calculation of the magnetic field of a uniformly magnetized ring-shaped magnet is reduced to computation of the magnetic field of two ring current layers that flow inside and outside the cylinder magnet boundaries. These currents have the same linear densities and opposite directions $\tau = \pm j_\theta \Delta S / \Delta z = \pm J$.

The current layers are divided for a number of small parts of extension Δz_k with elementary currents $\Delta I_k = \tau \Delta z_k$. Then, using (2.63) and (2.64) one obtains the following expressions for components of magnetic induction:

$$B_r = \frac{\mu_0}{2\pi} \sum_k \Delta I_k \frac{z - z_k}{r\left[(r+R_2)^2 + (z-z_k)^2\right]^{1/2}} f_1(r, R_2, z, z_k)$$

$$- \frac{\mu_0}{2\pi} \sum_k \Delta I_k \frac{z - z_k}{r\left[(z+R_1)^2 + (z-z_k)^2\right]^{12}} f_1(r, R_1, z, z_k)$$

$$B_z = \frac{\mu_0}{\pi} \sum_k \Delta I_k \frac{1}{\left[(r+R_1)^2 + (z-z_k)^2\right]^{1/2}} f_2(r, R_1, z, z_k)$$

$$- \frac{\mu_0}{2\pi} \sum_k \Delta I_k \frac{1}{\left[(z+R_2)^2 + (z-z_k)^2\right]^{1/2}} f_2(r, R_2, z, z_k) \,.$$

The value of magnetization in the operating points can be approximately found as [52]:

$$J \approx \frac{1}{2}(J_r + |H_{CB}|) \,, \tag{2.66}$$

where J_r – remanent magnetization, H_{CB}– coercive force on induction for the given magnetic materials.

Analysis of a ring-shaped magnet uniformly magnetized in the radial direction shows that the magnetic field of this magnet can be found as the field of the two ring current layers flowing at the end magnet surfaces, with linear current density being equal to $\tau = \pm J$.

2.6.3 Technique of Solution of Nonlinear Magnetic Problems

Designing of charge particle optical systems requires, as a rule, computation of magnetic fields in nonuniform magnetic media. A typical example of such systems is magnetic lines (Fig. 2.10) formed by ring-shaped permanent magnet (1) and soft iron magnetic pole pieces (2). The magnetic properties of magnetic media such as magnetic permeability, susceptibility and magnetization depend on the intensity of magnetic field. Because of this the problem of magnetic-field computation becomes a nonlinear one. For its solution the method of successive approximation is usually used. Its application reduces

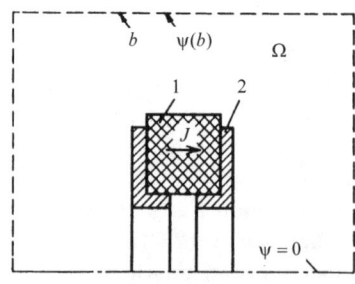

Fig. 2.10. Permanent magnet with shields

2.6 Numerical Calculation of Magnetic Field

the solution of the initial nonlinear problem to a sequence of linear problems with fixed properties of magnetic media. The method of successive approximation usually includes finding an initial approximation and its subsequent refinement by some iterative procedure.

As the concept of equivalent (virtual) current is used for magnetic-field computation, an iterative process in accordance with the following scheme is used for solution of (2.44):

$$\nabla^2 \vec{A}^{(n+1)} = -\mu_0 \left(\vec{\delta} + \vec{j}^{(n)} \right),$$

where $\vec{A}^{(n+1)}$ – vector magnetic potential of the current $(n+1)$-th approximation, $\vec{j}^{(n)}$ – density of virtual currents of the previous n-th approximation.

In the case, when magnetic field is described by (2.46), the following expression is used for the iterative procedure:

$$\text{rot}\left(v^{(n)} \text{ rot } \vec{A}^{(n+1)}\right) = \vec{\delta},$$

here $\nu^{(n)} = 1/\mu^{(n)}$ is determined from the field of the previous approximation.

In order to improve the convergence of the iterative process values of $\vec{j}^{(n)}$ and $\nu^{(n)}$ is to be computed with use of the relaxation equations:

$$\vec{j}^{(n)} = \omega \vec{\tilde{j}}^{(n)} + (1-\omega)\vec{j}^{(n-1)},$$

$$v^{(n)} = \omega \tilde{v}^{(n)} + (1-\omega) v^{(n-1)},$$

where $\vec{\tilde{j}}^{(n)}$ – value of $\vec{j}$ determined from the field of the n-th approximation, $\vec{j}^{(n-1)}$ – virtual current density of the $(n-1)$-th approximation, ω – relaxation factor, the superscript of ν has the same physical meaning.

The above process is known as a simple iteration process. It has relatively slow convergence.

A method of successive approximation, which provides for faster convergence, known as the Newton–Kontorovich method, is carried out with the following scheme [54]:

$$L\left(\vec{A}^{(n)}\right) + L'_{A^{(n)}}\left(\vec{A}^{(n+1)} - \vec{A}^{(n)}\right) = 0$$

where $L(\vec{A}) = \text{rot}(v \text{ rot } \vec{A}) - \vec{\delta} = 0$ – operator of the problem; $L'_{A^{(n)}}$ – Frechet derivative of the operator $L(\vec{A})$ in the point $A^{(n)}$.

The drawback of the method consists in that the iterative process converges only under conditions of a "good" initial approximation. To obtain such an initial approximation Koshelev's iteration process can be used [53, 55]:

$$\nabla^2 \vec{A}^{(n+1)} = \nabla^2 \vec{A}^{(n)} - \varepsilon L\left(\vec{A}^n\right),$$

where ε is a positive constant. This process converges starting from an arbitrary smooth approximation.

Let us consider the technique of solution of the nonlinear magnetic problem on computation of axially symmetric fields.

The following system of equations describes the magnetic field:

$$\frac{\partial^2 \psi}{\partial r^2} - \frac{1}{r}\frac{\partial \psi}{\partial r} + \frac{\partial^2 \psi}{\partial z^2} = -2\pi r \mu_0 (\delta + j), \qquad (2.67)$$

$$j = \operatorname{rot}_\theta \vec{J},$$

where ψ – magnetic flux enclosed by circle of radius r, $\vec{J}$ – vector of magnetization, $\operatorname{rot}_\theta \vec{J}$ – azimuth component of vector $\operatorname{rot} \vec{J}$.

For the solution the partial differential equation (2.67) is replaced by a finite-different equation, which, for a mesh with square cells, is written as

$$\psi_1 + \psi_2 + \psi_3 + \psi_4 - 4\psi_0 - \frac{h}{2r_0}(\psi_2 - \psi_1) = -\mu_0 h^2 2\pi r_0 (\delta + j). \qquad (2.68)$$

The solution is carried out at the region Ω (Fig. 2.10) with the following boundary conditions: $\psi = 0$ at the axis of symmetry and $\psi = \psi(b)$ at the rest of the boundary. In systems with an open boundary like that shown in Fig. 2.10 the value of $\psi = \psi(b)$ at the beginning of the computation is not known. If boundary "b" is removed from the system by a large distance the value $\psi = \psi(b)$ can be taken equal to zero. In the opposite case an approximate boundary value $\psi(b)$ can be first calculated through known δ and j with the help of equations of type (2.62).

The iteration process is performed in the following way. Let us suppose that the virtual current density $(n-1)$-th approximation $j^{(n-1)}$ is known. Then, it is possible to find the magnetic flux of n-th approximation ψ^n by using finite-difference equations of type (2.68) and calculate components of magnetic induction $B_r^{(n)}$ and $B_z^{(n)}$. These values are used to find the components and module of the vector of magnetic field strength $H_r^{(n)}$, $H_z^{(n)}$, $H^{(n)}$ for n-th approximation with help of magnetic B–H curves given in Figs. 2.11 and 2.12.

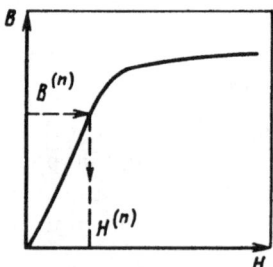

Fig. 2.11. Curve of magnetization of soft magnetic material

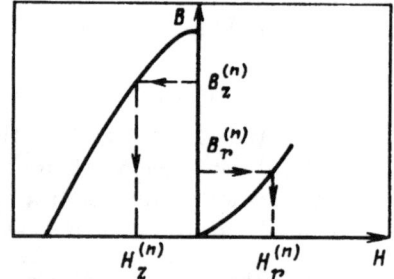

Fig. 2.12. Curves of magnetization and demagnetization of a permanent magnet

For soft magnetic materials the curves of type shown in Fig. 2.11 are used. The value of module $H^{(n)}$ is found through the module of $B^{(n)}$. Components of $H^{(n)}$ are calculated by the formulae: $H_r^{(n)} = B_r^{(n)} \frac{H^{(n)}}{B^{(n)}}$, $H_z^{(n)} = B_z^{(n)} \frac{H^{(n)}}{B^{(n)}}$.

For hard magnetic materials (permanent magnet) $B - H$ curves of the type shown in Fig. 2.11 are used. If a magnet was initially magnetized in the z-direction, then the ratio $B_z^{(n)}$ and $H_z^{(n)}$ is determined by the curve of the second quadrant, that is, demagnetization curve, while the ratio of $B_r^{(n)}$ and $H_r^{(n)}$ does so by the curve of the first quadrant, that is the magnetization curve. It should be noted that using B–H curves allows nonlinear properties of magnetic materials to be taken into account.

The found values of $H_r^{(n)}$, $H_z^{(n)}$ are used to find the magnetization vector $\vec{J}^{(n)}$ and the density of virtual currents of n-th approximation $\vec{j}^{(n)}$:

$$\vec{J}^{(n)} = \frac{1}{\mu_0} \vec{B}^{(n)} - \vec{H}^{(n)}$$

$$\vec{j}^{(n)} = \operatorname{rot} \vec{J}^{(n)} = \frac{1}{\mu_0} \operatorname{rot} \vec{B}^{(n)} - \operatorname{rot} \vec{H}^{(n)}\ .$$

Since

$$\frac{1}{\mu_0} \operatorname{rot} \vec{B}^{(n)} = \vec{\delta} + \vec{j}^{(n-1)}\ ,$$

then

$$\vec{j}^{(n)} = \vec{j}^{(n-1)} - \Delta \vec{j}\ ,$$

where $\Delta \vec{j} = \vec{\delta} - \operatorname{rot} \vec{H}^{(n)}$.

It should be noted that in the last equation $\operatorname{rot} \vec{H}^{(n)} \ne \vec{\delta}$, since at the given step of computation the value of $\vec{H}^{(n)}$ is an approximate one.

As iterations proceed $\operatorname{rot} \vec{H}^{(n)}$ will approach $\vec{\delta}$ and the discrepancy of current densities $\Delta \vec{j} = \vec{j}^{(n-1)} - \vec{j}^{(n)}$ approaches zero. This indicates the convergence of the iteration process.

It is necessary to note that computation of the field of the first (initial) approximation has some peculiarities. Permanent magnets are simply replaced by virtual currents according to the technique used for computation of ring-shaped magnets. Soft magnetic materials are assumed to be of a constant value of permeability $\mu = \operatorname{const}$ [56].

On the basis of the method and algorithm described above a computer code "Weber" has been developed and successfully used for designing various magnetic systems including systems based on permanent magnets [57, 58].

It should be noted that the method of integral equations is also used for magnetic-field computation, its detail description can be found in [44, 45].

3 Fundamentals of Charged-Particle Motion

3.1 Equations of Motion in Newton Form

Charged-particle motion in electric and magnetic fields is governed by the Lorenz equation:

$$m\frac{d\vec{v}}{dt} = e\vec{E} + e(\vec{v} \times \vec{B}) , \qquad (3.1)$$

m – mass of the charged-particle;
e – value of charge;
$\vec{v}$ – velocity vector;
$\vec{E}$ – electric field vector;
$\vec{B}$ – magnetic flux density vector;
$\frac{d\vec{v}}{dt}$ – acceleration vector.

The first term in the equation, $e\vec{E}$, expresses the force acting on a charged-particle due to the electric field. The second term, $e\left(\vec{v} \times \vec{B}\right)$, is the force acting on a charged-particle due to the magnetic field.

In general, $\vec{E}$ and $\vec{B}$ are functions of space coordinates and time. But, henceforth, these quantities shall be taken to be independent of time. Such an assumption is admissible in many practical problems, when the transit time is considerably less than the time period of the field variation. This assumption considerably simplifies the analysis of motion problems.

Usually, the Lorentz equation is used in the form of projections on coordinate axes. The form of equations depends on the choice of the coordinate system. In the rectangular coordinate system x, y, z it will be

$$\begin{aligned} m\ddot{x} &= eE_x + e\dot{y}B_z - e\dot{z}B_y ; \\ m\ddot{y} &= eE_y + e\dot{z}B_x - e\dot{x}B_z ; \\ m\ddot{z} &= eE_z + e\dot{x}B_y - e\dot{y}B_x , \end{aligned} \qquad (3.2)$$

where the values E and B with subscripts x, y, z, are the components of the vectors $\vec{E}$ and $\vec{B}$; $\dot{x}, \dot{y}, \dot{z}, \ddot{x}, \ddot{y}, \ddot{z}$ are components of the velocity and acceleration.

In the cylindrical coordinate system r, θ, z the equations of motion are expressed in the following form:

3 Fundamentals of Charged-Particle Motion

$$m(\ddot{r} - r\dot{\theta}^2) = eE_r + e(r\dot{\theta}B_z - \dot{z}B_\theta);$$

$$m\frac{1}{r}\frac{d}{dt}(r^2\dot{\theta}) = eE_\theta + e(\dot{z}B_r - \dot{r}B_z);$$

$$m\ddot{z} = eE_z + e(\dot{r}B_\theta - r\dot{\theta}B_r). \qquad (3.3)$$

Here $\dot{r}, \dot{\theta}, \dot{z}, \ddot{r}, \ddot{z}$ – first and second derivatives of coordinates with respect to time.

In generalized curvilinear coordinates q_1, q_2, q_3 the equations of motion can be written as follows [2]:

$$m\frac{d}{dt}(h_1\dot{q}_1) + mh_2\dot{q}_2\left(\frac{\dot{q}_1}{h_2}\frac{\partial h_1}{\partial q_2} - \frac{\dot{q}_2}{h_1}\frac{\partial h_2}{\partial q_1}\right) + mh_3\dot{q}_3\left(\frac{\dot{q}_1}{h_3}\frac{\partial h_1}{\partial q_3} - \frac{\dot{q}_3}{h_1}\frac{\partial h_3}{\partial q_1}\right) =$$
$$= eE_1 + e(h_2\dot{q}_2B_3 - h_3\dot{q}_3B_2);$$

$$m\frac{d}{dt}(h_2\dot{q}_2) + mh_1\dot{q}_1\left(\frac{\dot{q}_2}{h_1}\frac{\partial h_2}{\partial q_1} - \frac{\dot{q}_1}{h_2}\frac{\partial h_1}{\partial q_2}\right) + mh_3\dot{q}_3\left(\frac{\dot{q}_2}{h_3}\frac{\partial h_2}{\partial q_3} - \frac{\dot{q}_3}{h_2}\frac{\partial h_3}{\partial q_2}\right) =$$
$$= eE_2 + e(h_3\dot{q}_3B_1 - h_1\dot{q}_1B_3);$$

$$m\frac{d}{dt}(h_3\dot{q}_3) + mh_1\dot{q}_1\left(\frac{\dot{q}_3}{h_1}\frac{\partial h_{31}}{\partial q_1} - \frac{\dot{q}_1}{h_3}\frac{\partial h_1}{\partial q_3}\right) + mh_2\dot{q}_2\left(\frac{\dot{q}_3}{h_2}\frac{\partial h_3}{\partial q_2} - \frac{\dot{q}_2}{h_3}\frac{\partial h_2}{\partial q_3}\right) =$$
$$= eE_3 + e(h_1\dot{q}_1B_2 - h_2\dot{q}_2B_1). \qquad (3.4)$$

Here q_i ($i = 1, 2, 3$) – generalized coordinates; $\dot{q}_i$ ($i = 1, 2, 3$) – derivatives of generalized coordinates with respect to time, i.e. generalized velocities; E_i, B_i ($I = 1, 2, 3$) – components of electric and magnetic field; h_i ($i = 1, 2, 3$) – scale coefficients, or metrics, of the coordinate system.

Equations of motion in any particular coordinate system can be obtained with the help of the equations in generalized form by substitution in them of the corresponding coordinates and scale coefficients. For the more frequently used cylindrical and spherical coordinates the scale coefficients are expressed as

for cylindrical coordinates:

$$q_1 = r \quad q_2 = \theta \quad q_3 = z; \qquad h_1 = 1 \quad h_2 = r \quad h_3 = 1,$$

for spherical coordinates:

$$q_1 = r \quad q_2 = \theta \quad q_3 = \psi \qquad h_1 = 1 \quad h_2 = r \quad h_3 = r\sin\theta.$$

3.2 Law of Energy Conservation

Using Lorentz's equation (3.1) we obtain an important relationship expressing the kinetic energy and the velocity of a charged-particle in combined electric and magnetic fields in terms of the electric field potential U.

3.2 Law of Energy Conservation

Multiplying the left and right parts of the equation by $\vec{v}$ we find

$$m\vec{v}\frac{d\vec{v}}{dt} = e\vec{v}\vec{E} + e\vec{v}(\vec{v} \times \vec{B}).$$

Since

$$\vec{v}(\vec{v} \times \vec{B}) = 0 \quad \text{and} \quad \vec{E} = -\operatorname{grad} U,$$

this results in

$$\frac{d}{dt}\left(\frac{mv^2}{2}\right) = -e\vec{v}\operatorname{grad} U.$$

This gives

$$\frac{d}{dt}\left(\frac{mv^2}{2}\right) = -e\frac{dU}{dt},$$

and after integration

$$\frac{mv^2}{2} + eU = \text{const}. \tag{3.5}$$

This is the fundamental equation of energy conservation in stationary electric and magnetic fields.

It states that the total energy of a charged-particle, which is equal to the sum of kinetic energy $T = mv^2/2$ and potential energy $V = eU$, is conserved during its motion in electric and magnetic fields.

Alternatively (3.5) may be written as

$$\frac{mv^2}{2} - \frac{mv_0^2}{2} = -e(U - U_0).$$

In this form it states that the change in the particle kinetic energy while the particle moves from one point to another is determined by the difference of potentials of the two points. If the potential of the particle source U_0 and the initial velocity v_0 are known, the kinetic energy at an arbitrary position can be expressed as follows

$$\frac{mv^2}{2} = \frac{mv_0^2}{2} - e(U - U_0). \tag{3.6}$$

When electron motion is considered and a thermal cathode is used as the source of electrons, the initial energy is thermal in nature. In many practical cases its average value is negligible compared with the energy imparted to electrons by an accelerating field. Then, assuming the cathode potential $U_0 = 0$, we find from (3.6)

$$\frac{mv^2}{2} = -eU, \quad \text{or} \quad v = \left(2\frac{\hat{e}}{m}U\right)^{1/2}, \tag{3.7}$$

where $\hat{e}$ is the absolute value of electron charge.

Upon substitution of the numerical values this gives

$$v = 5.93 \times 10^7 U^{1/2} \quad [cm/s, V] . \tag{3.8}$$

Using the energy conservation law in its different versions it is possible to determine only the magnitude of velocity, while the direction of the velocity vector remains undetermined.

3.3 Lagrange Equations of Motion

The Newton equation is not the only way of formulating laws of motion. In many cases it is convenient to use Lagrange and Hamilton equations of motion. For the particular case of rectangular coordinates and conservative forces the Lagrange equation can be easily derived from Newtons second law

$$\frac{d}{dt}(m\dot{v}_x) = F_x .$$

Introducing a potential energy function $V(x, y, z)$ and kinetic energy $T = \frac{m}{2}(v_x^2 + v_y^2 + v_z^2)$, one can represent terms mv_x and F_x in the following form:

$$mv_x = \frac{\partial T}{\partial v_x} \quad \text{and} \quad F_x = -\frac{\partial V}{\partial x} .$$

This results in

$$\frac{d}{dt}\left(\frac{\partial T}{\partial v_x}\right) + \frac{\partial V}{\partial x} = 0 . \tag{3.9}$$

If we introduce the function L, known as the Lagrange function, which is equal to the difference between the kinetic and potential energy of a particle $L = T - V$, one finds:

$$\frac{\partial T}{\partial v_x} = \frac{\partial L}{\partial v_x} \quad \text{and} \quad \frac{\partial V}{\partial x} = -\frac{\partial L}{\partial x} .$$

Substitution of these expressions into (3.9) leads to

$$\frac{d}{dt}\left(\frac{\partial L}{\partial v_x}\right) - \frac{\partial L}{\partial x} = 0 . \tag{3.10}$$

A similar equation can be obtained for y and z, in this form they are called the Lagrange equations of motion.

The significance of these equations lies in the fact that they express motion in terms of kinetic and potential energy and hold their form for coordinates other than the rectangular ones.

In generalized curvilinear coordinates the equations are

$$\frac{d}{dt}\frac{\partial L}{\partial \dot{q}_i} - \frac{\partial L}{\partial q_i} = 0, i = 1, 2, 3 . \tag{3.11}$$

$L(q_i, q)$ is again the difference between kinetic and potential energy expressed in terms of generalized velocities and coordinates. They are usually derived from the principle of least action that states that action

$$\hat{S} = \int_{t_0}^{t} L(q_i, \dot{q}_i) dt \tag{3.12}$$

has its minimum values when it is taken along the real path of motion. Using the method of variation calculus it is possible to show that it results in (3.11) for charged-particle motion [60].

If we introduce the value known as the generalized momentum

$$p_i = \partial L / \partial \dot{q}_i , \tag{3.13}$$

then the Lagrange equation is written as:

$$\dot{p}_i = \frac{\partial L}{\partial q_i} .$$

An important connection exists between the generalized momentum and the action:

$$p_i = \frac{\partial \hat{S}}{\partial q_i} . \tag{3.14}$$

For motion in electrostatic fields the Lagrange function is set up in accordance with its definition:

$$L = T - V = \frac{m}{2} \sum_i h_i^2 \dot{q}_i^2 - eU , \tag{3.15}$$

where eU is the potential energy of a particle in an electrostatic field, $\frac{m}{2} \sum_i h_i^2 \dot{q}_i^2$ is the kinetic energy expressed in generalized coordinates.

For combined electric and magnetic fields the following form of the Lagrange equation is used:

$$L = T - V + e(\vec{A} \cdot \vec{v}) , \tag{3.16}$$

where $\vec{A}$ is the vector potential of the magnetic field.

In curvilinear coordinates it is expressed as

$$L = \frac{m}{2} \sum_i h_i^2 \dot{q}_i^2 - eU(q_i) + e \sum_i A_i h_i \dot{q}_i , \tag{3.17}$$

where A_i are projections of the vector potential on curvilinear coordinates.

The generalized momentum for the present case is:

$$p_i = \frac{\partial L}{\partial \dot{q}_i} = m h_i^2 \dot{q}_i + e A_i h_i . \tag{3.18}$$

3.4 Hamilton Equations

Introduction of a new function known as the Hamilton function

$$H = \sum_i p_i \dot{q}_i - L \tag{3.19}$$

allows the equation of motion to be obtained in Hamiltonian form [60]:

$$\dot{p}_i = -\frac{\partial H}{\partial q_i}, \quad \dot{q}_i = \frac{\partial H}{\partial p_i} \quad (i = 1, 2, 3). \tag{3.20}$$

Parameters q_i and p_i in these equations are considered to be undependable variables.

For motion in static electric and magnetic fields the Hamilton function has the simple physical meaning of the total energy E [71].

$$H = E = T + V. \tag{3.21}$$

Therefore, the Hamilton function can be set up to express the total energy in terms of the generalized moments and coordinates. For combined electric and magnetic fields the total energy of a charged-particle is

$$E = T + V = \frac{m}{2} \sum_i h_i^2 \dot{q}_i^2 + eU(q_i). \tag{3.22}$$

To obtain the Hamilton function it is necessary to express velocities $\dot{q}_i$ in terms of generalized moments:

$$\dot{q}_i = \frac{1}{mh_i^2}(p_i - eA_i h_i).$$

This results in:

$$H = \frac{1}{2m} \sum_i \frac{1}{h_i^2}(p_i - eA_i h_i)^2 + eU(q_i). \tag{3.23}$$

3.5 Hamilton–Jacobi Equation

The Hamilton–Jacobi equation of motion has the following general form [60]:

$$\frac{\partial \hat{S}}{\partial t} + H\left(q_1, q_2, q_3, \frac{\partial \hat{S}}{\partial q_1}, \frac{\partial \hat{S}}{\partial q_2}, \frac{\partial \hat{S}}{\partial q_3}, t\right) = 0, \tag{3.24}$$

where H is the Hamilton function, in which the generalized moments are expressed in the terms of the action function $\hat{S}$ as given by (3.14), that is $p_i = \partial \hat{S}/\partial q_i$.

For motion, in conservative fields, when the Hamilton function is independent of time the Hamilton–Jacobi equation is reduced to

$$H\left(q_1, q_2, q_3, \frac{\partial S}{\partial q_1}, \frac{\partial S}{\partial q_2}, \frac{\partial S}{\partial q_3}\right) = E, \tag{3.25}$$

where E is total energy $E = T + V$, being kept constant during the motion, $S(q_i)$ is the so-called shortened action function, which is connected with the total action (3.12) by the relation $S(q_i) = \hat{S}(q_i, t) + Et$.

For a charged-particle motion in electrostatic and combined fields the following equations are obtained for this function:

$$\frac{1}{2m}\sum_i \frac{1}{h_i^2}\left(\frac{\partial S}{\partial q_i}\right)^2 + eU(q_i) = E \qquad (3.26)$$

$$\frac{1}{2m}\sum_i \frac{1}{h_i^2}\left(\frac{\partial S}{\partial q_i} - eA_i h_i\right)^2 + eU(q_i) = E . \qquad (3.27)$$

The above equations are partial differential equations for the action function. The complete integral of these equations S contains a number of constants equal to the number of coordinates (degrees of freedom) $\alpha_1, \alpha_2, \alpha_3$; one of them is an additive constant; the energy E is involved in the solution as a parameter:

$$S = S(q_1, q_2, q_3, \alpha_1, \alpha_2, E) + \alpha_3 . \qquad (3.28)$$

If the complete integral of the equation is found, the motion of a charged particle is determined by the following equations:

$$\frac{\partial S}{\partial \alpha_i} = \beta_i , \qquad \frac{\partial S}{\partial E} = t - t_0 ,$$

where $\alpha_i, \beta_i, (i = 1, 2, 3)$ are primary and secondary constants, which determine the possible motion of a particle, t_0 is also considered to be an arbitrary constant.

The first three equations do not contain time, they define a trajectory of the particle. The last one determines particle position as a function of time.

The classical method of solution of the Hamilton–Jacobi equations is the method of separation of variables. In particular, it is applicable when the equation does not explicitly involve one or two coordinates. Suppose that it does not contain q_1 coordinate. Then a solution can be sought in the form

$$S = S_1(q_1) + S_2(q_2, q_3) .$$

Its substitution in (3.25) gives:

$$H\left(q_2, q_3, \frac{\partial S_1}{\partial q_1}, \frac{\partial S_2}{\partial q_2}, \frac{\partial S_2}{\partial q_3}\right) = E .$$

It is possible to solve this equation with respect to $\partial S_1/\partial q_1$ and write

$$\frac{\partial S_1}{\partial q_1} = F\left(q_2, q_3, \frac{\partial S_2}{\partial q_2}, \frac{\partial S_2}{\partial q_3}, E\right) .$$

The left part of this equation depends only on the q_1-coordinate, while the right part depends on q_1 and q_2 coordinates. This equality can exist only if each side is equal to some constant value α_1, that is

$$\frac{\partial S_1}{\partial q_1} = \alpha_1, \qquad F\left(q_2, q_3, \frac{\partial S_2}{\partial q_2}, \frac{\partial S_3}{\partial q_3}, E\right) = \alpha_1 .$$

The first of these equations has the integral $S_1 = \alpha_1 q_1 + \text{const}$, the last one contains fewer variables than the initial equation and usually allows us to find a simpler solution.

The Hamilton–Jacobi equation does not contain explicitly two coordinates, the solution can be sought in the form:

$$S = S_1(q_1) + S_2(q_2) + S_3(q_3).$$

Its substitution separates the partial differential equation into three ordinary differential equations that can be integrated directly.

The Hamilton–Jacobi method allows us to express the motion of a particle in terms of a single scalar function known as the action function S. This is an important property of the method, which has been successfully used for analysis of charged-particle motion in complicated electric and magnetic fields.

3.6 Equations of Motion in Axially Symmetric Fields

Study of charged-particle motion in this type of fields is of special interest, because focusing systems with axially symmetrical fields have found wide application in the majority of modern electronic devices.

It is natural to use the cylindrical coordinate system for analysis of charged-particle motion in such fields.

3.6.1 Equations of Motion in Newton Form

These are obtained from the general equations (3.3), if one substitutes in them zero values of azimuth field components ($E_\theta = B_\theta = 0$):

$$m(\ddot{r} - r\dot{\theta}^2) = eE_r + er\dot{\theta}B_z ; \tag{3.29}$$

$$m\frac{1}{r}\frac{d}{dt}(r^2\dot{\theta}) = e\dot{z}B_r - e\dot{r}B_z ; \tag{3.30}$$

$$m\ddot{z} = eE_z - er\dot{\theta}B_r . \tag{3.31}$$

Bush Theorem

There exists an important relation that characterizes motion in axially symmetrical fields, known as the Bush theorem.

Let us express in (3.30) the radial and axial magnetic field components in terms of a vector magnetic potential A_θ by means of expressions:

$$B_z = \frac{1}{r}\frac{\partial}{\partial r}(rA_\theta) ; \qquad B_r = -\frac{\partial A_\theta}{\partial z} .$$

Then we have

$$m\frac{1}{r}\frac{d}{dt}(r^2\dot{\theta}) = -e\dot{z}\frac{\partial A_\theta}{\partial z} - e\dot{r}\frac{1}{r}\frac{\partial}{\partial r}(rA_\theta) ,$$

or

$$m\frac{d}{dt}(r^2\dot{\theta}) = -e\dot{z}\frac{\partial(rA_\theta)}{\partial z} - e\dot{r}\frac{\partial}{\partial r}(rA_\theta) .$$

3.6 Equations of Motion in Axially Symmetric Fields

Since

$$\dot{z}\frac{\partial}{\partial z}(rA_\theta) + \dot{r}\frac{\partial}{\partial r}(rA_\theta) = \frac{dA_\theta}{dt},$$

we obtain

$$m\frac{d}{dt}(r^2\dot{\theta}) = -e\frac{d}{dt}(rA_\theta).$$

Its integration gives

$$mr^2\dot{\theta} + erA_\theta = C.$$

The value of the constant C can be found, if the values involved in the left part of the equation are known at any arbitrary point k of the article trajectory $C = mr_k^2\dot{\theta}_k + er_kA_{\theta k}$. Then,

$$mr^2\dot{\theta} - mr_k^2\dot{\theta}_k = -e(rA_\theta - r_kA_{\theta k}). \tag{3.32}$$

A_θ is a component of the vector magnetic potential at a radius r, which can be expressed in terms of magnetic flux ψ linked by a circumference of radius r:

$$A_\theta = \frac{\psi}{2\pi r} = \frac{2\pi \int_0^r B_z r dr}{2\pi r}. \tag{3.33}$$

The substitution of the last equation in (3.32) leads to

$$mr^2\dot{\theta} - mr_k^2\dot{\theta}_k = -\frac{e}{2\pi}(\psi - \psi_k), \tag{3.34}$$

where ψ and ψ_k are magnetic fluxes, passing through the surfaces bounded by the circles with radius r and r_k, respectively (Fig. 3.1).

This expression is known as the Bush theorem. It states that a change in the angular momentum is proportional to the magnetic-flux variation.

If a particle starts with zero initial velocity, the above expression for the angular momentum is written as

$$mr^2\dot{\theta} = -\frac{e}{2\pi}(\psi - \psi_k). \tag{3.35}$$

It follows that

$$r^2\dot{\theta} = -\frac{e}{2\pi m}(\psi - \psi_k). \tag{3.36}$$

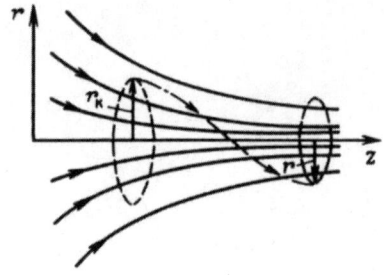

Fig. 3.1. Definition of magnetic fluxes ψ and ψ_k

Modified Equations of Motion

Using (3.36) we can transform (3.29)–(3.31) to the following form:

$$m\ddot{r} = -e\frac{\partial U}{\partial r} - \frac{e^2}{4\pi^2 m}\frac{\psi - \psi_k}{r}\left(2\pi B_z - \frac{\psi - \psi_k}{r^2}\right) ; \qquad (3.37)$$

$$m\ddot{z} = -e\frac{\partial U}{\partial z} - \frac{e^2}{4\pi^2 m}\frac{\psi - \psi_k}{r^2}\frac{\partial \psi}{\partial z} ; \qquad (3.38)$$

$$\dot{\theta} = -\frac{1}{2\pi}\frac{e}{m}\frac{\psi - \psi_k}{r^2} . \qquad (3.39)$$

The peculiarity of these equations lies in the fact that forces acting upon a charged-particle are expressed in terms of electric and magnetic fields parameters only. They allow some important properties of particle motion in axially symmetric fields to be deduced.

As follows from (3.37), the radial force acting upon a particle due to the magnetic field depends not only on the value of the magnetic induction at the particle position, but on the difference of magnetic fluxes $\psi - \psi_k$ as well. The force tends to zero when this difference approaches zero.

Equation (3.38) shows that the magnetic field varying in the z-direction produces a z-component of magnetic force. This results in a decrease of the z-component of the particle velocity when the magnetic field is increased in the z-direction and vice versa.

In the particular case of a uniform magnetic field directed along the z-axis ($B = B_z$) we have:

$$\psi = \pi r^2 B_z , \qquad \psi_k = \pi r_k^2 B_z , \qquad \frac{\partial \psi}{\partial z} = 0 .$$

Then system of equations (3.37)–(3.39) is reduced to

$$\ddot{r} = -\frac{e}{m}\frac{\partial U}{\partial r} - \frac{1}{4}\left(\frac{e}{m}\right)^2 B_z^2 r\left(1 - \frac{r_k^4}{r^4}\right);$$

$$\ddot{z} = -\frac{e}{m}\frac{\partial U}{\partial z};$$ (3.40)

$$\dot{\theta} = -\frac{1}{2}\frac{e}{m}B_z\left(1 - \frac{r_k^2}{r^2}\right).$$

In some cases it is expedient to represent (3.37)–(3.39) in the following form:

$$m\ddot{r} = -e\frac{\partial U}{\partial r} - \frac{e^2}{8\pi^2 m}\frac{\partial}{\partial r}\left(\frac{\psi - \psi_k}{r}\right)^2;$$

$$m\ddot{z} = -e\frac{\partial U}{\partial r} - \frac{e^2}{8\pi^2 m}\frac{\partial}{\partial z}\left(\frac{\psi - \psi_k}{r}\right)^2;$$ (3.41)

$$\dot{\theta} = -\frac{1}{2\pi}\frac{e}{m}\frac{\psi - \psi_k}{r^2}.$$

It follows from (3.41) that motion of a particle in axially symmetric fields can be considered as motion in fields with generalized potential Q:

$$Q = U + \frac{e}{8\pi^2 m}\left(\frac{\psi - \psi_k}{r}\right)^2,$$

$$m\ddot{r} = -e\frac{\partial Q}{\partial r} \qquad m\ddot{z} = -e\frac{\partial Q}{\partial z}.$$ (3.42)

The trajectory equation is obtained if we eliminate time from the above equations of motion:

$$\frac{d^2 r}{dz^2} = \left(\frac{\partial Q}{\partial r} - \frac{\partial Q}{\partial z}\frac{dr}{dz}\right)\left[1 + \left(\frac{dr}{dz}\right)^2\right]\frac{1}{2Q}.$$ (3.43)

Paraxial Equations of Motion

These equations are used for description of charged-particle motion in the vicinity of the axis of symmetry with small angles of trajectory slopes. They can be obtained from the general equations of motion if we present potentials and magnetic fluxes, including in them, as series expansions [see (2.25), (2.59), (2.60)] and to hold only the first terms of the series. Then one finds:

$$\frac{\partial U}{\partial z} \approx \frac{dU_0}{dz} = U_0', \qquad \frac{\partial U}{\partial r} \approx -\frac{r}{2}\frac{d^2 U_0}{dz^2} = -\frac{r}{2}U_0''$$

$$B_z \approx B_{z0} \qquad B_r \approx B_{z0}',$$

$$\psi \approx \pi r^2 B_{z0} \qquad \psi_k \approx \pi r_k^2 B_{z0k}.$$

In these equations the primes denote differentiation on the z-coordinate, zero in subscripts indicates the values at the axis of symmetry, B_{z0k} is the value of magnetic induction at the cathode.

Substitution of these expression in the general equations of motion (3.37)–(3.39) yields the following system of paraxial equations of motion:

$$\ddot{r} = \frac{1}{2}\frac{e}{m}rU_0'' - \frac{1}{4}\left(\frac{e}{m}\right)^2 rB_{z0}^2\left[1 - \left(\frac{r_k^2 B_{z0k}}{r^2 B_{z0}}\right)^2\right] \quad (3.44)$$

$$\ddot{z} = -\frac{e}{m}U_0' - \frac{1}{4}\left(\frac{e}{m}\right)^2 r^2 B_{z0}' B_{z0}\left(1 - \frac{r_k^2 B_{z0k}}{r^2 B_{z0}}\right) \quad (3.45)$$

$$\dot{\theta} = -\frac{1}{2}\frac{e}{m}B_{z0}\left(1 - \frac{r_k^2 B_{z0k}}{r^2 B_{z0}}\right). \quad (3.46)$$

Elimination of time from these equation allows us to obtain the paraxial trajectory equation:

$$\frac{d^2r}{dz^2} + \frac{U_0'}{2U_0}\frac{dr}{dz} + \frac{U_0''}{4U_0}r - \frac{e}{8m}r\frac{B_{z0}^2}{U_0}\left[1 - \left(\frac{r_k^2 B_{z0k}}{r^2 B_{z0}}\right)^2\right] = 0. \quad (3.47)$$

3.6.2 Lagrange and Hamilton Equations of Motion

One of the main properties of these equations is that they keep the same form for different coordinate systems. The only thing that is required is to express them in cylindrical coordinates and obtain expressions for the Lagrange and Hamilton functions.

This can be done, if one substitutes into general equations (3.11), (3.17), (3.20), (3.24) corresponding coordinates, scale coefficients, velocities, momentum and magnetic vector potential components:

$$\begin{aligned}
q_1 &= r & h_1 &= 1 & \dot{q}_1 &= \dot{r} & p_1 &= p_r & A_1 &= A_2 = 0 \\
q_2 &= \theta & h_2 &= r & \dot{q}_2 &= \dot{\theta} & p_2 &= p_\theta & A_2 &= A_\theta = A \\
q_3 &= z & h_3 &= 1 & \dot{q}_3 &= \dot{z} & p_3 &= p_z & A_3 &= A_z = 0.
\end{aligned}$$

Upon substitution we have:

$$\frac{d}{dt}\left(\frac{\partial L}{\partial r}\right) - \frac{\partial L}{\partial r} = 0; \quad \frac{d}{dt}\left(\frac{\partial L}{\partial \theta}\right) - \frac{\partial L}{\partial \theta} = 0; \quad \frac{d}{dt}\left(\frac{\partial L}{\partial z}\right) - \frac{\partial L}{\partial z} = 0;$$

$$L = \frac{m}{2}[\dot{r}^2 + r^2\dot{\theta}^2 + \dot{z}^2] + er\dot{\theta}A_\theta - eU(r,z);$$

3.6 Equations of Motion in Axially Symmetric Fields

$$\dot{p} = -\frac{\partial H}{\partial r}; \quad \dot{p}_\theta = -\frac{\partial H}{\partial \theta}; \quad \dot{p}_z = -\frac{\partial H}{\partial r};$$
$$\dot{r} = \frac{\partial H}{\partial p_r} \quad \dot{\theta} = \frac{\partial H}{\partial p_\theta}; \quad \dot{z} = \frac{\partial H}{\partial p_z}, \tag{3.48}$$

$$H = \frac{1}{2m}\left[p_r^2 + \frac{1}{r^2}(p_\theta - erA_\theta)^2 + p_z^2\right] - eU(r,z). \tag{3.49}$$

It is worth mentioning that the momenta involved in the Hamilton function, are generalized momenta:

$$p_r = \frac{\partial L}{\partial \dot{r}}; \quad p_\theta = \frac{\partial L}{\partial \dot{\theta}}; \quad p_z = \frac{\partial L}{\partial \dot{z}}.$$

It is interesting to note that the Bush theorem, obtained above, immediately follows from the Hamilton equations of motion. Since the Hamilton function does not depend explicitly on θ, we find:

$$\dot{p}_\theta = -\frac{\partial H}{\partial \theta} = 0; \quad p_\theta = \text{const}; \quad \dot{\theta} = \frac{\partial H}{\partial p_\theta} = \frac{1}{mr^2}(p_\theta - erA_\theta).$$

These equations give:

$$p_\theta = mr^2\dot{\theta} + erA_\theta = \text{const}.$$

The last equation leads to the Bush theorem.

3.6.3 Hamilton–Jacobi Equation of Motion

This can be found from (3.27) upon corresponding substitution and has the following form:

$$\frac{1}{2m}\left[\left(\frac{\partial S}{\partial r}\right)^2 + \frac{1}{r}\left(\frac{\partial S}{\partial \theta} - erA_\theta\right)^2 + \left(\frac{\partial S}{\partial z}\right)^2\right] + eU(r,z) = E = \text{const}. \tag{3.50}$$

It does not involve θ coordinate explicitly, and the method of separation of the variables can be applied for its solution.

3.6.4 Motion of Electrons in Cylindrical Magnetron

It is instructive to consider electron motion in a cylindrical magnetron as an example of application of the different equations of motion.

The cylindrical magnetron cosists of two concentric cylinders. The inner cylinder is the cathode having zero potential. The outer cylinder is the anode electrode with potential U_a. This results in a radial electric field E within the interelectrode space:

$$U = U(r) = U_a \frac{\ln r/a}{\ln b/a} \quad \text{and} \quad E_r = -\frac{U_a}{r \ln b/a}.$$

The magnetic field is assumed to be uniform and directed along the z–axis, which is normal to the page $B = B_z = $ const.

In this case the magnetic vector potential has only one θ-component, which can be expressed as:

$$A_\theta = \frac{\psi}{2\pi r} = \frac{\pi r^2 B}{2\pi r} = \frac{rB}{2} .$$

The initial velocity of electrons at the cathode is assumed to be equal to zero.

Equation of Motion in Newton Form

Since in the present case $\partial U/\partial z = \partial B/\partial z = 0$, $\psi = \pi r^2 B$ and $\psi_k = \pi a^2 B$ the equations of motion are reduced to

$$\ddot{r} = \frac{\hat{e}}{m}\frac{\partial U}{\partial r} - r\left(\frac{\hat{e}}{2m}B\right)^2\left(1 - \frac{a^4}{r^4}\right) ,$$

$$\dot{\theta} = \frac{1}{2}\frac{\hat{e}}{m}B\left(1 - \frac{a^2}{r^2}\right) \quad (3.51)$$

$$m\ddot{z} = 0, \quad \dot{z} = \text{const} = 0 .$$

Note that in the above equations and throughout this section $\hat{e}$ is the absolute value of electron charge.

The motion is performed in the $r\theta$-plane, the angular velocity $\dot{\theta}$ being a function of only the radial electron position.

The equation of trajectory in the $r\theta$ plane is found with the help of the energy conservation law:

$$\dot{r}^2 + r^2\dot{\theta}^2 = 2\frac{\hat{e}}{m}U .$$

This gives:

$$\dot{\theta} = \sqrt{2\frac{\hat{e}}{m}U}\left[\left(\frac{dr}{d\theta}\right)^2 + r^2\right]^{-\frac{1}{2}} .$$

Substitution of the last equation in (3.51) yelds

$$\sqrt{2\frac{\hat{e}}{m}U}\left[\left(\frac{dr}{d\theta}\right)^2 + r^2\right]^{-\frac{1}{2}} = \frac{1}{2}\frac{\hat{e}}{m}B\left(1 - \frac{a^2}{r^2}\right) ,$$

or, after some algebra:

$$\frac{d\theta}{dr} = \frac{\frac{1}{2}\left(\frac{\hat{e}}{m}\right)^{1/2}B\left(1 - \frac{a^2}{r^2}\right)}{\left[2U - \frac{1}{4}\frac{\hat{e}}{m}B^2 r^2\left(1 - \frac{a^2}{r^2}\right)^2\right]^{1/2}} . \quad (3.52)$$

3.6 Equations of Motion in Axially Symmetric Fields

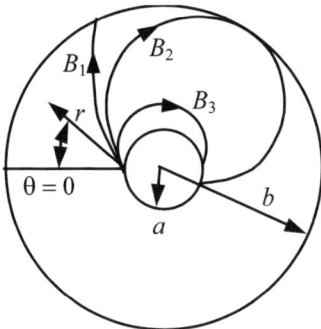

Fig. 3.2. Cylindrical magnetron

Its integration leads to:

$$\theta = \theta_0 + \int_a^r \frac{\frac{1}{2}\left(\frac{\hat{e}}{m}\right)^{1/2} B \left(1 - \frac{a^2}{r^2}\right) dr}{\left[2U(r) - \frac{1}{4}\frac{\hat{e}}{m}B^2 r^2 \left(1 - \frac{a^2}{r^2}\right)^2\right]^{1/2}}, \quad (3.53)$$

where

$$U(r) = U_a \frac{\ln r/a}{\ln b/a}.$$

These expressions are used for the calculation of electron trajectories $r = f(\theta)$ by a numerical method. Some electron trajectories corresponding to various values of magnetic field $B_3 > B_2 > B_1$ are shown in Fig. 3.2.

It is seen that beginning from a certain value of magnetic field $B = B_2$ electrons cannot reach the anode surface and return to the cathode and cutoff of the anode current takes place.

The cutoff condition can be determined without direct solution of (3.53). It is readily seen that, at the cutoff condition, electrons just graze the anode surface, their radial velocity at this point is equal to zero. Therefore, the cutoff condition can be written as follows:

$$\frac{dr}{dt} = 0, \quad \text{at} \quad r = b,$$

or

$$\frac{dr}{d\theta} = \frac{dr}{dt} \bigg/ \frac{d\theta}{dt} = 0.$$

Its substitution in (3.52) leads to the well-known cutoff relationship:

$$U_a = \frac{\hat{e}}{8m} B^2 b^2 \left(1 - \frac{a^2}{b^2}\right)^2.$$

Method of Hamilton Equations

For the present case the Hamilton function is given by:

$$H = \frac{1}{2m}\left[p_r^2 + \frac{1}{r^2}(p_\theta + \hat{e}rA_\theta)^2\right] - \hat{e}U(r) = 0,$$

or, since $A_\theta = rB/2$,

$$H = \frac{1}{2m}\left[p_r^2 + \frac{1}{r^2}\left(p_\theta + \frac{\hat{e}r^2B}{2}\right)^2\right] - \hat{e}U(r) = 0.$$

Then, using the Hamilton equations one obtains:

$$\dot{p}_\theta = -\frac{\partial H}{\partial \theta} = 0, \quad \text{therefore,} \quad p_\theta = \text{const},$$

$$\dot{\theta} = \frac{\partial H}{\partial p_\theta} = \frac{1}{mr^2}\left(p_\theta + \frac{\hat{e}r^2B}{2}\right).$$

This results in

$$p_\theta = mr^2\dot{\theta} - \hat{e}\frac{r^2B}{2} = \text{const}.$$

Using the initial condition $\dot{\theta} = 0$ at $r = a$ one obtains

$$p_\theta = -\hat{e}\frac{r^2B}{2} = \text{const}.$$

Therefore,

$$\dot{\theta} = \frac{1}{2}\frac{\hat{e}}{m}B\left(1 - \frac{a^2}{r^2}\right),$$

$$\dot{r} = \frac{\partial H}{\partial p_r} = \frac{p_r}{m}.$$

The value of p_r can be found directly from the Hamilton function:

$$p_r = \left[2m\hat{e}U - \frac{1}{r^2}\left(p_\theta + \frac{\hat{e}r^2B}{2}\right)^2\right]^{1/2}.$$

Substitution in this equation of $p_\theta = -\hat{e}\frac{a^2B}{2}$ yields:

$$p_r = \left[2m\hat{e}U - \frac{\hat{e}^2r^2B^2}{4}\left(1 - \frac{a^2}{r^2}\right)^2\right]^{1/2}.$$

The electron-trajectory equation is found in the following way:

$$\frac{d\theta}{dr} = \frac{\dot{\theta}}{\dot{r}} = \frac{m\dot{\theta}}{p_r} = \frac{\frac{1}{2}\left(\frac{\hat{e}}{m}\right)^{1/2}B\left(1 - \frac{a^2}{r^2}\right)}{\left[2U - \frac{1}{4}\frac{\hat{e}}{m}r^2B^2\left(1 - \frac{a^2}{r^2}\right)^2\right]^{1/2}}.$$

3.6 Equations of Motion in Axially Symmetric Fields

It is similar to (3.52) and has the same integral (3.53).

The formula for the current cutoff is obtained as follows. As has been shown the cutoff condition is

$$\dot{r} = \frac{dr}{dt} = 0 \quad \text{at} \quad r = b,$$

then, taking into account that $\dot{r} = \frac{p_r}{m}$, one finds

$$p_r|_{r=b} = \left[2m\hat{e}U - \frac{e^2 r^2 B^2}{4} \left(1 - \frac{a^2}{b^2} \right)^2 \right]^{1/2} = 0.$$

This yields

$$U_a = \frac{\hat{e}}{8m} B^2 b^2 \left(1 - \frac{a^2}{b^2} \right)^2.$$

Application of Hamilton–Jacobi Equation

In the case under cosideration this differential equation is written as:

$$\frac{1}{2m} \left[\left(\frac{\partial S}{\partial r} \right)^2 + \frac{1}{r^2} \left(\frac{\partial S}{\partial \theta} + \frac{\hat{e} r^2 B}{2} \right)^2 \right] - \hat{e} U(r) = E.$$

Since the cathode potential and initial electron velocities are assumed to be equal to zero the total energy will be also equal to zero, $E = 0$.

In accordance with the method of separation of variables a solution is presented in the following form:

$$S = S_1(r) + S_2(\theta).$$

Its substitution in the differential equation gives:

$$\left(\frac{dS_1}{dr} \right)^2 + \frac{1}{r^2} \left(\frac{dS_2}{d\theta} + \frac{er^2 B}{2} \right)^2 - 2\hat{e}mU(r) = 0.$$

It can be written as follows:

$$\frac{dS_2}{d\theta} = r \left[2\hat{e}mU(r) - \left(\frac{dS_1}{dr} \right)^2 \right]^{1/2} - \frac{\hat{e}r^2 B}{2}.$$

The right part of this equation depends only on the r-coordinate, while the left part depends on the θ-coordinate. This equality can exist only if each side is equal to some constant value α_2, that is

$$\frac{dS_2}{d\theta} = \alpha_2 \quad \text{and} \quad r \left[2\hat{e}mU(r) - \left(\frac{dS_1}{dr} \right)^2 \right]^{1/2} - \frac{\hat{e}r^2 B}{2} = \alpha_2.$$

The last equation can be written as

$$\frac{dS_1}{dr} = \sqrt{\varphi(r)},$$

where

$$\varphi(r) = 2\hat{e}mU(r) - \frac{1}{r^2}\left(\alpha_2 + \frac{\hat{e}r^2 B}{2}\right)^2.$$

Therefore, instead of the initial patial differential equation there are two ordinary differential equations for $S_1(r)$ and $S_2(\theta)$, which allow direct integration:

$$S_2 = \alpha_2 \theta + C_1, \qquad S_1 = \int \sqrt{\varphi(r)} dr + C_2,$$

where C_1 and C_2 are arbitrary constants.

Therefore, a solution of the initial patial differential equation can be written as:

$$S = S_1 + S_2 = \int \sqrt{\varphi(r)} dr + \alpha_2 \theta + C_3.$$

In accordance with the general method (see Sect. 3.5) the trajectory equation is determined by

$$\frac{dS}{d\alpha_2} = \beta_2,$$

where β_2 is a secondary arbitrary constant.

This gives

$$\theta + \int \frac{d}{d\alpha_2} \sqrt{\varphi(r)} dr = \beta_2.$$

Since

$$\frac{d}{d\alpha_2} \sqrt{\varphi(r)} = -\frac{\alpha_2 + \frac{\hat{e}r^2 B}{2}}{r^2 \sqrt{\varphi(r)}},$$

one obtains:

$$\theta - f(r) = \beta_2, \qquad f(r) = \int \frac{\alpha_2 + \frac{\hat{e}r^2 B}{2}}{r}^2 \sqrt{\varphi(r)} dr.$$

These equations determine a family of possible electron trajectories, a particular trajectory being determined by the values of the constants α_2 and β_2, which in turn depend on the initial conditions.

The constant β_2 can be found if the position of any trajectory point is known. Assuming that it is a point at the cathode surface with coordinates $r = a$ and θ_0, one finds:

$$\beta_2 = \theta_0 - f(a).$$

Then,

$$\theta = \theta_0 + \int_a^r \frac{\alpha + \frac{\hat{e}r^2 B}{2}}{r^2 \sqrt{\varphi(r)}} dr.$$

The arbitrary constant α_2 including in this equation can be easily found since it has the physical meaning of generalized momentum $\alpha_2 = \partial S/\partial \theta = p_\theta$. At the point occurring on the cathode surface $(r = a)$ $p_\theta = -ea^2 B/2$, therefore, $\alpha_2 = p_\theta = -\hat{e}a^2 B/2$.

The ultimate result will be

$$\theta = \theta_0 + \int_a^r \frac{\frac{1}{2}\left(\frac{\hat{e}}{m}\right)^{1/2} B\left(1 - \frac{a^2}{r^2}\right) dr}{\left[2U(r) - \frac{1}{4}\frac{\hat{e}}{m}B^2 r^2 \left(1 - \frac{a^2}{r^2}\right)^2\right]^{1/2}}.$$

This is the equation of a particular electron trajectory. It agrees with those obtained by other methods.

3.7 Motion in Planar Two-Dimensional Fields

3.7.1 Equations of Motion in Newton Form

For the rectangular coordinate system x, y, z taking the x-components of electric and magnetic fields equal to zero, $E_x = B_x = 0$, we find the following equations of motion:

$$m\ddot{x} = e\dot{y}B_z - e\dot{z}B_y, \quad m\ddot{y} = eE_y - e\dot{x}B_z, \quad m\ddot{z} = eE_z + e\dot{x}B_y .$$
(3.54)

3.7.2 Lagrange and Hamilton Equations

These are obtained from the equation of motion in generalized curvilinear coordinates, if we introduce into them the expressions for coordinates, impulses, and components of the vector magnetic potential. Taking into account that the vector magnetic potential has only one component $A_1 = A_x = A$, we obtained the following expressions for the Lagrange and Hamilton functions:

$$L = \frac{m}{2}(\dot{x}^2 + \dot{y}^2 + \dot{z}^2) + eA\dot{x} - eU(x,y) ;$$

$$H = \frac{1}{2m}[(p_x - eA)^2 + p_y^2 + p_z^2] + eU(x,y) .$$

Analog of Bush Theorem

The Hamilton function does not involve the x-coordinate explicitly, therefore,

$$\dot{p}_x = -\frac{\partial H}{\partial x} = 0 \quad \text{and} \quad p_x = \text{const} .$$

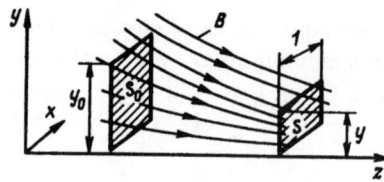

Fig. 3.3. Definition of magnetic fluxes ψ and ψ_0

Since $p_x = \frac{\partial L}{\partial \dot{x}} = m\dot{x} + eA$, it gives $m\dot{x} + eA = \text{const}$ or in equivalent form:

$$m\dot{x} - m\dot{x}_0 = -e(A - A_0) \, ,$$

where $\dot{x}_0$ and A_0 are the values of the velocity and the vector potential at some initial point of the trajectory.

When a magnetic field has the plane of symmetry ($y = 0$) the magnetic potentials can be expressed by the magnetic field flows, ψ and ψ_0, passing through the surfaces of unit width S and S_0 (Fig. 3.3).

This leads to the following expression, which is analogues to the Bush theorem:

$$m\dot{x} - m\dot{x}_0 = -e(\psi - \psi_0) \, . \tag{3.55}$$

3.8 Numerical Calculation of Charged-Particle Trajectories

Motion of the charged-particles is described by the ordinary differential equation of the second order. Several methods are used for their integration: standard methods such as Runge–Kutta and Adams–Stormer methods, methods of the Taylor series and special ones, which take into account the specificity of the integrated equations.

3.8.1 Method of Taylor Series

We consider the application of the method for the calculation of charged-particle trajectories in a two-dimensional electrostatic field. In the rectangular coordinate system y, z the equations of motion are written as follows:

$$\frac{d^2y}{dt^2} = \frac{e}{m}E_y \, , \qquad \frac{d^2z}{dt^2} = \frac{e}{m}E_z \, .$$

Let us suppose that at some initial time t_0 the charged-particle coordinates y_0, z_0 and the components of velocities $\dot{y}_0, \dot{z}_0$ are known. Then, using the Taylor series it is possible to find its position and the velocity in the next time interval $t_1 = t_0 + \Delta t$:

$$y_1 = y_0 + \dot{y}_0 \Delta t + \frac{1}{2}\ddot{y}_0 \Delta t^2 + \frac{1}{6}\dddot{y}_0 \Delta t^3 + \cdots \, ,$$

$$z_1 = z + \dot{z}_0 \Delta t + \frac{1}{2}\ddot{z}_0 \Delta t^2 + \frac{1}{6}\dddot{z}_0 \Delta t^3 + \cdots ,$$

where the dots above the coordinates ($\dot{y}_0, \ddot{y}_0$, etc.) indicate derivatives of the coordinates with time.

Assuming that time step Δt is small, it is possible to neglect the terms of the series containing time steps of degree higher then the second. Then the following system of equations is obtained for determination of coordinates and components of velocity at the end of the time step Δt:

$$y_1 \simeq y_0 + \dot{y}_0 \Delta t + \frac{1}{2}\ddot{y}_0 \Delta t^2 , \qquad \dot{y}_1 \simeq \dot{y}_0 + \ddot{y}_0 \Delta t + \frac{1}{2}\dddot{y}_0 \Delta t^2 ,$$

$$z_1 \simeq z_0 + \dot{z}_0 \Delta t + \frac{1}{2}\ddot{z}_0 \Delta t^2 , \qquad \dot{z}_1 \simeq \dot{z}_0 + \ddot{z}_0 \Delta t + \frac{1}{2}\dddot{z}_0 \Delta t^2 .$$

The values $\ddot{y}_0, \dddot{y}_0, \ddot{z}_0, \dddot{z}_0$ appearing in these equation are found from equations of motion:

$$\ddot{y}_0 = \frac{e}{m}(E_y)_0 , \qquad \dddot{y}_0 = \frac{e}{m}\left[\left(\frac{\partial E_y}{\partial y}\right)_0 \dot{y}_0 + \left(\frac{\partial E_y}{\partial z}\right)_0 \dot{z}_0\right] ,$$

$$\ddot{z}_0 = \frac{e}{m}(E_z)_0 \qquad \dddot{z}_0 = \frac{e}{m}\left[\left(\frac{\partial E_z}{\partial y}\right)_0 \dot{y}_0 + \left(\frac{\partial E_z}{\partial z}\right)_0 \dot{z}_0\right] ,$$

the components of the electrostatic field as well as their derivatives being determined at the initial point having coordinates y_0, z_0.

The found values $y_1, \dot{y}_1, z_1, \dot{z}_1$ are used for the determination of coordinates and components of velocities at the next time step and so on. This gives the following algorithm of trajectory computation:

$$y_{n+1} \simeq y_n + \dot{y}_n \Delta t + \frac{1}{2}\ddot{y}_n \Delta t^2 , \qquad \dot{y}_{n+1} \simeq \dot{y}_n + \ddot{y}_n \Delta t + \frac{1}{2}\dddot{y}_n \Delta t^2 ,$$

$$z_{n+1} \simeq z_n + \dot{z}_n \Delta t + \frac{1}{2}\ddot{z}_n \Delta t^2 , \qquad \dot{z}_{n+1} \simeq \dot{z}_n + \ddot{z}_n \Delta t + \frac{1}{2}\dddot{z}_n \Delta t^2 ,$$

$$\ddot{y}_n = \frac{e}{m}(E_y)_n , \qquad \dddot{y}_n = \frac{e}{m}\left[\left(\frac{\partial E_y}{\partial y}\right)_n \dot{y}_n + \left(\frac{\partial E_y}{\partial z}\right)_n \dot{z}_n\right] ,$$

$$\ddot{z}_n = \frac{e}{m}(E_z)_n \qquad \dddot{z}_n = \frac{e}{m}\left[\left(\frac{\partial E_z}{\partial y}\right)_n \dot{y}_n + \left(\frac{\partial E_z}{\partial z}\right)_n \dot{z}_n\right] .$$

3.8.2 Runge–Kutta Method of Fourth Order of Accuracy

Let us consider the application of the method for numerical solution of the paraxial trajectory equation for a particle moving in an axially symmetrical electrostatic field:

$$\frac{d^2r}{dz^2} + \frac{U_0'}{2U_0}\frac{dr}{dz} + \frac{U_0''}{4U_0}r = 0,$$

where U_0, U_0', U_0'' are, respectively, axial potential and its first and second derivatives on the z-coordinate.

Its solution yields the function $r = f(z)$, which describes the particle trajectory. The equation can be rewritten in the following form:

$$\frac{d^2r}{dz^2} = F(r, z, r'),$$

where

$$F(r, z, r') = -\frac{U_0'}{2U_0}r' - \frac{U_0''}{4U_0}r \quad \text{and} \quad r' = \frac{dr}{dz}.$$

The solution of this equation by the Runge–Kutta method is produced by the following standard algorithm [61]:

$$r_{n+1} = r_n + h\left[r_n' + \frac{1}{6}(k_1 + k_2 + k_3)\right],$$

$$r_{n+1}' = r_n' + \frac{1}{6}(k_1 + 2k_2 + 2k_3 + k_4).$$

In these equations h is the step of integration, the values k_1, k_2, k_3, k_4 are determined by the equations:

$$k_1 = hF(z_n, r_n, r')$$

$$k_2 = hF\left(z_n + \frac{h}{2},\ r_n + \frac{h}{2}r_n' + \frac{h}{8}k_1,\ r_n' + \frac{1}{2}k_1\right),$$

$$k_3 = hF\left(z_n + \frac{h}{2},\ r_n + \frac{h}{2}r_n' + \frac{h}{8}k_1,\ r_n' + \frac{1}{2}k_2\right),$$

$$k_4 = hF\left(z_n + h,\ r_n + hr_n' + \frac{h}{2}k_3,\ r_n' + k_3\right).$$

The initial values z_0, r_0, r_0' are used for the first integration step. The magnitude of the error of this method is of order of $\varepsilon \approx (h^5)$.

3.8.3 Spatial Method of Trajectory Tracing

This method is based on the finite-difference equations, which directly follow from the motion equations. For instance, for the equation of motion in a two-dimensional electrostatic field they are:

$$y(t + \Delta t) = y(t) + \dot{y}\left(t + \frac{1}{2}\Delta t\right)\Delta t,$$

$$\dot{y}\left(t + \frac{1}{2}\Delta t\right) = \dot{y}\left(t - \frac{1}{2}\Delta t\right) + \frac{e}{m}E_y(t)\Delta t$$

$$z(t + \Delta t) = z(t) + \dot{z}\left(t + \frac{1}{2}\Delta t\right)\Delta t,$$

$$\dot{z}\left(t + \frac{1}{2}\Delta t\right) = \dot{z}\left(t - \frac{1}{2}\Delta t\right) + \frac{e}{m}E_z(t)\Delta t.$$

The equations have a simple physical interpretation. The increments of the coordinates are equal to the product of the corresponding components of velocity on the integration time step Δt. The increments of velocity components are determined as the product of the acceleration on the time step.

It should be noted that the values of velocity and accelerated components, which are used in the above equations, correspond to the middle of the time step.

3.9 Electrostatic Charged-Particle Lenses

Some types of specially shaped electric fields possess the property to focus electron beams passing through them. The fields are known as electron lenses. Amongst the number of lens types the lenses of axial symmetry are most widely used in electron devices.

3.9.1 Types of Axially Symmetric Electrostatic Lenses

Axially symmetric electrostatic spatially shaped fields of a finite extension can create the electrostatic lenses, which possess the property to focus or defocus charged-particle beams.

The lenses can be classified by the potential distribution along the lens axis and by the design of electrodes that produce the lens field.

When on both sides of a lens potential is constant and has the same value it is called a unipotential (or single) lens (Fig. 3.4a,b,c). If potentials are constant but have different values on the two sides of the lens, it is known as an immersion lens (Fig. 3.4d,e,f).

As is seen, lens electrode systems consist of different combinations of cylinders and diaphragms.

A special type of lens is formed by a diaphragm electrode with regions of uniform field at one or both sides of it (Fig. 3.5a,b,c). They are known as diaphragm lenses. Diaphragm lenses are usually components of more complicated electron-optical systems.

3.9.2 Focusing Properties of Electrostatic Lenses

Focusing properties can be investigated by analysis of the charged-particle trajectory equation, which is obtained by the elimination of time from the equations of motion.

3 Fundamentals of Charged-Particle Motion

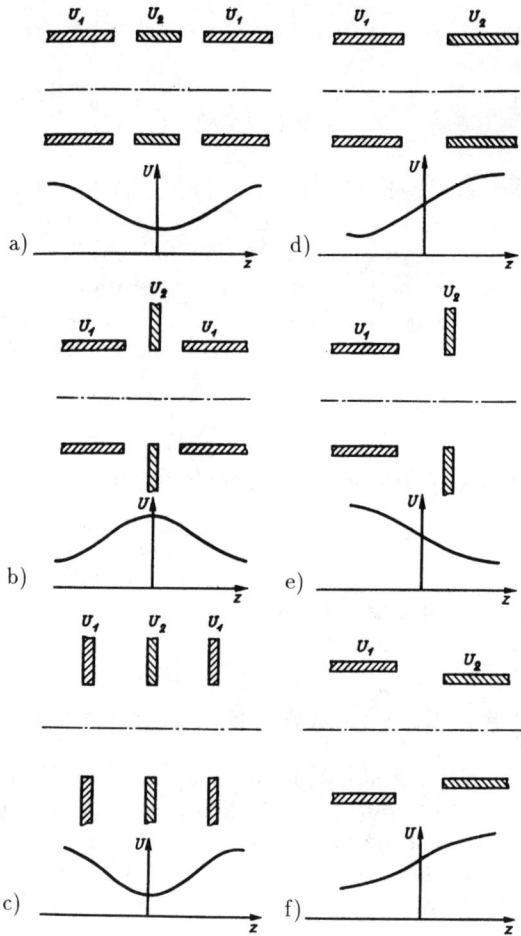

Fig. 3.4. Types of electrostatic lenses with corresponding potential distribution

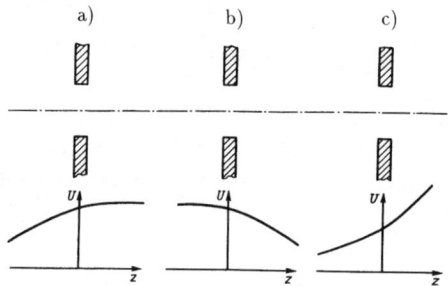

Fig. 3.5. Potential distribution in the diaphragm lenses

3.9 Electrostatic Charged-Particle Lenses

$$m\frac{d^2r}{dt^2} = -e\frac{\partial U}{\partial r}$$

$$m\frac{d^2z}{dt^2} = -e\frac{\partial U}{\partial z}.$$

The procedure of time elimination leads to the following trajectory equation

$$\frac{d^2r}{dz^2} = \left(\frac{\partial U}{\partial r} - \frac{\partial U}{\partial z}\frac{dr}{dz}\right)\left[1 + \left(\frac{dr}{dz}\right)^2\right]\frac{1}{2U}.$$

This can be simplified, if we consider only paraxial trajectories, assuming that r and $\frac{dr}{dz}$ are small. In this case:

$$1 + \left(\frac{dr}{dz}\right)^2 \approx 1, \qquad U \approx U_0(z),$$

$$\frac{\partial U}{\partial r} \approx -U_0''(z)\frac{r}{2}, \qquad \frac{\partial U}{\partial z} \approx U_0'(z),$$

where $U_0(z)$, $U_0'(z)$, U_0'' – potential and its derivatives at the axis of symmetry.

Substitution of these results in the above equations gives the paraxial trajectory equation:

$$\frac{d^2r}{dz^2} + \frac{1}{2}\frac{dr}{dz}\frac{U_0'}{U_0} + \frac{r}{4}\frac{U_0''}{U_0} = 0.$$

It is easy to verify that this equation can be also written in the form:

$$\frac{d}{dz}\left(\sqrt{U_0}\frac{dr}{dz}\right) = -\frac{r}{4}\frac{U_0''}{\sqrt{U_0}}. \tag{3.56}$$

Let us apply it to the investigation of charged-particle motion in the lens region, which is supposed to be limited by the interval $z_1 - z_2$ of the z-axis as is shown in Fig. 3.6.

The integration of this equation within these limits of $z_1 - z_2$ gives:

$$\left(\sqrt{U_0}\frac{dr}{dz}\right)_{z_1}^{z_2} = -\int_{z_1}^{z_2}\frac{r}{4}\frac{U_0''}{U_0}dz,$$

or

$$\sqrt{U_{02}}\left(\frac{dr}{dz}\right)_2 - \sqrt{U_{01}}\left(\frac{dr}{dz}\right)_1 = -\int_{z_1}^{z_2}\frac{r}{4}\frac{U_0''}{\sqrt{U_0}}dz.$$

This equation permits determination of the slope of a trajectory at the exit of the lens if the entrance slope $(dr/dz)_1$ is known.

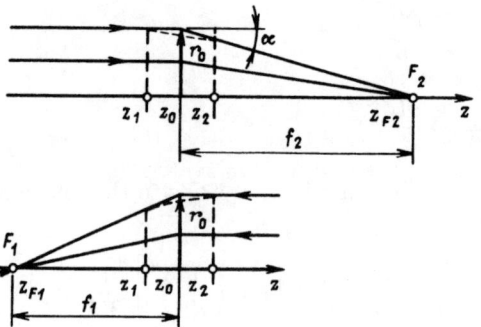

Fig. 3.6. Definition of focusing distances of a thin lens

Thin Lens

A lens is called "thin" or "weak", if the extent of the lens field is short and the field is weak. The change of charged-particle radial position within the thin lens is rather small. With a fair degree of accuracy we can assume it to be constant $r \approx r_0 = \text{const}$.

Then, (3.56) can be written as

$$\sqrt{U_{02}}\left(\frac{dr}{dz}\right)_2 - \sqrt{U_{01}}\left(\frac{dr}{dz}\right)_1 = -\frac{r_0}{4}\int_{z_1}^{z_2}\frac{U_0''}{\sqrt{U_0}}dz \; .$$

The integration by parts of the integral leads to:

$$\sqrt{U_{02}}\left(\frac{dr}{dz}\right)_2 - \sqrt{U_{01}}\left(\frac{dr}{dz}\right)_1 = -\frac{r_0}{4}\left(\frac{U_0'}{\sqrt{U_0}}\right)_{z_1}^{z_2} - \frac{r_0}{8}\int_{z_1}^{z_2}\frac{(U_0')^2}{U_0^{3/2}}dz \; ,$$

or

$$\sqrt{U_{02}}\left(\frac{dr}{dz}\right)_2 - \sqrt{U_{01}}\left(\frac{dr}{dz}\right)_1 = \frac{r_0 U_{01}'}{4\sqrt{U_{01}}} - \frac{r_0 U_{02}'}{4\sqrt{U_{02}}} - \frac{r_0}{8}\int_{z_1}^{z_2}\frac{(U_0')^2}{U_0^{3/2}}dz \; ,$$

where U_0' and U_{02}' are the gradients of the electric potential at the left and the right sides of the lens.

For the charged-particle entering the lens parallel to the z-axis with zero slope $(dr/dz)_1 = 0$ we obtain the following equation for the trajectory slope at the exit of the lens:

$$\left(\frac{dr}{dz}\right)_2 = \frac{r_0}{4\sqrt{U_{02}}}\left(\frac{U_{01}'}{\sqrt{U_{01}}} - \frac{U_{02}'}{\sqrt{U_{02}}}\right) - \frac{r_0}{8}\int_{z_1}^{z_2}\frac{(U_0')^2}{U_0^{3/2}}dz \; .$$

If the regions at the left and at the night sides of a lens are free of field $U_{01}' = U_{02}' = 0$, we obtain:

$$\left(\frac{dr}{dz}\right)_2 = -\frac{r_0}{8\sqrt{U_{02}}} \int_{z_1}^{z_2} \frac{(U_0')^2}{U_0^{3/2}} dz \ . \tag{3.57}$$

Since the value of $(U_0')^2$ is certainly positive, the slope of a trajectory $(dr/dz)_2$ will be negative for $r > 0$ and positive for $r < 0$ so that trajectories at the exit of the lens will be inclined towards the axis (Fig. 3.6). So we can state that electric fields of finite length create lenses that possess the convergent action. Single and immersion lenses mentioned above are, therefore, convergent.

Focal Length of a Thin Lens

The focal length of the thin lens is the distance from the central plane of the lens to the point where the trajectory, which is parallel to the $z-$ axis at lens entrance, crosses the axis (Fig. 3.6).

Assuming that the trajectory slope is changed by a step at the middle plane of the lens, we find the following trajectory equation after the lens:

$$r = r_0 + \left(\frac{dr}{dz}\right)_2 (z - z_0) \ .$$

From this follows the expression for the focal length:

$$f_2 = z_{F_2} - z_0 = -\frac{r_0}{\left(\dfrac{dr}{dz}\right)_2} \ .$$

Using (3.57) we can obtain the focal length of the thin electrostatic lens. For charged-particles moving from the left to the right (Fig. 3.6) we find:

$$\frac{1}{f_2} = -\frac{1}{r_0}\left(\frac{dr}{dz}\right)_2 = \frac{1}{8\sqrt{U_{02}}} \int_{z_1}^{z_2} \frac{(U_0')^2}{U_0^{3/2}} dz \ . \tag{3.58}$$

By analogy with the above analysis we find that the lens is also convergent for charged-particles moving from the right to the left. Its focal length will be

$$\frac{1}{f_1} = -\frac{1}{8\sqrt{U_{01}}} \int_{z_1}^{z_2} \frac{(U_0')^2}{U_0^{3/2}} dz \ , \tag{3.59}$$

the distance to the left of the lens central plane z_0 is supposed to be negative.

From (3.58) and (3.59) we find:

$$\frac{f_1}{f_2} = -\frac{\sqrt{U_{01}}}{\sqrt{U_{02}}} \ .$$

It is easy to see that the values of focal lengths f_1 and f_2 are equal for single lenses and different for immersion lenses.

In contrast to the considered cases the diaphragm lenses can be convergent or divergent depending on the field gradients at the left and the right sides of the diaphragm.

Using the definition of focal length $\frac{1}{f_2} = -\frac{1}{r_0}\left(\frac{dr}{dz}\right)_2$ given above, we find the following equation for the focal length of diaphragm lenses:

$$\frac{1}{f_2} = \frac{1}{4\sqrt{U_{02}}}\left(\frac{U'_{02}}{\sqrt{U_{02}}} - \frac{U'_{01}}{\sqrt{U_{01}}}\right) + \frac{r_0}{8}\int_{z_1}^{z_2}\frac{(U'_0)^2}{U_0^{3/2}}dz .$$

In the case of the diaphragm lens the interval of varying field $z_2 - z_1$ approaches zero and the second term including the integral vanishes, which results in:

$$\frac{1}{f_2} \approx \frac{1}{4\sqrt{U_{02}}}\left(\frac{U'_{02}}{\sqrt{U_{02}}} - \frac{U'_{01}}{\sqrt{U_{01}}}\right) .$$

Taking into account that in the case under consideration $U_{01} \approx U_{02} \approx U_0$ we obtained

$$\frac{1}{f_2} \approx \frac{1}{4\sqrt{U_0}}(U'_{02} - U'_{01}) , \quad \text{or} \quad \frac{1}{f_2} \approx \frac{1}{4U_0}(E_1 - E_2) , \qquad (3.60)$$

where E_1 and E_2 are electric field strength on the left and on the right of the diaphragm.

Matrix Equation for the Thin Unipotential Lens

The paraxial equation of the charged-particle trajectory is the linear differential equation. This allow us to consider an electrostatic lens as a system that produces the linear transformation of the entrance trajectory parameters r_1 and $\left(\frac{dr}{dz}\right)_1$ to the final ones r_2 and $\left(\frac{dr}{dz}\right)_2$, and use the matrix method for the description of the lens action.

As mentioned above, the thin lens does not change the radial position of charged-particles passing through it but it changes the trajectory slope by a step. Therefore, we can put $r_2 = r_1 = r_0$. The transformation of the slope is governed by the following equation

$$\left(\frac{dr}{dz}\right)_2 = \left(\frac{dr}{dz}\right)_1 - \frac{r_0}{8\sqrt{U_{02}}}\int_{z_1}^{z_2}\frac{(U'_0)^2}{U_0^{3/2}}dz .$$

Since

$$\frac{1}{8\sqrt{U_{02}}}\int_{z_1}^{z_2}\frac{(U'_0)^2}{U_0^{3/2}}dz = \frac{1}{f_2} ,$$

we obtain the following expression for the slope transformation:

$$\left(\frac{dr}{dz}\right)_2 = \left(\frac{dr}{dz}\right)_1 - \frac{r_0}{f_2} ,$$

or, denoting $\left(\frac{dr}{dz}\right)_1 = r'_1$ and $\left(\frac{dr}{dz}\right)_2 = r'_2$, we get $r'_2 = r'_1 - \frac{r_0}{f}$.

Taking into account the last expression and the equality $r_2 = r_1 = r_0$ we can find the equivalent matrix lens equation:

$$\begin{vmatrix} r_2 \\ r'_2 \end{vmatrix} = M \begin{vmatrix} r_1 \\ r'_1 \end{vmatrix}. \tag{3.61}$$

where M is the matrix of the thin lens:

$$M = \begin{vmatrix} 1 & 0 \\ -1/f & 1 \end{vmatrix}. \tag{3.62}$$

Angle of Trajectory Refraction

The angle of trajectory refraction is an important parameter that is convenient to use in the theory and design of the focusing systems for intense charged-particle beams. In accordance with the drawing shown in Fig. 3.6 the angle of trajectory refraction is determined as:

$$\alpha \approx tg\alpha = \frac{r_0}{f_2} = \frac{r_0}{8\sqrt{U_{02}}} \int_{z_1}^{z_2} \frac{(U'_0)^2}{U_0^{3/2}} dz. \tag{3.63}$$

Thick Particle Lens

The extension of the active field region in the thick lens is compared with the lens focal distance. The focusing action of a thick lens can be described by refraction of the charged-particle trajectories, but in contrast to the thin lens, the refractions are attached not to the middle plane of the lens but to the principal planes H_1 and H_2 as is shown at Fig. 3.7.

The focusing action is described by the following four parameters: f_1 and f_2 – the focal lengths and distances h_1 and h_2 that determine the position of the principal planes. Knowing these parameters it is possible to find the charged-particle trajectory behind the lens if a entrance parameters are known.

The matrix equation of the thick lens has the same form as for a thin lens:

$$\begin{vmatrix} r_2 \\ r'_2 \end{vmatrix} = M \begin{vmatrix} r_1 \\ r'_1 \end{vmatrix},$$

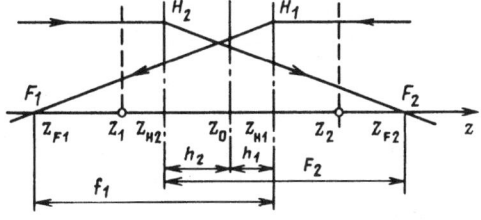

Fig. 3.7. Geometrical model of the thick lens and its main dimensions.

where M is the matrix of transformation that depends on the lens type.

For instance, the unipotential symmetric lens has the following matrix of transformation (for the region limited by the two principal planes H_1 and H_2):

$$M = \begin{vmatrix} 1 - s/d & 2s - s^2/f \\ -1/f & 1 - s/f \end{vmatrix}, \tag{3.64}$$

where $f = f_2 = |f_1|$, $s = 2h$, $h = h_1 = |h_2|$.

Spherical Aberration of Lenses

As applied to the calculation and design of focusing systems for intense charged-particle beams the lens spherical aberration is most important as compared with other lens aberrations. It appears in the nonlinear dependence of the angle of trajectory refraction on its radial position r_0 at the middle plane of the lens.

As a result of this peripheral trajectories are refracted more strongly than predicated by (3.63). A peripheral trajectory crosses the lens axis before the point of the paraxial focus (Fig. 3.8), its radial coordinate δ in the focal plane is equal to

$$\delta = C_s \alpha^3 ,$$

where C_s is the coefficient of spherical aberration.

For the case of intense charged-particle beams the effect of the spherical aberration can be taken into account by the correction of the angle of refraction which is expressed as [10]:

$$\Delta\alpha = \left(\frac{C_s}{f}\right) \alpha^3 \cos\alpha . \tag{3.65}$$

Then for the total angle of refraction we find:

$$\alpha' = \alpha + \Delta\alpha = \alpha + \left(\frac{C_s}{f}\right) \alpha^3 \cos^2\alpha .$$

For rather small angles of refraction taking $\alpha \approx r_0/f$, $\cos^2\alpha \approx 1$ and introducing the angle coefficient of spherical aberration $C_\alpha = C_s/f^3$, we obtain:

$$\alpha' = \left(\frac{r_0}{f}\right)\left(1 + C_\alpha r_0^2\right) . \tag{3.66}$$

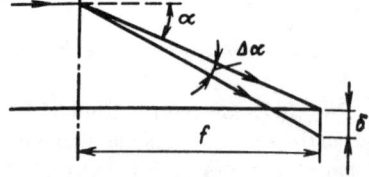

Fig. 3.8. Trajectory refraction in the lens possessing the spherical aberration. f – the paraxial focal length, α – the paraxial angle of refraction, $\Delta\alpha$ – the correction for the refraction angle

3.9 Electrostatic Charged-Particle Lenses

Therefore, correction for the angle of trajectory refraction is expressed as:

$$\Delta\alpha/\alpha = C_\alpha r_0^2 .$$

Parameters of Some Electrostatic Lenses

The two-electrode immersion lens consists of two coaxial cylinders separated by a gap, the cylinders being kept at different potentials U_1 and U_2. Depending on the relation of the potentials U_2/U_1 the lens can be accelerating or decelerating. The parameters of the lens are presented in Fig. 3.9. Focal lengths f_1 and f_2 and the distances $p_1 = z_0 - z_{F1}$ and $p_2 = z_{F2} - z_0$, which determine the position of the focal points F_1 and F_2, are plotted as a function of the potential relation U_2/U_1.

The parameters of a three-electrode unipotential lens are given in Fig. 3.10 when the potential of the middle electrode is equal to the cathode potential [62].

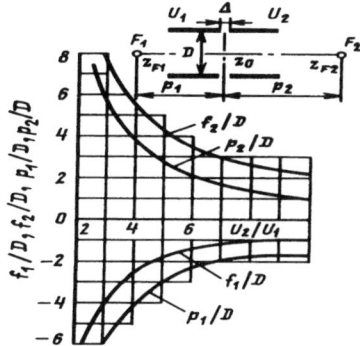

Fig. 3.9. Parameters of an immersion lens

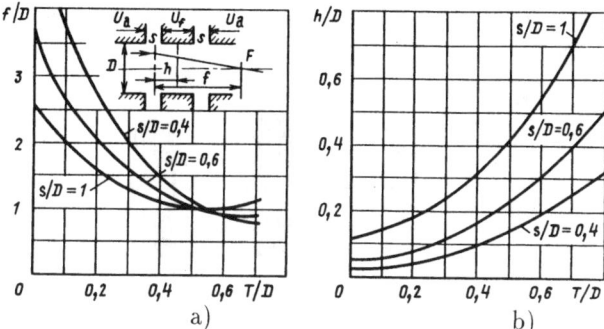

Fig. 3.10. Parameters of a unipotential lens as function of its geometry. **a** focal length f as function the middle electrode thickness T. **b** distance of the principal plane from the middle plane of the lens h as a function of the middle electrode thickness T

3.10 Magnetic Lenses

Magnetic lenses are created by axially symmetric magnetic fields of a finite extent, which are produced by electromagnetic coils or ring-shaped permanent magnets (Figs. 3.11–3.16).

The properties of magnetic lenses can be investigated by using the paraxial trajectory equation, which is obtained from the general paraxial trajectory equation (3.47), if we assume that there is no axis potential variation ($U_0 =$ const) and the charged-particle emitter is shielded from the magnetic field ($B_{z0k} = 0$):

$$\frac{d^2 r}{dz^2} - \frac{e}{8m} \frac{B_0^2(z)}{U_0} r = 0 ,$$

where $B_0(z)$ is the axial component of magnetic induction U_0 – axis potential.

Assuming that the magnetic field exists inside the finite interval of the z-axis $z_2 - z_1$ we find the after integration:

$$\left(\frac{dr}{dz}\right)_2 - \left(\frac{dr}{dz}\right)_1 = \frac{e}{8mU_0} \int_{z_1}^{z_2} r B_0^2(z) dz .$$

Or, after rearranging of the right part of this equation:

$$\left(\frac{dr}{dz}\right)_2 - \left(\frac{dr}{dz}\right)_1 = -\frac{e^2}{8m |eU_0|} \int_{z_1}^{z_2} r B_0^2(z) dz ,$$

where $\left(\frac{dr}{dz}\right)_1$ and $\left(\frac{dr}{dz}\right)_2$ are the slopes of the charged-particle trajectory at the entrance and the exit of the lens.

3.10.1 Thin Magnetic Lens Approximation

For the case of the thin or weak lens the radial coordinate r is considered to be unchanged within the interval $z_2 - z_1$, ($r = r_0 =$ const) and the only

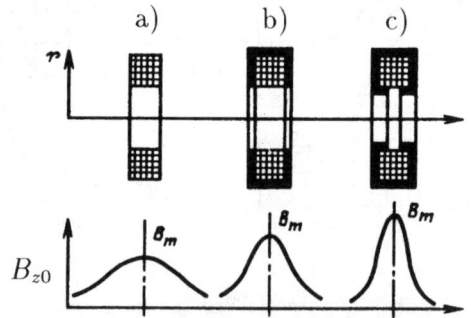

Fig. 3.11. Solenoid-type magnetic lenses and corresponding axial magnetic distribution. **a** – coil without shield, **b** and **c** – shielded coils

trajectory slope is changed abruptly. This allows us to take the r-coordinate from the integral and write:

$$\left(\frac{dr}{dz}\right)_2 - \left(\frac{dr}{dz}\right)_1 = -\frac{e^2 r_0}{8m|eU_0|} \int_{z_1}^{z_2} B_0^2(z)dz .$$

For a charged-particle moving parallel to the axis of symmetry and entering the lens from the left to the right, so that $\left(\frac{dr}{dz}\right)_1 = 0$, we find:

$$\left(\frac{dr}{dz}\right)_2 = -\frac{e^2 r_0}{8m|eU_0|} \int_{z_1}^{z_2} B_0^2(z)dz .$$

Evidently, the electron trajectory is deflected toward the axis.

The same result is obtained for the electron trajectory passing the lens from the right to the left.

$$\left(\frac{dr}{dz}\right)_1 = \frac{e^2 r_0}{8m|eU_0|} \int_{z_2}^{z_1} B_0^2(z)dz .$$

A thin magnetic lens has a convergent action. Its focal lengths are:

$$\frac{1}{f_2} = \frac{e^2}{8m|eU_0|} \int_{z_1}^{z_2} B_0^2(z)dz , \quad \frac{1}{f_1} = -\frac{e^2}{8m|eU_0|} \int_{z_2}^{z_1} B_0^2(z)dz . \quad (3.67)$$

The angle of trajectory refraction is expressed as

$$tg\alpha = \frac{e^2 r_0}{8m|eU_0|} \int_{z_1}^{z_2} B_0^2(z)dz . \quad (3.68)$$

Azimuthal Charged-Particle Drift

In the case of the magnetic lens a charged-particle passing the lens field is affected by the azimuthal force that leads to the charge azimuthal drift expressed by the formula

$$\theta_2 - \theta_1 = -\frac{e}{(8m|eU_0|)^{1/2}} \int_{z_1}^{z_2} B_0(z)dz .$$

As is seen, the azimuth drift depends on the charged-particle sign and the direction of the axis of magnetic induction $B_0(z)$.

3.10.2 Thick Magnetic Lens

In the case when the extension of the lens magnetic field is compared with the lens focal length, the lens should be considered as a thick one. The geometrical model used for description of its focusing action is similar to the model applied for the thick electrostatic lens with the same set of main lens parameters.

In general, the lens parameters are determined numerically by calculation of charged-particle trajectories.

The matrix equation for the thick symmetrical magnetic lens coincides with the one used for the thick symmetrical unipotential electrostatic lens.

3.10.3 Parameters of Some Magnetic Lenses

Thin Coil

The simplest magnetic lens is created by the thin magnetic coil (Fig. 3.12). When the sizes of the coil cross section are considerably less than its average radius, the axial magnetic field distribution can be approximately expressed by the formula obtained for the single ring turn with the current being equal to the thin coil excitation IN (where I is coil current, N is the number of coil turns):

$$B_{z0} = \frac{1}{2}\mu_0 R^2 IN \left(z^2 + R^2\right)^{-3/2}.$$

Substitution of this equation in the formula for focus length of the thin lens and integration in the limits $z_1 = -\infty$ and $z_2 = +\infty$ yields

$$\frac{1}{f} = \frac{3\pi e^2 \mu_0^2}{16^2 m |eU_0|} \frac{(IN)^2}{R}. \tag{3.69}$$

Magnetic Coil with a Shield

The geometry of the coil is presented in Fig. 3.13. The value of magnetic induction and its axial distribution depends on the coil excitation IN and on the magnetic-shield geometry.

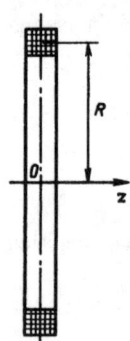

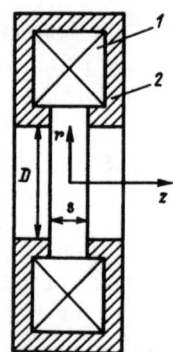

Fig. 3.12. Thin magnetic coil

Fig. 3.13. Shielded coil. 1 – winding, 2 – shield

3.10 Magnetic Lenses

For the case when the relation $0.5 \leq D/s \leq 2$ is fulfilled and the saturation of the shield material is absent, the axial distribution of the magnetic induction is expressed by the formula [63]:

$$B_{z0} = \frac{1.257 \times 10^{-4} IN}{2s} \left[\frac{z + \frac{s}{2}}{\sqrt{\left(\frac{D}{3}\right)^2 + \left(z + \frac{s}{2}\right)^2}} - \frac{z - \frac{s}{2}}{\sqrt{\left(\frac{D}{3}\right)^2 + \left(z - \frac{s}{2}\right)^2}} \right],$$

where magnetic induction B_{z0} is expressed in Tesla, excitation IN in Amperes, the linear dimensions in centimeters.

In general, the optical parameters of the lens are found by numerical trajectory computation. When the inequality $\frac{s}{D} \leq 1$ is fulfilled, it is possible to approximate the induction distribution by the equation:

$$B_{z0} = \frac{B_m}{1 + \left(\frac{z}{d}\right)^2}.$$

This is known as Glaser's approximation. Then, substituting this formula in the equation for the focal length of the thin lens we find:

$$\frac{1}{f} = \frac{e^2}{8m|eU_c|} \int_{-\infty}^{+\infty} \frac{B_m^2}{\left[1 + \left(\frac{z}{d}\right)^2\right]^2} dz = \frac{e^2}{8m|eU_0|} \frac{\pi d B_m^2}{2} = \frac{\pi k^2}{2d},$$

where $k^2 = e^2 B_m^2 d^2 / 8m|eU_0|$ is a parameter that defines the refractive lens action, for electrons $k^2 = 2.2 \times 10^4 B_m^2 d^2 / U_0$.

Parameters B_m and d appearing in the above equations are connected with the lens parameters by the equalities:

$$d = 0.48\sqrt{s^2 + 0.45 D^2}, \quad B_m = \frac{\mu_0 IN}{\pi d}.$$

The estimation of the angle spherical aberration of the lens leads to the equation

$$C_\alpha \approx \frac{1.25}{Ds},$$

where D and s are lens dimensions expressed in cm.

Long Magnetic Lens

The lens is created by a long coil with uniform winding (Fig. 3.14). The magnetic field of the coil has the regular region, in which the field is near uniform with only a longitudinal component of magnetic induction B_z, and edge regions, where the magnetic field has both field components B_r and B_z.

Let us consider the motion of a charged-particle, which is emitted and accelerated by a source located outside of the magnetic field.

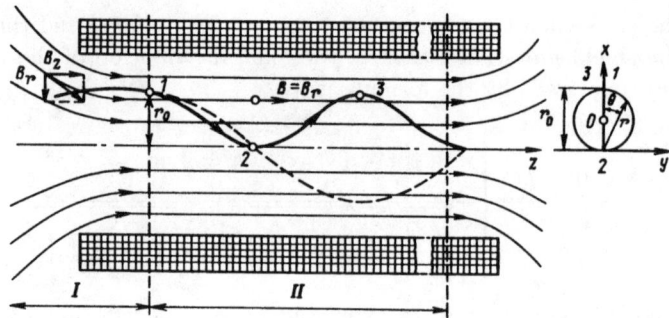

Fig. 3.14. Long magnetic lens: Magnetic field lines and charged-particle trajectories

When moving through the edge region the charged-particle interacts with B_r and B_z components of magnetic induction and as a result of that obtains the azimuthal component of velocity $v_\theta = r\dot\theta$, its value being found from the Bush theorem. Then the motion of the charged-particle in the regular region will be described by the following systems of equations:

$$\ddot r = -\frac{1}{4}\left(\frac{e}{m}\right)^2 B_z^2, \qquad \dot\theta = \dot\theta_0 = -\frac{1}{2}\frac{e}{m}B_z, \qquad \ddot z = 0.$$

Their integration with time gives:

$$r = r_0 \cos\omega_L t + \left(\frac{\dot r_0}{\omega_L}\right)\sin\omega_L t,$$

$$z = z_0 + \dot z_0 t, \qquad \theta = \theta_0 + \dot\theta t = \theta_0 - \frac{1}{2}\frac{e}{m}B_z t,$$

where $r_0, \theta_0, z_0, \dot r_0, \dot\theta_0, \dot z_0$ are the initial coordinates and velocity components at the entrance to the uniform field region, $\omega_L = \frac{|e|}{2m}B_z$ is a parameter known as the Larmor frequency.

Assuming for the simplification of further analysis $z_0 = 0$, $\theta_0 = 0$, $\dot r_0 = 0$ and excluding time t with the help of the expressions $t = z/\dot z_0$, $t = \theta/\dot\theta_0$, we obtain:

$$r = r_0 \cos\left(\frac{\omega_L z}{\dot z_0}\right), \qquad r = r_0 \cos\left(\frac{\omega_L \theta}{\dot\theta_0}\right).$$

The first equation describes the projection of the particle trajectory on the meridian plane rz rotating with angular velocity $\dot\theta$. The second one gives the trajectory projection on the azimuthal plane $r\theta$.

Introducing the rectangular coordinates x, y, z in the way shown in Fig. 3.14, we have $x = r\cos\theta$, $y = r\sin\theta$. Taking into account the above two equations we obtain the projection of particle trajectory on xy and xz planes:

$$\left(x - \frac{r_0}{2}\right)^2 + y^2 = \frac{r_0^2}{2}$$

$$x = \frac{r_0}{2}\left(1 + \cos\omega_L \frac{z}{\dot{z}_0}\right).$$

The trajectory projection on the xy plane is the circle with the center at the point O having the coordinates $x = r_0/2$, $y = 0$ and radius of the circle equal to $r_0/2$. The trajectory projection xz plane is the curve shown in Fig. 3.14 by the solid line. The dashed line presents the trajectory projection on the meridian rz plane rotating with angular velocity $\dot{\theta}$.

The particle trajectory passes through the $z-$axis at the points for which the value x became zero. The positions of the points are found from the equations:

$$2\omega_L \frac{z}{\dot{z}} = n\pi, \quad \text{or} \quad z = \frac{n\pi}{2\omega_L}\dot{z}_0,$$

where n are integer odd numbers.

At fixed values of ω_L the z-coordinates of these points depend only on the z-component of velocity $\dot{z}_0$ and does not depend on the initial radial coordinate r_0. This means that charged-particles entering the regular region of the long magnetic coil the equal z-component of velocity but at a different radial coordinate r_0 will be focused at the same points of the z-axis. Therefore, a narrow parallel charged-particle beam being injected in the long magnetic lens will also be focused at these points.

Magnetic Lens with Divergent Action

This lens can be realized if the emitter of charged-particles is placed in the sharply falling magnetic field, produced, for instance, by the shielded solenoid (Fig. 3.15).

A charged-particle moving from the emitter toward the field-free region obtains the azimuthal component of velocity; its value is found from the Bush theorem:

$$v_\theta = r\dot{\theta} = -\frac{1}{2\pi}\frac{e}{m}\frac{\psi - \psi_k}{r}.$$

Substitution in this equation $\psi = 0$ gives the azimuthal velocity at the entrance to the field-free region.

Subsequently, the particle will move away from the $z-$axis with transverse velocity

$$v_\perp = \frac{1}{2\pi}\frac{e}{m}\frac{\psi_k}{r_0},$$

where r_0 is the radial particle coordinate at the entrance to the field-free region.

3.10.4 Magnetic Lenses with Permanent Magnets

Several types of lenses with the corresponding curves of the axial distributions of magnetic induction are shown in Fig. 3.16: a – ring-shaped magnet (barium

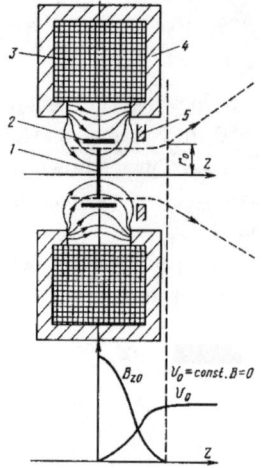

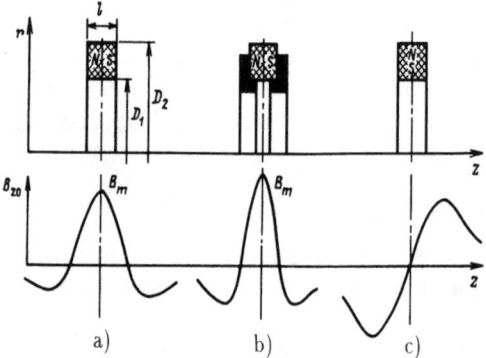

Fig. 3.15. Magnetic lens with divergent action: 1 – emitter, 2 – focusing electrode, 3 – coil, 4 – shield, 5 – accelerating electrode; dashed lines with arrows shows charged-particle trajectories. Axial distributions of magnetic induction B_{z0} and accelerating potential U_0 are plotted on the lower drawing

Fig. 3.16. Types of magnetic lenses: **a** – ring-shaped magnet magnetized in axial direction, **b** – magnet with the soft iron shield, **c** – ring-shaped magnet magnetized in radial direction

ferrite) magnetized in the axial direction, b – ring-shaped magnet magnetized in the axial direction with the iron shield, c – ring-shaped magnet magnetized in the radial direction.

Ring-Shaped Magnet Magnetized in Axial Direction

The axial distribution of magnetic induction and configuration of magnetic force lines are presented in Figs. 3.16a and 3.17a.

When the magnetization of the magnet J is uniform the axial distribution of the magnetic induction is found from the expression [52]

$$B_{z0} = \frac{\mu_0 J}{2} F(z) \ . \tag{3.70}$$

The function $F(z)$ depends on the axial coordinate z, the ring dimensions D_1, D_2, l and can be calculated in the following way:

$$F(z) = \frac{f_1}{F_1} + \frac{f_2}{F_2} - \frac{f_3}{F_3} - \frac{f_4}{F_4} \ . \tag{3.71}$$

3.10 Magnetic Lenses

The functions $f_1, f_2, f_3, f_4, F_1, F_2, F_3, F_4$, including in the above formula, are determined by the equations:

$$F_1 = (1+f_1)^{1/2}, \quad F_2 = (1+f_2)^{1/2}, \quad F_3 = (1+f_3)^{1/2},$$

$$F_4 = (1+f_4)^{1/2},$$

$$f_1 = \frac{l-2z}{D_1}, \quad f_2 = \frac{l+2z}{D_1}, \quad f_3 = \frac{l+2z}{D_2}, \quad f_4 = \frac{l-2z}{D_2}.$$

The value of magnetization J in the operational point is found from the formula

$$J \approx \frac{1}{2}(J_r + H_{CB}),$$

where J_r and H_{CB} are the residual magnetization and the coercive force of the magnet.

When conditions $l/D_2 \leq 0.2$ and $D_1/D_2 \geq 0.3$ are fulfilled, the axial distribution of the magnetic induction is approximated by the expression [64]:

$$B_{z0} = B_m \frac{1-2Z^2}{(1+Z^2)^{5/2}}, \qquad (3.72)$$

where Z is the normalized axial coordinate $Z = z/R$, $R = \frac{D_1+D_2}{4}$ is the average beam radius.

Substitution of this expression in the formula for focal length of the thin lens and numerical integration gives:

$$\frac{1}{f} = \frac{1.43 \times 10^6 B_m^2 R}{U_0},$$

the values B_m, R and U_0 being expressed in Tesla, centimeters and Volts, respectively.

Using an approximate formula (3.72) it is possible to compute the angle coefficient of spherical aberration:

$$C_\alpha \approx \frac{2.5}{R^2}.$$

Ring-Shaped Magnet Magnetized in Radial Direction

The axial magnetic induction distribution and configuration of magnetic force lines are presented in Figs. 3.16c and 3.17b.

The axial induction distribution is expressed by the equation:

$$B_{z0} = \frac{\mu_0 J}{2} F(z),$$

where J is radial magnetization of the magnet, $F(z)$ is the function of coordinate z and the ring dimensions.

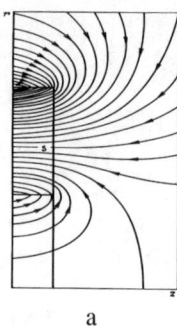 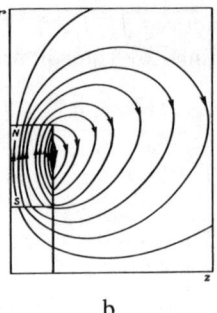

a b

Fig. 3.17. Pictures of magnetic force lines for ring-shaped magnets magnetized in axial direction (**a**) and radial direction (**b**)

The function $F(z)$ is calculated in the following way:

$$F(z) = \frac{1}{F_1} - \frac{1}{F_2} - \frac{1}{F_3} + \frac{1}{F_4} + ln\,|F_5|\ .$$

The functions F_1, F_2, F_3, F_4, included in this formula, are the same as in (3.71), F_5 is found from the expression:

$$F_5 = \frac{(1+f_3)(1+f_2)}{(1+f_1)(1+f_4)}\ .$$

Numerical computation is used for determination of the lens parameters.

Ring-Shaped Magnet with Soft Iron Shield

Application of the soft iron shield leads to concentration of the magnetic field at the center of the lens and increasing of the maximum magnetic induction B_m (Fig. 3.16b). This results in the increasing of the refraction action of the magnetic lens.

Magnetic-field computation as well as calculation of the lens parameters are produced by numerical methods using corresponding computer program (codes).

4 Motion of Intense Charged-Particle Beams

4.1 Peculiarities of Intense Particle-Beam Motion

4.1.1 Effects of Self-Fields and Velocity Spreading

Motion of intense charge particle beams is considerably affected by particle interactions. There exist two types of interaction: particle to particle collisions and collective interaction. In the latter case the charge particles interact through total self-fields created all the particles of the beam.

As has been shown by Vlasov [65], mainly the collective interactions are important, when intense beam motion is treated.

There are collective self-fields of different nature: electrostatic field and magnetic field. The former is the field creating by continuously distributed space charge of density $\rho(x, y, z)$. It obeys the Poisson equation:

$$\nabla^2 U = -\rho/\varepsilon_0. \tag{4.1}$$

The magnetic field is created by a convection current with density $\vec{j} = \rho \vec{v}$.

In this chapter we consider mainly effects of electric self-fields, leaving the effects of magnetic fields for Chap. 10, where intense relativistic beams are analyzed, since the effects of a self-magnetic field essentially appears at relativistic velocities of beam motion.

The most evident effects of electrostatic self-fields are spreading of the beams, enlargement of their cross section.

Another effect of the space charge is the changing (depression) of the potential of the space, where beams are moving. This effect leads to limitation of the value of the beam current, which can be transmitted through interelectrode spaces and tube channels.

The essential feature of the intense beam motion lies in the fact that the motion occurs in the fields, which in turn depends on this motion. Such fields are known as self-consistent fields.

Particles of real beams leave the emitter surface with different values and directions of velocities. This results in nonlaminar charge particle flows, when at an arbitrary point of flow there exist particles with different velocities. The effects of initial velocities leads to spreading of edges in extended beams and limitation of the attainable minimum beam cross section in point focus beams.

4.1.2 Mathematical Description of Multiple-Velocity Beams

The concept of phase space, that is the six-dimensional space of coordinates and velocities x, y, z, v_x, v_y, v_z is used for the mathematical description of multiple-velocity beam motion.

The elementary volume of phase space is $\Delta \tau = dxdydzdv_xdv_ydv_z$. The number of charged-particles occurring in the volume is equal:

$$dN = f(x, y, z, v_x, v_y, v_z)\, dx, dy, dz, dv_x, dv_y, dv_z ,$$

where $f(x, y, z, v_x, v_y, v_z)$ is the phase-space particle density.

In the case when collisions of charge particles are absent the Liouville theorem states that the phase-space particle density satisfies the equation [66, 67]:

$$\frac{d}{dt} f(x, y, z, v_y, v_z) = 0 . \tag{4.2}$$

This result has the following physical sense: the phase-space particle density in the vicinity of a particle, remains a constant, when the particle is moving.

The total derivative of phase space density can be expressed as follows [66]:

$$\frac{\partial f}{\partial t} + \vec{v}\nabla f + \frac{d\vec{v}}{dt}\nabla_v f = 0 , \tag{4.3}$$

where $\Delta \frac{\partial f}{\partial x} + \vec{i}_y \frac{\partial f}{\partial y} + \vec{i}_z \frac{\partial f}{\partial z}$, $\Delta_v f = \vec{i}_x \frac{\partial f}{\partial v_x} + \vec{i}_y \frac{\partial f}{\partial v_y} + \vec{i}_z \frac{\partial f}{\partial v_z}$ are gradients of density in coordinate and velocity spaces, respectively.

The particle acceleration appearing in the above equation can be expressed in terms of the Lorenz force

$$\frac{d\vec{v}}{dt} = \frac{e}{m}(\vec{E} + \vec{v} \times \vec{B}) . \tag{4.4}$$

Substitution of (4.4) in (4.3) gives

$$\frac{\partial f}{\partial t} + v\nabla f + \frac{e}{m}(\vec{E} + \vec{v} \times \vec{B})\nabla_v f = 0 . \tag{4.5}$$

If the electric field $\vec{E}$ includes the space-charge field, this equation is called the kinetic equation with self-consistent field, or the Vlasov equation.

In the case of steady-state flow the value of $\partial f/\partial t$ in (4.3) and (4.5) is taken to be equal zero, $\partial f/\partial t = 0$.

A partial solution of (4.2) and, therefore, solutions of (4.3) and (4.5) can be arbitrary function of integrals of motion, which are the values characterizing of charge particle motion, being invariant of motion $J(t) = \text{const}$.

4.1 Peculiarities of Intense Particle-Beam Motion 97

When motion proceeds in static fields an integral of motion is total energy

$$mv^2/2 + eU = \text{const},$$

which remains a constant throughout the particle motion.

So, if there is a function $f(J)$, with J being invariant of t, this function will satisfy (4.2):

$$\frac{df}{dt} = \frac{\partial f}{\partial J}\frac{dJ}{dt} = 0, \quad \text{since} \quad \frac{dJ}{dt} \equiv 0. \tag{4.6}$$

Law of Phase-Space Volume Conservation

This law states that the value of the elementary phase volume $\Delta\tau_0$ occupied by a group of particles at some initial moment t_0 will remain constant throughout the motion, that is

$$\Delta\tau = \Delta\tau_0, \tag{4.7}$$

where $\Delta\tau$ is the space volume occupied by the group of particles at the current moment.

This result can be obtained if we express the number of particles dN in the group in question through the phase density $dN = f_0\Delta\tau_0 = f\Delta\tau$. Since $f = f_0$, (4.7) follows.

Invariance of phase volume is fulfilled and for a finite phase volume [67]:

$$\tau = \int dx\,dy\,dz\,dv_x\,dv_y\,dv_z = \text{const}.$$

An important feature of charge particle motion in phase space is the absence of particle trajectories intersections. This means that particles occurring inside a closed hypersurface do not cross it throughout the motion. This allows us to consider only motion of boundary particles, when motion of a group of particles is analyzed.

Transverse Phase Space

In most charge particle focusing systems charge particle motions in longitudinal and transverse directions can be considered as independent. And what is more, in many cases motion in the transverse plane in two reciprocally perpendicular directions can also be considered as independent. For the description of the transverse motions two phase plane are used the xv_x and yv_y. Frequently, instead of phase coordinates v_x and v_y the normalized coordinates $x' = dx/dz = v_x/v_z$ and $y' = dy/dz = v_y/v_z$ are introduced.

When axially symmetric beams are considered a phase plane with phase coordinates r and $r' = dr/dz$ is used for beam-motion analysis, r and r' being radial position and slope of particle trajectory, respectively. A set of trajectories forming a charge particle beam are presented at the phase plane as a number of points, which create some phase pattern at the plane. It may be a straight line, a curve or a figure of finite area.

98   4 Motion of Intense Charged-Particle Beams

The phase pattern of a finite area corresponds to a multiple-velocity beam, in which several trajectories can pass through a certain point of coordinate space (Fig. 1.2). The phase pattern allows us to estimate the degree of order of a beam and its quality.

Emittance

For evaluation of the beam degree of order and its quality the value known as emittance is introduced. The emittance is defined as the phase-pattern area A divided by π:

$$\varepsilon = A/\pi \ .$$

The phase-pattern area is conserved (like six-dimensional phase-space volume) if the beam moves with unchanged velocity v_z through a charge particle optical system, which focusing forces depend linearly on the radial coordinates. This leads to conservation of beam emittance.

It is important to note that a charge particle beam with a finite value of emittance can not be focused to a point.

More detailed consideration of the emittance and the beam brightness is given in the Appendix.

System of Equation of Beam Motion in Self-Consistent Fields

The system can include the above equations in different combinations. But most frequently it consists of the following: Poisson and Vlasov equations (4.1) and (4.5), which are complemented by the equations expressing space charge and current densities in terms of the phase-space density:

$$\rho(x,y,z) = e \int v dv \tag{4.8}$$

$$\vec{j} = e \int \vec{v} f dv \ . \tag{4.9}$$

The system of equations (4.1), (4.5), (4.8), (4.9) most completely describes motion of charge particles in self-consistent stationary fields, but its practical application involve considerable difficulties [68–70]. So that frequently some simplified models of charge particle beams are applied for their analysis and design.

4.2 Simplified Models of Charged-Particle Beams

4.2.1 Monovelocity Models

The models are based on the assumption that particles originate from an emitter with zero velocities or with finite ones being directed normal to the emitter surface and equal for all particles. This assumption is valid when the

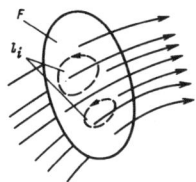

Fig. 4.1. For definition of the normal congruence F – an arbitrary surface crossing the congruence, l_i – contours of integration

initial energy of a particle is considerably less than the energy resulting from particle acceleration.

As shown in classical mechanics and electron optics [71, 72], on the basis of general laws of particle motion in conservative force fields, if particles leave a source with negligible energy, their trajectories create a congruence. The latter is such a set of trajectories that in general only a single trajectory passes though any given point, except some peculiar points through which there can pass all trajectories of the set. Apart from trajectories, particle motion is characterized by vector fields of particle velocities $\vec{v}$ and total momentum, $\vec{P} = m\vec{v} + \vec{A}$, where $\vec{A}$ is the vector magnetic potential.

The orderliness of trajectories provides the orderliness of these vector fields with $\vec{v}$ and $\vec{P}$ being single-valued functions of position, except for some peculiar points.

The congruence of trajectories is referred to as a normal one if the vectors of total momentum $\vec{P}$ can be expressed as the gradient of a scalar function S known as the action function (see Chap. 3). In this case, vector $\vec{P}$ is normal to the surfaces of constant action ($S = \text{const}$). This explains the term " normal congruence" [72].

In order for the trajectory congruence to be a normal one the normal component of vector $\text{rot}\vec{P}$ must be equal to zero at an arbitrary surface F crossing the congruence (Fig. 4.1). This condition is equivalent to the requirement for circulation of vector $\vec{P}$ to be equal to zero any contour l_i belonging to this surface, $\oint_{l_i} \vec{P} d\vec{l} = 0$.

It is possible to take as the surface in question the surface of emitter of charge particles. Then, assuming that charge particles are emitted with zero velocity one finds:

$$\oint_{l_k} \vec{P} d\vec{l} = \oint_{l_k} \vec{A} d\vec{l},$$

or applying Stokes theorem:

$$\oint_{l_k} \vec{A} d\vec{l} = \int_{F_k} \text{rot}\, \vec{A} d\vec{S} = \int_{F_k} \vec{B} d\vec{S} = \psi_k,$$

where l_k – contour of integration belonging to the emitter surface, F_k – emitter surface, closed by this contour, ψ_k magnetic flux through this surface.

From the last equation it follows that congruence is a normal one, if the magnetic flux through the emitter surface is equal to zero. In the opposite case $\vec{P} \neq \operatorname{grad} S$ and the set of trajectories creates a so-called skew congruence.

Evidently, that in pure electrostatic fields $\left(\vec{A} \equiv 0\right)$ starting from the emitter with zero initial velocity charged particle trajectories create the normal congruence, the field of velocities being a gradient field, which is determined as $m\vec{v} = \operatorname{grad} S$.

The above-noted peculiarities of motion of one-velocity beams allows the hydrodynamic approximation to be used for their description. This consists of replacement of the discrete charge particle flow with that of charged-liquid flow.

4.2.2 Hydrodynamic (Laminar) Model of Flow

According this model charge particle flow as a whole is considered to be the laminar flow of charged liquid. Such flow, in accordance with definition accepted in hydrodynamics, is characterized by absence of crossing its layers, the field of velocities being single valued function of coordinates.

The following system of equations are used for description of this flow.

The equation of motion of elementary volume of charge liquid, which coincides with that for a single charge particle:

$$\frac{d\vec{v}}{dt} = \frac{e}{m}(\vec{E} + \vec{v} \times \vec{B}) \,. \tag{4.10}$$

Field equations:

$$\nabla^2 U = -\frac{\rho}{\varepsilon_0}, \qquad E = -\operatorname{grad} U = -\nabla U \,, \tag{4.11}$$

$$\vec{B} = \operatorname{rot} \vec{A} = \nabla \times \vec{A} \,. \tag{4.12}$$

Equation of continuity:

$$\operatorname{div} \rho\vec{v} = \nabla \left(\rho\vec{v}\right) = 0 \,. \tag{4.13}$$

In the vector analysis and hydrodynamics there exists the following expression for the field of velocities:

$$\frac{d\vec{v}}{dt} = (\vec{v}\nabla)\vec{v} = \frac{1}{2}\nabla\left(v^2\right) - \vec{v} \times (\nabla \times \vec{v}) \,.$$

Its substitution in the left part of (4.10) gives:

$$\frac{1}{2}\nabla\left(v^2\right) - \vec{v} \times (\nabla \times \vec{v}) = -\frac{e}{m}\nabla U + \frac{e}{m}\left(\vec{v} \times \vec{B}\right) \,,$$

or

$$\vec{v} \times \left(\nabla \times \vec{v} + \frac{e}{m}\vec{B}\right) = \nabla\left(\frac{v^2}{2} + \frac{e}{m}U\right) \,.$$

This equation can be written as:

4.2 Simplified Models of Charged-Particle Beams

$$\vec{v} \times \left(\operatorname{rot} \vec{v} + \frac{e}{m}\vec{B}\right) = \operatorname{grad}\left(\frac{v^2}{2} + \frac{e}{m}U\right). \quad (4.14)$$

The value in the brackets of the right part of this equation is the total energy of a "particle", which is considered to be an elementary volume of the charged liquid.

For monoenergetic flows this value is the same for all particles. This results in $\operatorname{grad}\left(v^2/2 + (e/m)U\right) = 0$.

The equation of motion is reduced to $\vec{v} \times \left(\operatorname{rot}\vec{v} + (e/m)\vec{B}\right) = 0$, substitution in it $\vec{B} = \operatorname{rot}\vec{A}$ gives $v \times \operatorname{rot}\left(m\vec{v} + e\vec{A}\right) = 0$, or

$$\vec{v} \times \operatorname{rot}\vec{P} = 0. \quad (4.15)$$

This equality will be fulfilled if we put $\operatorname{rot}\vec{P} = 0$. Flows in which this condition is satisfied are known as normal or regular flows.

When an emitter of particles is unipotential, the necessary and sufficient condition of existence of normal flows is a zero value of the magnetic flux through the emitter surface. In pure electrostatic fields flows are normal ones, and, besides this, they have velocity fields of a gradient type

$$m\vec{v} = \operatorname{grad} S = \nabla S. \quad (4.16)$$

In this case the system of equations (4.10), (4.11), (4.13) can be replaced by one partial differential equation in the following way. Assuming that the potential of the emitter is equal to zero, from the integral of the total energy one finds $U = -mv^2/(2e)$. Substitution in this equation the value of $v^2 = \vec{v}\vec{v} = m^{-2}(\nabla S)^2$ gives:

$$U = -\frac{1}{me}\frac{(\nabla S)^2}{2}. \quad (4.17)$$

In view of this formula the Poisson equation can be written as

$$\rho = \frac{\varepsilon_0}{2me}\nabla^2(\nabla S)^2.$$

Then, substituting in the continuity expression (4.13) the found values of velocity vector from (4.16) and space-charge density from (4.17), one finds the required equation [73]:

$$\nabla\left[\nabla^2(\nabla S)^2 \nabla S\right] = 0. \quad (4.18)$$

This is sometimes referred to as Spangenberg's equation.

The hydrodynamic model seems to be correct when applied to the synthesis of charge particle optical systems. As in this case the laminar structure of flow is postulated and external conditions are found to realize this laminar flow. Sometimes the hydrodynamic model is used for analysis of charge particle beam motion in exterior fields. However, the correctness of the obtained results can not be guaranteed in this case, as within the framework of the laminar model it is impossible to take into account the crossing of separate layers that can take place in beams moving in real external fields.

4.2.3 Quasihydrodynamic Flow Model

The charge particle flow is divided for separate layers (tubes). Charge particle velocities of a given layer are assumed to be a single-valued function of coordinates. In this connection motion of particles belonging to a separate layer can be considered in hydrodynamic approximation using (4.10)–(4.13), or equivalent ones. The model allows the separate layers to be crossed and provides the correct analysis of real beams in real fields [10, 15].

4.3 Methods of Solution of Motion Equations

4.3.1 Method of Curvilinear Coordinates

In this method the trajectories of charged particles (or, more exactly, the elementary volumes of charged liquid) are supposed to coincide with coordinate lines of a curvilinear coordinate system [74, 75]. In this case, the system of motion equations is reduced to an ordinary differential equation of second order.

For simplicity, the further analysis will be restricted by consideration of axially symmetric flows. In a curvilinear orthogonal coordinate system q_1, q_2, q_3 trajectories of charge particle flow are supposed to coincide with the coordinate line q_1.

All flow parameters are considered to be independent of the coordinate q_3 ($\partial/\partial q_3 = 0$) to meet the condition of the axial symmetry.

In this coordinate system the charge particle velocity has only a single component $\vec{v} = \vec{i}_1 v_1 = \vec{i}_1 h_1 \dot{q}_1$ ($\vec{i}_1$ and h_1 are unit vector and scaling parameter of the coordinate system, respectively). This velocity can be expressed in terms of the action function S:

$$v = v_1 = \frac{1}{mh_1}\frac{\partial S}{\partial q_1} \left(\frac{\partial S}{\partial q_2} = \frac{\partial S}{\partial q_3} = 0\right).$$

The following equations are used to described charge particle flow:

Integral of equation of motion

$$U = -\frac{m}{2e}\left(\frac{1}{mh_1}\frac{\partial S}{\partial q_1}\right)^2, \qquad (4.19)$$

Poisson equation

$$\frac{1}{h_1 h_2 h_3}\left[\frac{\partial}{\partial q_1}\left(\frac{h_2 h_3}{h_1}\frac{\partial U}{\partial q_1}\right) + \frac{\partial}{\partial q_2}\left(\frac{h_3 h_1}{h_2}\frac{\partial U}{\partial q_2}\right)\right] = -\frac{\rho}{\varepsilon_0}, \qquad (4.20)$$

Continuity equation

$$\frac{\partial}{\partial q_1}(h_2 h_3 \rho v_1) = 0.$$

As is seen, the value in brackets of the last equation does not depend on q_1 and therefore can be a function of coordinate q_2 only:

$$h_2 h_3 \rho v_1 = F(q_2) . \tag{4.21}$$

This gives

$$\rho = F(q_2)/h_2 h_3 v_1 .$$

Substitution of this and equation (4.19) in the Poisson equation (4.20) results in

$$\frac{\varepsilon_0 m}{2eh_1^2} Q \left\{ \frac{\partial}{\partial q_1} \left[\frac{h_2 h_3}{h_1} \frac{\partial}{\partial q_1} \left(\frac{Q}{h_1} \right)^2 \right] + \frac{\partial}{\partial q_2} \left[\frac{h_3 h_1}{h_2} \frac{\partial}{\partial q_2} \left(\frac{Q}{h_1} \right)^2 \right] \right\} = F , \tag{4.22}$$

where $Q = Q(q_1) = 1/m \, (\partial S / \partial q_1)$.

After some algebra (4.22) is transformed to

$$Q \left[f_1 \frac{d^2 Q^2}{dq_1^2} + \frac{\partial f_1}{\partial q_1} \frac{dQ^2}{dq_1} + f_2 Q^2 \right] = \frac{2e}{\varepsilon_0 m} F , \tag{4.23}$$

where $f_1 = \frac{h_2 h_3}{h_1^5}$, $f_3 = \frac{h_2 h_3}{h_1} \nabla^2 \left(\frac{1}{h_1^2} \right)$.

Thus, using orthogonal curvilinear coordinates allows us to reduce the initial system of equations, describing motion in self-consistent fields, to one ordinary differential equation of second order.

Charge particle flows, moving along one coordinate of a curvilinear coordinate system, are referred to as single-component flows [75]. However, as has been shown in [76] only a limited number of orthogonal curvilinear coordinate systems allow us to realize such flows.

A simple example of the application of curvilinear coordinates is given below.

Example 4.1. Let us apply the method of curvilinear coordinates for calculation of the radial electron flow in a spherical diode. The spherical coordinates $\bar{r}, \theta, \varphi$ are used for this, with the flow being directed along the r-coordinate.

In this case: $q_1 = \bar{r}$, $q_2 = \theta$, $h_1 = 1$, $h_2 = \bar{r}$, $h_3 = \bar{r} \sin \theta$, $v_1 = v_r$, $v_2 = v_3 = 0$.

Coefficients, f_1, $\partial f_1/\partial q$, f_3 and F included in (4.23), are expressed as:

$$f_1 = \bar{r}^2 \sin \theta , \quad \partial f_1/\partial q_1 = 2\bar{r} \sin \theta , \quad f_3 = 0 , \quad F = \rho v_r \bar{r}^2 \sin \theta$$

Substitution of this expressions in (4.23) leads to the equation of single-component motion in this coordinate system:

$$Q \left[\bar{r}^2 \frac{d^2 Q^2}{d\bar{r}^2} + 2\bar{r}^2 \frac{dQ^2}{d\bar{r}} \right] = \frac{2e}{m} \bar{r}^2 \rho v_r .$$

The parameter $\bar{r}^2 \rho v_r$ including in the right part of this equation can be rewritten as

$$\bar{r}^2 \rho v_r = \frac{4\pi \bar{r}^2 \rho v_r}{4\pi} = \frac{I}{4\pi},$$

where I is the total current of spherical flow.

Furthermore, from (4.19) and (4.22) we find $Q^2 = -(2e/m)U$. The equation of flow motion can be written as follows:

$$\frac{d}{d\bar{r}}\left(\bar{r}^2 \frac{dU}{d\bar{r}}\right) = -\frac{I}{4\pi\varepsilon_0\sqrt{2|eU|/m}}.$$

In the case of an electron beam, when $U \geq 0$ and $I = -4\pi\bar{r}^2\rho v_r$, we obtain:

$$\frac{d}{d\bar{r}}\left(\bar{r}^2 \frac{dU}{d\bar{r}}\right) = \frac{I}{4\pi\varepsilon_0\sqrt{2|e|U/m}}.$$

A solution of this equation with initial conditions, given at the cathode surface $\bar{r} = \bar{r}_k$ $U = 0$, $dU/dr = 0$, are well known:

$$I = \frac{16\pi\varepsilon_0}{9}\sqrt{2\frac{|e|}{m}}\frac{U^{3/2}}{(-\alpha^2)}, \quad U = U_a \frac{(-\alpha)^{4/3}}{(-\alpha_a)^{4/3}},$$

where $(-\alpha)^2$ and $(-\alpha)^{4/3}$ are tabulated functions of the ratio of the radius of cathode curvature $\bar{r}_k$ to the current radial coordinate $\bar{r}$; $(-\alpha_a)^{4/3}$ is the value of the function $(-\alpha)^{4/3}$ calculated for ratio, $\bar{r}_k/\bar{r}_a$, where r_a is the radius of the anode curvature.

4.3.2 Method of Expansion in Series. Paraxial Trajectory Equation

Let us consider the essence of the method for the case of an axially symmetric intense charge particle beam. The initial system of equations, describing the motion of such beams in self-consistent fields, includes the following equations:

$$\frac{dv_r}{dt} = \frac{e}{m}E_r, \quad \frac{dv_z}{dt} = \frac{e}{m}E_z$$

$$\frac{\partial^2 U}{\partial r^2} + \frac{1}{r}\frac{\partial U}{\partial r} + \frac{\partial^2 U}{\partial z^2} = \frac{\rho}{\varepsilon_0}, \quad \operatorname{div}\vec{j} = \operatorname{div}\rho\vec{v} = 0. \quad (4.24)$$

Excluding time from the equations of motions one obtains the equation of trajectory:

$$\frac{d^2 r}{dz^2} = \left(\frac{\partial U}{\partial r} - \frac{\partial U}{\partial z}\frac{dr}{dz}\right)\left[1 + \left(\frac{dr}{dz}\right)^2\right]\frac{1}{2U}. \quad (4.25)$$

Potential U and space charge density ρ in the Poisson equation can be presented in the following series expansions:

$$U(r,z) = U_0(z) + U_2(z)r^2 + U_4(z)r^4 + \ldots + U_n(z)r^n$$

$$\rho(r,z) = \rho_0(z) + \rho_2(z)r^2 + \rho_4(z)r^4 + \ldots + \rho_n(z)r^n .$$

The two series contain only the terms in even powers of variable r to meet the potential symmetry $U(r) = U(-r)$.

Substitution of these series in the Poisson equation gives, after transformation, the following series for the potential distribution [77]:

$$U(r,z) = U_0 - \frac{1}{4}\left(U_0'' + \frac{\rho_0}{\varepsilon_0}\right)r^2 + \frac{1}{16}\left(\frac{1}{4}U_0^{IV} + \frac{1}{4}\frac{\rho_0''}{\varepsilon_0} - \frac{\rho_2}{\varepsilon_0}\right)r^4 - \ldots .$$

Here the Roman numeral superscript after potential U_0 denotes the fourth derivative with respect to the z variable.

Further analysis is unstrained by consideration of paraxial beams that trajectories form very small angles to the axis and whose distance from the axis also remains small throughout the motion.

If we substitute the above series expansions in the equation of trajectory (4.25) and omit all terms of higher order than the first one in r and r', the following paraxial equation of trajectory will be obtained:

$$\frac{d^2r}{dz^2} + \frac{1}{2U_0}\frac{dU_0}{dz}\frac{dr}{dz} + \frac{1}{4U_0}\frac{d^2U_0}{dz^2}r = \mp\frac{I}{4\pi\varepsilon_0 r\sqrt{2|eU_0|/m}\,U_0} ,$$

where I is the beam current; the upper and lower signs in the right part of the equation are applied for the positive charge particle beams and for the negative particle beams accordingly.

It is important to note that the axial potential in this equation is the potential of the self-consistent field, which takes into account the space charge field.

4.4 Approximate Method of Space-Charge Account

The total electric field can be presented as the sum of external field, $\vec{E}_L$, which is created by electrodes, and the self-field of space charge $\vec{E}_\rho$:

$$\vec{E} = \vec{E}_L + \vec{E}_\rho .$$

In an extended beam, which length considerably exceeds the beam transverse dimensions, the transverse component of the space-charge field is considerably greater than the longitudinal one $E_{\rho r} \gg E_{\rho z}$ and mainly impacts on particle-beam motion.

This can be found by use of the Gauss theorem, which states that the total electric field flow through a closed surface is equal to the charge inside the volume enclosed by this surface. Let us consider an axially symmetric extended beam and apply the Gauss theorem to an elementary cylindrical volume $V = \pi r^2 \Delta z$ located at the beam axis (Fig. 4.2). This gives:

$$\frac{1}{\varepsilon_0}\int_V \rho 2\pi r\, dr\, dz = \int_{S_0} E_{\rho r}\, dS + \int_{S_1} E_{\rho z}\, dS + \int_{S_2} E_{\rho z}\, dS .$$

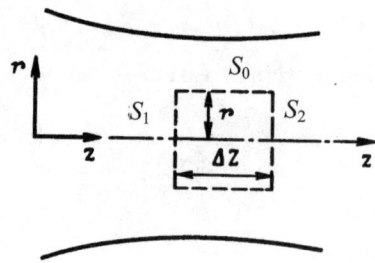

Fig. 4.2. For calculation of space-charge field by use of Gauss theorem

The total boundary surface of the cylinder is equal to the sum of the lateral face S_0 and the two end faces S_1 and S_2, V being the volume of the cylinder.

For approximate calculation, taking into account that $E_{\rho z} \ll E_{\rho r}$ one obtains:

$$\int_{\Delta z} E_{\rho r} 2\pi r dz = \frac{1}{\varepsilon_0} \int_V \rho 2\pi r dr dz ,$$

or, assuming $E_{\rho r} = \text{const}$ on Δz and ρ is independent of z,

$$E_{\rho r} = \frac{1}{\varepsilon_0 r} \int_0^r \rho r dr = \frac{q_1}{2\pi \varepsilon_0 r} , \qquad (4.26)$$

where q_1 is the value of space charge per unit beam length.

Supposing that the electron velocity is constant over the cross section of the elementary volume, it is possible to express q_1 in terms of current ΔI passing through the volume cross section and the axial velocity of electrons v_z:

$$q_1 = 2\pi \int_0^r \rho r dr = \frac{2\pi}{v_z} \int_0^r v_z \rho r dr ,$$

or

$$q_1 = \pm \frac{2\pi}{v_z} \int_0^r j_z r dr = \pm \frac{\Delta I}{v_z} .$$

In this formula, as well as in these given below, the upper sign corresponds to the beams of positive charge particles, while the lower one corresponds to the negative particle beams.

Then, for the transverse component of the space-charge field $E_{\rho r}$ the following expression is obtained

$$E_{\rho r} = -\frac{\partial U_\rho}{\partial r} = \pm \frac{\Delta I}{2\pi \varepsilon_0 r v_z} . \qquad (4.27)$$

For the case of a laminar flow, when there are no trajectories crossing, this equation can be modified and used for determination of the beam-boundary trajectory of a laminar beam:

$$E_{\rho r} = -\frac{\partial U_\rho}{\partial r} = \pm \frac{I}{2\pi\varepsilon_0 r v_z}, \quad (4.28)$$

where I is total current of the beam.

The expressions for the transverse components of the space-charge field presented above can be introduced in the equations of motion and trajectory equations. For instance, (3.43) and (3.47) can be rewritten in such a way:

$$\frac{d^2 r}{dz^2} = \left(\frac{\partial Q}{\partial r} \mp \frac{I}{2\pi\varepsilon_0 r v_z} - \frac{\partial Q}{\partial r}\frac{dr}{dz} \right) \left[1 + \left(\frac{dr}{dz} \right)^2 \right] \frac{1}{2Q}, \quad (4.29a)$$

$$2U_0 \frac{d^2 r}{dz^2} + \frac{dU_0}{dz}\frac{dr}{dz} + \frac{1}{2}\frac{d^2 U_0}{dz^2} - \frac{e}{4m} B_{z0}^2 r \left[1 - \left(\frac{r_k^2 B_{z0k}}{r^2 B_{z0}} \right)^2 \right] =$$

$$= \mp \frac{I}{2\pi\varepsilon_0 r v_z}. \quad (4.29b)$$

In these equations I – beam current, Q – generalized potential, U_0 – axis potential.

The similar approximate expressions for the transverse component of the space-charge field can be obtained for two-dimensional planar space charge flows.

4.5 Motion of Charge-Beams in Channels Free from External Fields

In electron devices (microwave tubes, etc.), installations for charge particle beam technology and physical experiments, charge particle beams pass a certain part of their travel in channels, which are free of any external fields. Therefore, analysis of such motion counts for much in practice.

According to the results obtained in two previous sections, an approximate equation (laminar flow is assumed) can be written as

$$d^2 r/dz^2 = K/r, \quad K = \frac{I}{4\pi\varepsilon_0 \sqrt{|e|/m} U_0^{3/2}}, \quad (4.30)$$

where U_0 is the absolute value of the axial potential, which can be approximately taken equal to the potential of the metallic channel surrounding the beam $U_0 \approx U_a$.

Introducing the normalized variables $R = r/r_0$, $Z = (2K)^{1/2} z/r_0$ we obtain:

$$d^2 R/dZ^2 = 1/2R\,.$$

A first integration of this equation gives:

$$\left(\frac{dR}{dZ} \right)^2 = \ln \frac{R}{R_0} + \left(\frac{dR}{dZ} \right)_0^2, \quad (4.31)$$

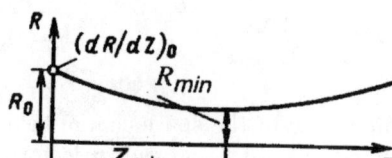

Fig. 4.3. Beam-boundary trajectory (beam envelope)

where R_0 and $(dR/dZ)_0$ are the initial radius and trajectory slope at the entrance of the channel (at $Z = 0$) (Fig. 4.3).

The value of beam radius at the plane of minimum beam cross section ($Z = Z_{\min}$) will be found if we put in (4.31) $\left(\frac{dR}{dZ}\right) = 0$:

$$R = R_{\min} = R_0 \exp - \left(\frac{dR}{dZ}\right)_0^2.$$

A second integration of (4.31) results in the following equation for the beam-boundary trajectory (or beam envelope):

$$Z = \int_{R_0}^{R} \frac{dR}{\sqrt{\ln \frac{R}{R_0} + \left(\frac{dR}{dZ}\right)_0^2}},$$

with coordinate Z being measured from the plane of the minimum beam cross-section ($Z = Z_{\min} = 0$, $R_0 = R_{\min}$, $(dR/dZ)_0 = 0$) the above equation is reduced to

$$Z = \pm \int_{R_{\min}}^{R} \frac{dR}{\sqrt{\ln R}}$$

This shows that the beam boundary is symmetric with regard to the coordinate $Z = Z_{\min}$.

Introducing the new normalization of variables, with radius $r_{\min}$ being used as the scaling coefficient, one finds the following system of equations, describing the beam boundary in an equipotential channel:

$$Z = \pm \int_{R_{\min}}^{R} \frac{dR}{\sqrt{\ln R}}, \qquad \frac{dR}{dZ} = \sqrt{\ln R}, \qquad R_{\min} = 1.$$

The results of solutions of this equation are presented in Fig. 4.4, where universal curves for calculation of the beam-boundary trajectory and its slope are shown.

For $0 \leq Z \leq 3$ the curve $R = R(Z)$ can be approximated by the cubic spline:

$$R \approx 1 + 0.25 Z^2 - 0.01 Z^3. \qquad (4.32)$$

4.5 Motion of Charge-Beams in Channels Free from External Fields

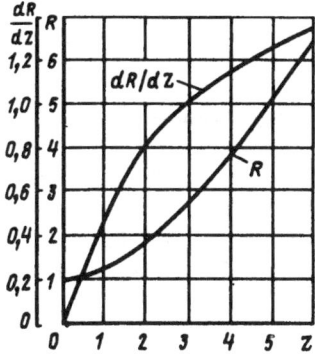

Fig. 4.4. Universal curves for calculation of the beam-boundary trajectory

The above expressions and curves can be used for calculation of negative and positive charge particle beam motion as well.

For the case of electron beams the normalized coordinate Z can be written as $Z = 0.174\sqrt{P}z/r_{\min}$, the value of beam perveance P being expressed in $\mu A/V^{3/2}$.

It is worth mentioning that the curves and expressions can be applied not only for calculation of initially parallel beams $((dR/dZ)_0 = 0)$, but also for initially convergent or divergent beams $(dR/dZ)_0 \neq 0$.

In the previous analysis the potential inside the channel was supposed to be constant and equal to the potential of the metallic cylinder surrounding the beam. This approximation is valid for relatively small values of beam perveance $p \leq 1$ $\mu A/V^{3/2}$. High values of beam perveance lead to potential depression inside the channel and at some value of perveance, known as the critical one, the axis potential can drop to zero and a virtual cathode is created. This effect results in collapse of beam transmission. Calculation gives the following approximate value of critical perveance P_{cr} for an election beam in a cylindrical channel [79]:

$$P_{cr} = \frac{I_{cr}}{U_a^{3/2}} \simeq 32.4q\left(\frac{b}{a}\right) .$$

In this equation, P_{cr} – critical perveance in $\mu A/V^{3/2}$, I_{cr} is critical electron beam current in μA, U_a – accelerating voltage in V, q is a function of the beam to channel radii ratio b/a as given by:

b/a	1	0.8	0.6	0.4	0.2
q	1	0.5	0.33	0.22	0.15

Detailed analysis of the effect of the virtual cathode is given in Sect. 10.2.

4.6 Influence of Residual Gases on Electron-Beam Motion

4.6.1 Effect of Space-Charge Neutralization by Positive Ions

Positive ions arise as a result of collisions of beam electrons with molecules of residual gases. They can be accumulated in the region occupied by the electron beam. This results in partial or full neutralization of the electron space charge. This effect in turn, influences the electron motion and beam geometry.

The rate of formation of ions depends on the nature and pressure of the residual gases, and the density and velocity of beam electrons. It is expressed by the formula:

$$n'_i = k_p B_i p \left(j/|e|\right) = k_p B_i p n_e v \,, \tag{4.33}$$

where n'_i is the number of ions creating in unity of volume 1 cm^3 per second, B_i is specific ionization, that is the number of ions created by one electron on distance equal 1 cm at residual gas pressure equal $p = 133.3$ Pa, n_e – electron concentration (number of beam electrons per unit volume), v is beam velocity, $k_p = 1/133,3$

The specific ionization depends on the type of residual gas and electron energy. For instance, for nitrogen (N$_2$) it is equal to 10 for energy 100 eV, 4 – at 1000 eV, 2.8 – at 2000 eV.

Assuming, for instance, $U = 1000\ V, p = 1.33 \times 10^{-5}$ Pa $B_i = 4$ and using (4.33) one obtains $n'_i = 750 n_e$. Therefore, even a small part of ion charge created per second is enough to neutralize the electron space charge.

However, the effect of ion neutralization is not always observed in electron beams, because of leakage of ions from the volume occupied by the electron beam. There are several reasons leading to the ion losses. One of them is extraction of ions due to the action of external electric fields. For example, the ions, formed in the region of an electron gun, are accelerated toward the cathode of the gun and can not be accumulated in the gun region.

Another cause of ion leakage is their thermal motion. The thermal velocities of ions allow them to leave the potential trap, which was created by the electron beam space charge. This is why the effect of space-charge neutralization is observed only when there are conditions for capture and accumulation of ions.

4.6.2 Process of Neutralization of the Electron Beam in a Long Channel

When an electron beam is propagating in an equipotential channel, the potential minimum is created at the axis of the channel. It serves as a trap for ions, which are created due to collisions of the beam electrons with molecules of residual gases.

4.6 Influence of Residual Gases on Electron-Beam Motion

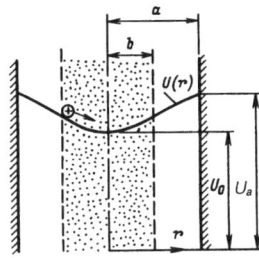

Fig. 4.5. Potential well in the channel, created by electron beam

In the process of accumulation of ions their charge will neutralize the electron beam space charge. This results in decreasing the potential minimum and increasing the number of ions leaving the region occupied by the electron beam.

The process of neutralization is established when there is a dynamic balance of the number of ions creating in the beam and leaving it for the channel wall.

The number of ions creating per unit of beam length per second can be expressed with the help of (4.33)

$$N'_i = n'_i \pi b^2 = k_p B_i p n_e v \pi b^2 = k_p B_i p n_e \sqrt{2\,(|e|/m)\,U_a}\,\pi b^2. \quad (4.34)$$

where U_a is the potential of the channel wall, which determines the energy and velocity of electrons of the beam, b is beam radius.

The number of ions leaving the electron beam per unit beam length per second due to their thermal velocities can be estimated by the expression [80]:

$$L'_i = 2\pi a n_{i0} \left(e_i U_i / 2\pi m_i\right)^{1/2} e^{\frac{U_0 - U_a}{U_i}} \quad (4.35)$$

where a – radius of cylindrical channel, n_{i0} – ion concentration on the channel axis, m_i – ion mass, $e_i U_i$ – average thermal energy of a single ion, U_0 – axial potential.

Equating (3.34) and (4.35) and assuming $|e| = e_i$ one finds:

$$p \frac{a n_{i0}}{\sqrt{\pi} k_p b^2 B_i n_e} \left(\frac{m}{m_i} \frac{U_i}{U_a}\right)^{1/2} e^{\frac{U_0 - U_a}{U_i}}. \quad (4.36)$$

Assuming in this equation $U_0 = U_a$ and $n_e = n_{i0}$ it is possible to determine the value of pressure of residual gases, which provides full neutralization of the electron beam space charge. For example, at $U_a = 1000\,\text{V}$, $a = b = 1$ cm, $m_i/m = 52\,000$ (N_2 as residual gas) and $e_i U_i = 0.26$ eV, which corresponds to a gas temperature of 300 K, calculation gives the following value of residual gas pressure that provides full space-charge neutralization:

$$p = \frac{133.3}{\sqrt{\pi}4} \left(\frac{1}{52\,000} \times \frac{0.026}{1000}\right)^{1/2} = 42 \times 10^{-5}\,\text{Pa}\,.$$

The previous analysis is valid for infinitely long channels. In channels of a finite length the process of neutralization can be essential other. For instance, if at the end of the channel there exists an electric field, that is accelerating for ions, they can be drawn from the channel, and the effect of neutralization will be absent.

4.6.3 Effect of Ion Background on Electron-Beam Motion

Full or partial neutralization of the electron space charge by ions can produce a marked influence on the electron-beam motion. In particular, in the absence of external fields, the effect of the magnetic field of the beam becomes significant. Assuming for simplicity that the degree of space-charge neutralization is uniform and introducing the coefficient of neutralization $f = n_i/n_e$, it is possible to obtain the following equation for the beam-boundary trajectory [81]:

$$\frac{d^2 r}{dz^2} = \frac{K}{r}\left(1 - f - \beta^2\right), \qquad (4.37)$$

where $K = \dfrac{I}{4\pi\varepsilon_0 \sqrt{2|e|/m}\, U_a^{3/2}} = 1.5 \times 10^{-2} P$, P is the perveance of the electron beam expressed in $\mu A/V^{3/2}$, the term $\beta^2 = v^2/c^2$ (v and c being beam and light velocities respectively) accounts for the self-magnetic field action.

For the full neutralization, $f \approx 1$, the self-magnetic field defines the form of the beam-boundary trajectory.

Equation (4.37) can be rewritten as follows:

$$d^2 r/dz^2 = K_1/r, \qquad \text{where} \qquad K_1 = K\left(1 - f - \beta^2\right).$$

The results of integration of the last differential equation are presented in normalized form in the diagram of Fig. 4.6. In this diagram $R = r/r_0$ – normalized radial coordinate, $Z = 2|K_1|^{1/2}\, z/r_0$ – normalized longitudinal coordinate, r_0 – initial beam radius. As is seen, under the action of its own magnetic field a fully neutralized beam ($K_1 < 0$) is focused to a point at a distance of $z/r_0 = 1.27/|K_1|^{1/2}$.

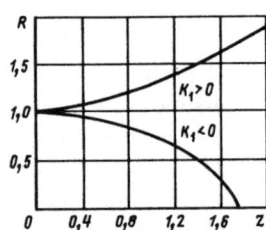

Fig. 4.6. Beam-boundary trajectories for different degrees of space-charge compensation

4.6.4 Ion Focusing

The effect of ion focusing arises when the value of residual gas pressure is increased above the value at which full beam neutralization is observed. In

4.6 Influence of Residual Gases on Electron-Beam Motion

this case, a column of positive-ion space charge is created at the axis of the channel, resulting in the effect of ion focusing.

The theory of ion focusing has been developed in [82, 83]. Using the results of [83] it is possible to obtain the following expressions for calculation of ion-focusing parameters:

$$\frac{n_i - n_e}{n_e} = \frac{C}{\sqrt{U_2}} - 1, \qquad \frac{\Delta U}{U_a} = K\frac{n_i - n_e}{n_e}, \qquad (4.38)$$

where n_e and n_i are electron and ion concentrations, $(n_i - n_e)/n_e$ – degree of ion overcompensation. ΔU is the difference in potentials at the beam axis and beam boundary.

Parameters K, C, U_2 included in these formulas are found from the following expressions:

$$K = 1.5 \times 10^4 P, \qquad\qquad C = 0.75\sqrt{m_i/m}B_i p\sqrt{U_a}$$

$$U_2 = \frac{2}{3}\chi_1 \text{sh}^2\left(\sqrt{\psi/3}\right), \qquad \chi_1 = 1.5 \times 10^{10}\frac{I}{b^2\sqrt{U_a}},$$

$$\text{sh}\psi = \frac{15.8 \times 10^{-6}}{\sqrt{P}}bB_i p\sqrt{\frac{m_i}{m}}.$$

where P – perveance of electron beam, $A/V^{3/2}$; I – beam current, A; U_a – beam accelerating voltage, V; b – average beam radius, cm; B_i – specific ionization, cm^{-1}; p – residual gas pressure, Pa; m_i/m – relation of ion and electron masses.

In the theory of ion focusing developed in [83] the secondary electron, created in acts of ionization, is supposed to leave the beam region and not have an effect on the process of ion focusing. This is valid if the potential difference ΔU does not exceed the average velocities of secondary electrons expressed in volts U_e. In the opposite case, the secondary electrons can not leave the beam region and take part in the processes of space-charge compensation.

On the basis of results obtained in [84] it is possible to find the following values of average velocities of the secondary electrons: $U_e = 15$ V for nitrogen N_2 as residual gas and $U_e = 20$ V for oxygen O_2. Assuming in (4.38) $\Delta U = U_e$ one can estimate the possible degree of ion overcompensation [85]

$$\frac{n_i - n_e}{n_e} = \frac{U_e}{KU_a} = \frac{U_e}{1.5 \times 10^4 PU_a} = \frac{66 \times 10^{-6}}{P}\frac{U_e}{U_a}.$$

It follows from this equation that the effect of ion overcompensation can be significant for electron beams with low voltage and perveance. For beams having $P \geq 10^{-6}$ A/V$^{3/2}$ and $U_a \geq 10^4$ V, $(n_i - n_e)/n_e \ll 1$ and the effect of ion background is reduced to compensation of the electron beam space charge.

4.6.5 Scattering of Electron Beam by Molecules of Residual Gas

At a pressure of residual gas of about 10^3 Pa there exists a clear effect of electron-beam scattering by molecules (atoms) of residual gas.

Scattering by Single Elastic Collisions

An elastic collision does not change electron energy but leads to deflection of its trajectory in the electric field of the atom nucleus (Fig. 4.7). The angle of trajectory deflection θ is determined by:

$$\mathrm{tg}\theta/2 = \frac{|e|(|e|Z)}{mv^2 b} = \frac{e^2 Z}{mv^2 b},$$

where $|e|Z$ – nucleu charge, $|e|$ – absolute value of electron charge, Z – atomic number, v – electron velocity, b – impact parameter, that is, minimum distance between the nucleus and the undeflected electron trajectory (dashed line in Fig. 4.7).

Let us consider scattering of an electron beam on a thin gas layer. The thickness of the layer Δz is assumed to be small enough to admit only single-electron collisions with residual gas atoms. If the area of beam cross section is taken to be S, then the number of atoms in the elementary volume of the layer $\Delta V = S\Delta z$ will be expressed by the formula:

$$n_S = n\Delta V = nS\Delta z,$$

where n – concentration of atoms of residual gas.

The number of electrons passing through this volume per unit time is expressed as $N' = I/|e|$, where I is beam current. The electrons will be deflected for different angles θ, the value of θ being dependent on the value of the impact parameter b. It is possible to calculate the number of electrons having the value of impact parameter lying in the range "$b \div b + db$". It is necessary for this to surround every scattering center (nucleus) with a ring with radii b and $b + db$. The area of it is equal $2\pi b db$. Then, the number of beam electrons having the values of collision parameters in the range $b \div b + db$ will be:

$$dN'/N' = 2\pi b db n_S / S,$$

where n_S – number of atoms in elementary volume ΔV, $n_S = nS\Delta z$, S – area of electron beam cross section, $2\pi b db n_S$ – total area of all elementary rings with boundary radii b and $b + db$.

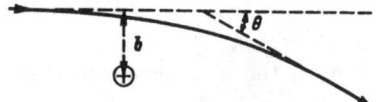

Fig. 4.7. Electron-trajectory deflection in the field of atomic nucleus

4.6 Influence of Residual Gases on Electron-Beam Motion

Substitution of $n_S = nS\Delta z$ in the above equation gives:

$$dN'/N' = 2\pi b\, db\, n\Delta z = n_1 2\pi b\, db , \qquad (4.39)$$

where $n_1 = n\Delta z$ – number of atoms per unit area of beam cross section.

Taking into account the relation between the angle of deflation θ and the impact parameter b it is possible to find:

$$\frac{dN'}{N'} = \pi n_1 \left(\frac{e^2 Z}{mv^2}\right)^2 \frac{\cos\theta/2}{\sin^3\theta/2} d\theta .$$

This formula gives the relative number of electrons deflected in the range of angles $\theta \div \theta + d\theta$.

As applied to motion of a single electrons the relation dN'/N' expresses the probability of deflection in this range of angles. The equation shows that the probability of deflection for small angles θ is prevailing.

Let us introduce the solid angle created by two conical surfaces with angles of inclination equal, respectively, to θ and $\theta + d\theta$

$$d\Omega = 2\pi \sin\theta\, d\theta = 4\pi \sin\theta/2 \cos\theta/2\, d\theta .$$

Then,

$$\frac{dN'}{N'} = \frac{n_1}{4} \left(\frac{e^2 Z}{mv^2}\right)^2 \frac{d\Omega}{\sin^4\theta/2} . \qquad (4.40)$$

This equation is known the as Rutherford equation. It shows the relative number of electrons deflected in the solid angle $d\Omega$.

Usually, (4.40) is written in the following form:

$$dN'/N' = n_1 \sigma(\theta)\, d\Omega ,$$

where value $\sigma(\theta) = \frac{1}{4}\left(\frac{e^2 Z}{mv^2}\right)^2 \frac{1}{\sin^4\theta/2}$ refers as the effective cross section of scattering.

There exist more precise equations for the calculation of scattering, which account for the effect of shielding of the nucleus field by orbital electrons and the finite size of the nucleus [16]. The shielding effect limits the maximum value of the collision parameter by the radius of an atom $b \approx r_a = 7.4 \cdot 10^{-9} Z^{1/3}$ and determines the minimum value of the angle of deflection θ [16]:

$$\mathrm{tg}\frac{\theta_{\min}}{2} = \frac{e^2 Z}{mv^2 r_a} = \frac{10^{-4} Z^{4/3}}{2\gamma\beta^2} ,$$

where $\beta = v/c$, c – velocity of light, γ – relativistic factor.

The finite value of radius of a nucleus r_n determines the minimum of the impact parameter, and, therefore the maximum value of the angle of deflection:

$$\mathrm{tg}\frac{\theta_{\max}}{2} = \frac{e^2 Z}{mv^2 r_n} .$$

4 Motion of Intense Charged-Particle Beams

Scattering by Multiple Collisions

An example of an approximate electron trajectory at multiple interactions is shown in Fig. 4.8 [86].

The points A, B, C, D are the points of collisions of an electron with atoms of residual gas. The collisions lead to abrupt changes of the electron-trajectory direction by an angle θ_i. The average square of the deflection angles is used as a measure of electron scattering:

$$\bar{\theta}^2 = \frac{1}{k}\sum_{i=1}^{k}\theta_i^2 \ .$$

The following equation is recommended for estimation of electron-beam scattering in a gas at relative high pressure [87]:

$$\bar{\theta}^2 = 8\pi r_0^2 n Z^2 \left(\frac{1}{\beta^4 \gamma^2}\right) z \ln\frac{\theta_{\max}}{\theta_{\min}} , \qquad (4.41)$$

where r_0 – classical radius of electron, $r_0 = 2.83 \times 10^{-15} m$; n – gas density, m^{-3}; $\theta_{\min}$ and $\theta_{\max}$ – minimum and maximum values of angles of deflection, which are determined by the equations given above; z is distance passed by electrons in gas.

If an electron beam of small radius is injected into a gas (at plane $z = 0$) it will be dissipated because of the multiple collisions. Its cross section is increased. The current-density distribution is described by [87]:

$$j(r,z) = j(0,z)\exp\left[-\left(r/\bar{r}^2\right)\ln 2\right] , \qquad (4.42)$$

where $j(0,z)$ – current density at the beam axis, $\bar{r}$ – value of radius at which current density is half as large as $j(0,z)$.

The value of $\bar{r}$ can be expressed in terms of $\bar{\theta}^2$ and the distance z passed by the beam in a gas:

$$\bar{r} = \left[(1/3)(\ln 2) z^2 \bar{\theta}^2\right]^{1/2} \ .$$

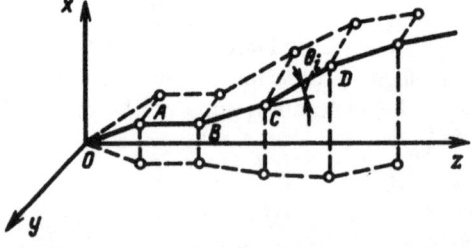

Fig. 4.8. Electron trajectory resulting from multiple collisions

4.7 Estimation of Effect of Thermal Electron Velocities

The effect of thermal (initial) velocities leads to spreading of beam edges, increasing beam cross section and redistribution of current density over the beam cross section [10, 88].

The following simplifications are used in the analysis of the effects of thermal velocities:

- configuration and structure of a laminar beam in the absence of thermal velocities are supposed to be known
- only transverse components of thermal velocities are considered
- analysis of motion of electrons possessing transverse velocities is carried out in the background of the electric field of the laminar flow
- linear variations of this field with transverse coordinate are additionally assumed.

4.7.1 Calculation of Transverse Particle Displacement

Equation of motion of "thermal" electrons, that is electrons possessing transverse thermal velocities, can be written as:

$$\frac{d^2r}{dt^2} = \left(\frac{e}{m}\right) E_r , \qquad (4.43)$$

where E_r is transverse (radial) component of electric field assumal to be created by the original laminar flow, r is the radial coordinate of a thermal electron.

The value of E_r can be expressed in terms of a dynamic parameter of laminar flow, which is taken to be the radial acceleration of its particles:

$$E_r = q \frac{m}{e} \frac{d^2R}{dt^2} , \qquad (4.44)$$

where R is the radial coordinate of a boundary particle of the laminar flow, d^2R/dt^2 – its radial acceleration, $q = r/R$ is the relative radial coordinate of the thermal electron under consideration.

Substitution of (4.44) into (4.43) gives:

$$\frac{d^2r}{dt^2} = q \frac{d^2R}{dt^2} . \qquad (4.45)$$

Bearing in mind that $r = qR$, the left part of this equation can be transformed thus:

$$\frac{d^2r}{dt^2} = \frac{d^2}{dt^2}(qR) = R\frac{d^2q}{dt^2} + 2\frac{dq}{dt}\frac{dR}{dt} + q\frac{d^2R}{dt^2} .$$

Substitution of this equation into (4.45) gives the following equation of "thermal" electrons in the background of the field of the laminar flow:

$$\frac{d^2q}{dt^2} = -\frac{2}{R}\frac{dR}{dt}\frac{dq}{dt} . \qquad (4.46)$$

It is essential to note that this equation describes the motion of the thermal electron in terms of its relative coordinate $q = r/R$ and parameters of the laminar flow R and dR/dt. Equation (4.46) can be rewritten as:

$$\left(\frac{dq}{dt}\right)^{-1} \frac{d}{dt}\left(\frac{dq}{dt}\right) = -\frac{2}{R}\frac{dR}{dt}.$$

Since

$$\frac{1}{R}\frac{dR}{dt} = \frac{d}{dt}\ln R \quad \text{and} \quad \left(\frac{dq}{dt}\right)^{-1}\frac{d}{dt}\left(\frac{dq}{dt}\right) = \frac{d}{dt}\ln\frac{dq}{dt},$$

then,

$$\frac{d}{dt}\ln\frac{dq}{dt} = -2\frac{d}{dt}\ln R.$$

The first integration of this equation gives:

$$\ln\frac{dq}{dt} - \ln\left(\frac{dq}{dt}\right)_k = -\ln R^2 + \ln R_k^2 \tag{4.47}$$

where R_k – the initial coordinate of the boundary trajectory of the laminar flow, $(dq/dt)_k$ – initial relative transverse velocity of the thermal electron.

Equation (4.47) can be presented in the form:

$$\frac{dq}{dt} = \left(\frac{dq}{dt}\right)_k \frac{R_k^2}{R^2}.$$

Its integration gives:

$$q - q_k = \left(\frac{dq}{dt}\right)_k \int_{t_k}^{t} \left(\frac{R_k}{R}\right) dt,$$

where q_k – initial relative velocity of the thermal particle, t_k – initial time.

The last equation can be written as:

$$(q - q_a)R = R_k \left(\frac{dq}{dt}\right)_k \frac{R}{R_k} \int_{t_k}^{t} \left(\frac{R_k}{R}\right)^2 dt,$$

or

$$\Delta r = r - r_0 = \left(\frac{dr}{dt}\right)_k \Delta t_e, \tag{4.48}$$

where Δr is the transverse displacement of the thermal particle resulting from the initial thermal velocities; $r = qR$ – current transverse coordinate of the particle; $r_0 = q_k R$ – transverse coordinate of the same particle in the laminar (undisturbed) flow; $(dr/dt)_k$ – initial transverse particle velocity; Δt_e – equivalent time of particle motion:

$$\Delta t_e = \left(\frac{R}{R_k}\right) \int_{t_k}^{t} \left(\frac{R_k}{R}\right)^2 dt.$$

4.7.2 Calculation of Current-Density Redistribution

For this calculation all the parameters of the initial (undisturbed) laminar flow are supposed to be known.

The elementary current, dI_Q passing through the surface $\Delta S = r_0 dr_0 d\theta$ of cross section of the laminar beam is $dI_Q = j_0 r_0 dr_0 \theta$, where j_0 is current density of laminar beam (Fig. 4.9).

As a result of thermal velocities spreading, the particles that earlier passed through elementary surface ΔS will be distributed over the cross section plane $r\theta$. A part of them can pass through the elementary surface $\Delta F = dxdy$ located in the vicinity of a point P. To calculate it we suppose that distribution of transverse particle velocities obeys the Maxwell law at an emitter. Then, the part of the current dI_Q created by particles having transverse velocity in the range from v_x to $v_x + dv_x$ and from v_y to $v_y + dv_y$ is expressed as:

$$dI_{v_x, v_y} = dI_Q \frac{m}{2\pi kT} \exp\left[-\left(\frac{mv_x^2}{2kT} + \frac{mv_y^2}{2kT}\right)\right] dv_x dv_y . \quad (4.49)$$

Using this equation it is possible to calculate the part of the current, dI_Q that comes to the elementary surface ΔF. It is created by particles with initial velocities lying in the ranges: from $v_x = x/\Delta t_e$ to $v_x + dv_x = x/\Delta t_e + dx/\Delta t_e$ and from $v_y = y/\Delta t_e$ to $v_y + dv_y = y/\Delta t_? + dy/\Delta t_?$.

Substitution of these value in (4.49) leads to the equation for the part of dI_Q, that reaches the surface ΔF:

$$dI_P = \frac{1}{2\pi} dI_Q \exp\left[-\left(x^2 + y^2\right)\frac{1}{(2kT/m)\Delta t_e^2}\right]\frac{dxdy}{(kT/m)\Delta t_e^2} . \quad (4.50)$$

In this equation, Δt_e is the equivalent transit time from the emitter to the transverse plane under consideration.

It is convenient to introduce a parameter $\sigma = \Delta t_e (kT/m)^{1/2}$, which has dimensions of length and can be considered as a measure of the thermal

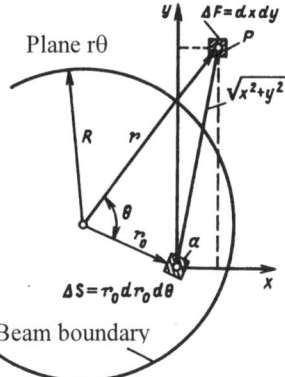

Fig. 4.9. For calculation of current-density distribution

electron deviation from the laminar trajectory (note that thermal energy per one degree of freedom is equal to $kT/2$).

Taking into account that $dI_Q = j_0 r_0 dr_0 d\theta$ we can write (4.50) in the following form:

$$dI_P = \frac{1}{2\pi} j_0 r_0 dr_0 d\theta \exp\left(-\frac{x^2+y^2}{2\sigma^2}\right) \frac{dxdy}{\sigma^2} .$$

Then, the partial current density at the surface ΔF is found as:

$$dj_P = \frac{dI_P}{dxdy} = \frac{j_0}{2\pi\sigma^2} \exp\left(-\frac{x^2+y^2}{2\sigma^2}\right) r_0 dr_0 d\theta .$$

The total current density at ΔF is determined by integration of the above equation over the cross section of the undisturbed (laminar) beam:

$$j_P = \int_0^R \int_0^{2\pi} \frac{j_0}{2\pi\sigma^2} \exp\left(-\frac{x^2+y^2}{2\sigma^2}\right) r_0 dr_0 d\theta .$$

Assuming that the current density of the laminar flow j_0 is constant and using the relation $x^2 + y^2 = r_0^2 + r^2 - 2r_0 r \cos\theta$ we can write:

$$j_P = \int_0^R \int_0^{2\pi} \frac{j_0}{2\pi\sigma^2} \exp\left(-\frac{x^2+y^2}{2\sigma^2}\right) r_0 dr_0 d\theta .$$

Since

$$\int_0^{2\pi} e^{\frac{r_0 r \cos\theta}{\sigma^2}} d\theta = 2\pi I_0\left(r_0 r/\sigma^2\right) ,$$

the ultimate result will be

$$j(r) = j_0 e^{-r^2/(2\sigma^2)} \int_0^{R/\sigma} \frac{r_0}{\sigma} e^{-r_0^2/(2\sigma^2)} I_0\left(\frac{r_0 r}{\sigma^2}\right) d\frac{r_0}{\sigma} .$$

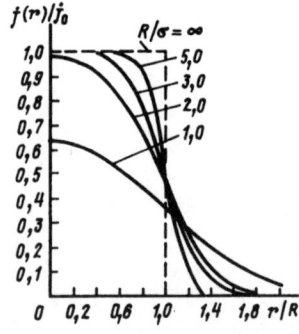

Fig. 4.10. Curves showing current-density distribution resulting from the thermal-velocity spreading: J_0 – current density of undisturbed beam, R – radius of undisturbed beam, σ – parameter characteristic effect of thermal-velocity spreading (4.50)

In these equation, $I_0(r_0 r/\sigma)$ is the modified Bessel function of zero order.

The curves showing the current-density distribution over the beam cross section are presented in Fig. 4.10, with R/σ being taken as a parameter.

As is seen, at $R/\sigma \leq 2$ there exists an essential redistribution of current density over the beam cross section as compared with the laminar beam, in which the current density is supposed to be uniform ($j_0 = $ const).

4.8 Methods of Solution of Self-Consistent Problems

Computation and design of charge particle optical systems (for formation, transport and focusing of intense charge particle beams) require solution of the self-consistent space-charge problem. Its essential feature lies in the fact that particle motion occurs in the fields, which in turn depends on this motion. The problem includes the joint solution of the equations of motion and field equations. The method of successive approximation is widely used for solution of this problem [89].

4.8.1 Method of Successive Approximation

An electrostatic field without a space-charge field is taken as the first approximation and charged particle trajectories of the first approximation are calculated in this field. They are used to determine the distribution of space charge. This allows calculation of the electric field of the second approximation with the space-charge field, which is used to calculate the trajectories of the second approximation. The space-charge distribution is corrected and the new field is calculated.

This procedure is repeated several times until the results of the subsequent n-th approximation is close enough to the results of the previous one, the $(n-1)$-th. As a criterion of convergence of the process of successive approximation, for instance, the slopes of charged particle trajectories in some chosen plane of the analyzed electrode system can be applied. Usually, 5–10 approximations are necessary to get the accuracy required.

A discrete model of a charged particle beam is used for the above procedure. For this purpose, the beam, being analyzed, is divided over its cross section on several elementary layers or tubes of current. The partial current ΔI_k, which a tube carries, is calculated from the area or tube cross section and current-density distribution over the beam cross section, which is supposed to be known. This partial tube current is attributed (attached) to one "central" trajectory of the tube in equation. This trajectory, bringing current ΔI_k, is calculated in the procedure of successive approximation.

Let us briefly consider the technique of account of the beam self-field. A mesh, which is used for electric-field calculation by finite-difference or discrete Green function methods, is shown at Fig. 4.11. The line with the arrow, which crosses the "$ABCD$" cell of the mesh is the charged particle trajectory being

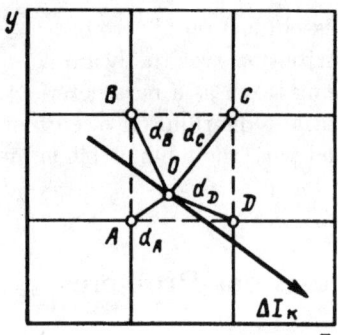

Fig. 4.11. Distribution of space charge on nodes of a mesh cell

found in the previous $(n-1)$-th approximation. It carries the value of current ΔI_k. The elementary space charge, which is contributed in this cell, will be equal to $q_k = \Delta I_k \tau_k$, with τ_k being the transit time of a charge particle in the cell under consideration. It is found as $\tau_k = l_k/v_k$, where l_k is a part of the trajectory lying within the given cell, v_k – average particle velocity.

For calculation of the field of the next n-th approximation the elementary charge Δq is supposed to be concentrated at the middle point 0 of the trajectory segment l_k and distributed between the cell nodes A, B, C, D in accordance with the law of inverse distances. For instance, the part of q_k being distributed in node A is determined as:

$$q_A = \frac{q_k}{d_A} \frac{1}{1/d_A + 1/d_B + 1/d_C + 1/d_D}.$$

If several trajectories cross a cell, then charges of the cell nodes are found by summation of the contributions of the separate trajectories.

A self-magnetic field is usually taken into account approximately. In the case of axially symmetric beams, as a rule, only the θ component of self-magnetic field B_θ is considered. It is found by use of the Ampere law:

$$B_\theta = \frac{\mu_0}{2\pi r} \sum_{k=i}^{k=n} \Delta I_k.$$

This equation determines the azimuthal component of magnetic induction B_θ at a distance r from the axis of symmetry, with summation being performed over the n trajectories passing inside the cycle of radius r.

It is necessary to note that among self-consistent problems there are problems that can not be solved by the process of successive approximation, for instance, the problem of virtual-cathode formation, as convergence of the process is absent. In this case an other method described below is used.

4.8.2 Method of "Step by Step" (or Algorithm Modeling of Transient Process)

A discrete beam model known as the model of "large particle" is used for this method of modeling. In the case of an axially symmetric beam the "large particle" is a particle of a ring form. The model is realized in the following way. The charge particle beam at the entrance of an electrode system to be analyzed is divided into several elementary tubes with partial currents ΔI_k. The space charge brought by these tubes is injected into the system by a discrete portion with time interval Δt. Then, the charge of this discrete portion, that is the large particle, is determined as $q_k = \Delta I_k \Delta t$. The first portion of large particles injected in an analyzing system meets the electric field free of the space-charge field, that is the Laplace field. The second one will move in the Laplace field and the partial space-charge field created by the first group of particles and so on. Step by step the electrode space is filled by space charge and steady-state beam motion is established.

5 Electron Guns

5.1 The Problem of Electron-Beam Formation

The formation of electron beams is provided by special electron-optical systems – electron guns. It can be fulfilled both in pure electrostatic fields and in combined electrostatic and magnetic fields. The problem of electron-beam formation is formulated as follows: there are known electrical and geometrical parameters of the beam, such as current, velocity and form and size of the beam cross section, it is required to determine the form of electrodes and magnetic-field configuration under which the formation of the beam with known parameters is provided.

At the present time, to solve the problem of formation two methods are used: the method of analysis (the method of trials and corrections) and the method of synthesis.

The method of analysis consists of successive changes of the gun electrode's geometry and magnetic-field distribution until the parameters of the beam become close to the required ones. This process includes the following basic steps: the choice of the original geometry of the gun and magnetic-field configuration, trajectory analysis, on the results of which the parameters of the beam are defined, making changes to the original geometry, and the successive trajectory analysis for the new electrode systems and so on. It is evident that the procedure of gun design by this method is quite a laborious one.

In the method of synthesis the electrode geometry and magnetic-field configuration providing formation of a beam with the required parameters are found in a direct way without using the successive approximation process. The classical example of synthesis is the design of electron guns with rectilinear trajectories suggested by Pierce [7]. The design is based on using the relations that describe the one-dimensional electron flow in rectangular, cylindrical and spherical coordinate systems (see Sect. 4.3). According to the Pierce method, from an electron flow the beam of finite transverse sizes is "being cut" and the rest of the flow is thrown out, its action being replaced by the equivalent action of the focusing electrodes. These electrodes must create, along the beam boundary, the same distribution of the electric potential and its derivative as that of the original flow.

The method of synthesis includes solution of the two problems. One of them, known as the inner (inside) problem, consists of the solution of the equation describing electron flows in the hydrodynamic approach, with the purpose of determining the flow with the required electrical and geometrical parameters. The other, known as the outer (outside) problem, supposes the determination of the gun-electrode shapes, it is reduced to the solution of Cauchy's problem for the Laplace equation, with initial conditions given at the beam boundary.

At present, the two variants of synthesis are used in practice. The first of these uses some known particular solution of the flow equations giving a flow with certain geometrical and electrical characteristics (for example, the flow with rectilinear trajectories in Pierce's gun). In this case the characteristics of the flow are known, though, perhaps, they do not always completely meet the requirements of the practical problem being solved. The second variant of the synthesis supposes finding a spatial solution of the inside problem, which would completely meet the requirements regarding electrical and geometrical beam parameters.

In both cases, the shapes of the focusing electrodes require the solution of the outer (outside) problem, that is Cauchy's problem for the Laplace equations, with initial conditions taken from the solution of the inner (inside) problem. The Cauchy problem for the Laplace equations is known to be an improperly set mathematical problem. Its solution is unstable with regard to small changes of the initial conditions, the spatial approach being used for its approximate numerical solution.

5.2 Guns for Formation of Strip Beams

5.2.1 Parallel Strip Beam

The electron gun forming a parallel strip beam may be created by using a part of a parallel rectilinear flow, which is characterized by the following relations:

$$U = Az^{3/4} \tag{5.1}$$

$$j = 2.33 \times 10^{-6} U^{3/2}/z^2 , \tag{5.2}$$

where z – is the longitudinal coordinate (Fig. 5.1); $A = U_a/d^{4/3}$, U_a – is potential of the anode electrode.

In accordance with the general synthesis procedure a layer of thickness $2y_b$ is "being cut", the rest of the flow being thrown out. To keep the nature of electron motion in the layer the following conditions are to be fulfilled on its boundaries:

$$U = Az^{4/3}, \qquad dU/dy = 0 . \tag{5.3}$$

The above relations are used as boundary conditions for solution of Cauchy's problem for Laplace's equation. The solution found by the method

5.2 Guns for Formation of Strip Beams   127

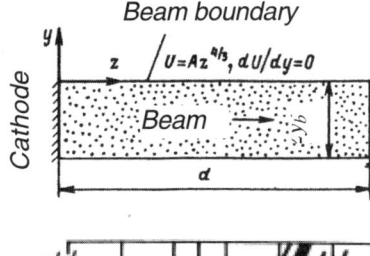

Fig. 5.1. Parallel strip beam, boundary conditions

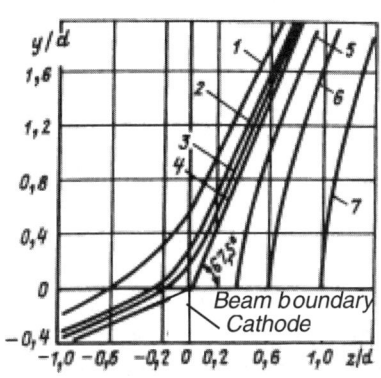

Fig. 5.2. Shapes of equipotential lines, 1 – $-0.25U_a$; 2 – $-0.1U_a$; 3 – $-0.05U_a$; 4 – 0; 5 – $0.25U_a$; 6 – $0.5U_a$; 7 – U_a

of analytical continuation of the function $U = Az^{4/3}$ (see Sect. 2.3) is as follows:

$$U = Re\left[A(z+iy)^{4/3}\right] = ARe\left[r^{4/3}e^{4/3i\theta}\right] = Ar^{4/3}\cos\frac{4}{3}\theta. \quad (5.4)$$

The shapes of equipotential lines found from this relation are presented in Fig. 5.2. The zero equipotential line is the straight one inclined to the beam boundary by the angle of 67.5°. The obtained results are strictly true for the flow of infinite extent in the direction of the x-axis that is perpendicular to the plane of the figure. With a certain approximation they can be used for beams of finite width x_b, under the condition that the beam width $x_b \gg 2y_b$, when edge effects are negligible.

The perveance of such a gun calculated per unit flow width is found from the 3/2 power law:

$$P_1 = I_1/U_a^{3/2} = 2.33 \times 10^{-6} 2y_b/d^2.$$

The anode electrode of a real gun usually has the aperture not covered with a grid (Fig. 5.3). The aperture disturbs the electric-field, leads to the apperance of the y-component of the field near the anode electrode and to decreasing its z-component at the cathode surface. This, in turn, yields the y-components of electron velocities at the exit of the gun and leads to the decrease of the beam perveance.

The defocusing action of the anode aperture can be taken into account by considering it as a slit lens. The focal distance of the latter is determined by the relation $f = 2U/(E_1 - E_2)$, where U – potential at the center of the

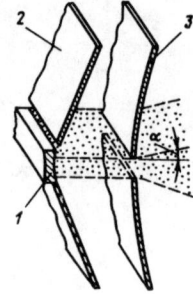 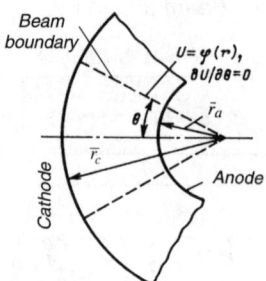

Fig. 5.3. Electron gun for strip-beam formation; 1 – cathode, 2 – focusing electrode, 3 – anode

Fig. 5.4. Wedge-shaped beam, boundary conditions

electrode slit, which is taken to be equal to the anode electrode potential ($U = U_a$); E_1 and E_2 are the electric-field strength to the left and to the right from the anode. Their values are calculated under the condition that there is no aperture in the anode electrode. In the considered case $E_2 = 0$, E_1 is calculated as follows:

$$E_1 = -\left.\frac{dU}{dz}\right|_{z=d} = -\frac{U_a}{d^{4/3}}\left.\frac{d}{dz}\left(z^{4/3}\right)\right|_{z=d} = -\frac{4}{3}\frac{U_a}{d}.$$

In this case we obtain $f = -3/2d$ The negative value of the focal length indicates the defocusing lens action.

The angles of an electron trajectory slopes at the gun exit is expressed by the following approximate formula: $\alpha \simeq \tan\alpha = y/|f| = 2/3(y/d)$ and for the boundary electron: $(y = y_b)$ $\alpha \simeq 2/3(y_b/d)$.

5.2.2 Wedge-Shaped Beam

The electron gun for formation of a wedge-shaped beam can be designed by making use a part of radial cylindrical flow (Fig. 5.4). Such a flow is characterized by the following basic equations:

Equation for potential distribution:

$$U(r) = U_a \left[\frac{\frac{r}{r_c}(-\beta)^2}{\frac{r_a}{r_c}(-\beta_a)^2}\right]^{2/3}.$$

The value of current in the sector of unit width (the size in the direction normal to the plane of the figure) and half-angle θ:

$$I_1 = 14'.66 \times 10^{-6}\frac{2\theta}{360}\frac{U^{3/2}}{r(-\beta)^2},$$

5.2 Guns for Formation of Strip Beams   129

Table 5.1. Values of function $(-\beta)^2$

r_c/r	$(-\beta)^2$	r_c/r	$(-\beta)^2$	r_c/r	$(-\beta)^2$
1.00	0.00000	1.15	0.02186	1.8	0.5572
1.01	0.00010	1.2	0.03849	1.9	0.6947
1.02	0.00040	1.3	0.08504	2.0	0.8454
1.04	0.00159	1.4	0.14856	2.1	1.0086
1.06	0.00356	1.5	0.22820	2.2	1.1840
1.08	0.00630	1.6	0.32330	2.3	1.3712

$(-\beta_a)^2$ is the value of $(-\beta)^2$ calculated for relation $r_c/r = r_c r_a$, where r_a – is radius of gun anode.

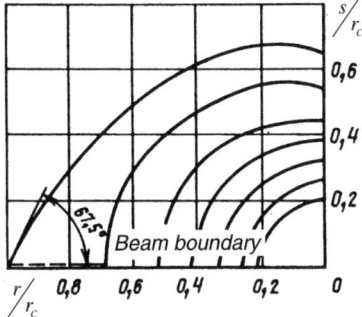

Fig. 5.5. Map of equipotential lines for wedge-shaped beam

here $(-\beta)^2$ – is a function relating cathode curvature radius r_c to the current radius r given in Table 5.1, half-angle θ is expressed in degrees.

As in the previous case, for keeping the character of electron motion in the sector with half-angle θ, the action of the thrown part of the electron flow is replaced with the action of focusing electrodes, which should provide, along the beam boundary, the following boundary condition: $U = U(r)$, $\partial U/\partial \theta = 0$. Forms of focusing electrodes are found as a result of solution of the outer (outside) problem. Figure 5.5 shows the family of equipotential lines analytically found in [90].

The aperture in the anode electrode produces a certain decrease of cathode current and beam defocusing. The latter effect can be taken into account by a method similar to that used in the previous case.

5.2.3 Strip-Beam Forming in Crossed Fields

Analytical calculations of guns that provide formation of a strip beam in crossed electric and magnetic fields is based on a partial solution of the system of equations describing the electron flow in the hydrodynamic approximation [91]. This solution assumes that potential velocity components, current and space charge densities depend on one coordinate, for instance, the y-coordinate.

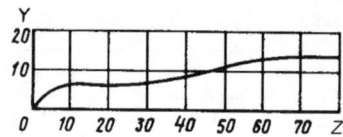

Fig. 5.6. An electron trajectory in crossed fields (case $p \ll 1$)

In normalized variables, the system of equations describing the electron flow is written as [91]:

$$Z = Z_0 + u^2/2 - (1-p)(1-\cos u) \ ;$$
$$Y = u - (1-p)\sin u \ ;$$
$$\Phi = u^2/2 - (1-p)(u\sin u + \cos u - 1) \ ,$$

where Z and Y – are normalized longitudinal and transversal coordinates that are defined by relations: $Z = \varepsilon_0 \omega_c^3 z/(\eta j_y)$; $Y = \varepsilon_0 \omega_c^3 y/(\eta j_y)$; $\Phi = \varepsilon_0 \omega_c^4 U/(\eta j_y^2)$ – normalized potential; $u = \omega_c t$ – is dimensionless time of electron motion from the cathode to a current point of coordinate space; Z_0 – coordinate of electron emission on the cathode ($u = 0$, $Y = 0$); $p = \varepsilon_0 \omega_c^2 \dot{y}_0/(\eta j_y)$ – parameter of initial conditions.

The values included in the above relations have the following sense: ω_c – cyclotron frequency $\omega_c = \eta B$; $\eta = |e|/m$; j_y – y-component of the current density; $\dot{y}_0$ – initial electron velocity on the cathode.

Depending on the value of parameter p the following particular cases of an electron-beam motion are possible:

a) $p \ll 1$; $Z = Z_0 + u^2/2 + \cos u - 1$, $Y = u - \sin u$, $\Phi = u^2/2 - u\sin u - \cos u + 1$.
b) $p \approx 1$; $Z = Z_0 + u^2/2$, $Y = u$, $\Phi = u^2/2$.
c) $p \gg 1$; $Z = Z_0 + u^2/2 + p(1-\cos u)$, $Y = u + p\sin u$, $\Phi = u^2/2 + p(u\sin u + \cos u - 1)$.

Cases a) and b) are of practical interest, they are taken as the basis for designing of guns providing the formation of strip beams. In the former case, electron trajectories appear to be relatively complicated curves, one of them is shown in Fig. 5.6. In points $u = 2\pi n$ ($n = 1, 2, 3, \ldots$) trajectories prove to be parallel to the Z-axis. In the latter case, electron trajectories are parabolas, which are approaching the lines parallel to the Z-axis, with the Z-coordinate being increased.

The former case is used for design of the so-called short-focus gun, while the latter is applied for design of the long-focus gun (Figs. 5.7 and 5.8). For designing the gun the part of the flow limited to the two trajectories starting from the cathode points $Z = 0$ and $Z = Z_0$, is "being cut", the rest of the flow is removed and its action is replaced by the equivalent action of focusing electrodes, which are determined as the result of the solution of outer (outside) problem.

Let us consider the calculation of forms of electrodes for the long-focus gun with a parabolic form of trajectories. Supposing that the beam boundary

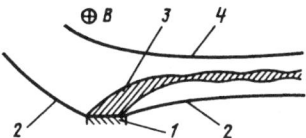
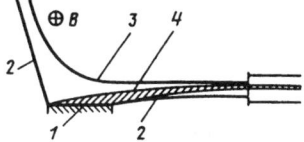

Fig. 5.7. Short-focus gun; 1 – cathode, 2 – focusing electrodes, 3 – strip beam, 4 – anode

Fig. 5.8. Long-focus gun; 1 – cathode, 2 – focusing electrodes, 3 – strip beam, 4 – anode

lies in the complex plane $\zeta = Z + iY$, we will map it onto the real axis of the auxiliary complex plane $\psi = u + iv$. Since the parametrical equation of the boundary in the initial plane is $Y = u$ and $Z = Z_0 + u^2/2$, the following function is used for this transformation:

$$\zeta = Z(\psi) + i\, Y(\psi) = Z_0 + \psi^2/2 + j\psi \;.$$

Substitution of $\zeta = Z + iY$ and $\psi = u + iv$ into this equation gives

$$Z + iY = iu - v + (1/2)\left(u^2 + 2iuv - v^2\right) + Z_0 \;,$$

or, after separation of the imaginary and real parts:

$$Z = Z_0 + \frac{1}{2}\left(u^2 - v^2\right) - v; Y = u(1+v) \;. \tag{5.5}$$

The initial conditions on the transformed beam boundary in the complex plane ψ are expressed as:

$$f_1(u) = \Phi|_{v=0} = \frac{u^2}{2}; \qquad f_2(u) = \left.\frac{\partial \Phi}{\partial v}\right|_{v=0} = u^2;$$

$$f_3 = \left.\frac{\partial \Phi}{\partial u}\right|_{v=0} = u; \qquad f_3(u) = \left.\frac{\partial \Phi}{\partial u}\right|_{v=0} = u \;.$$

Then, the complex potential W in the plane ψ is determined by the equations (see (2.18)):

$$\frac{dW}{d\psi} = f_3(\psi) - if_2(\psi) \;, \qquad \frac{dW}{d\psi} = \psi - i\psi^2 \;.$$

This results in $W = \psi^2/2 - i\psi^3/3 + C$. The potential Φ, being the real part of W, is found as:

$$\Phi = Re\,[W] = Re\left[\frac{\psi^2}{2} - i\frac{\psi^3}{3} + C\right]$$

$$= \frac{1}{2}\left(u^2 - v^2\right) + u^2 v - \frac{v^3}{2} + Re\,C \;.$$

Since the point $u = 0$, $v = 0$ corresponds to the point $Z = Z_0$ and $Y_0 = 0$, located on the cathode, then its potential is equal to zero. Assuming in the previous equation $u = v = \Phi = 0$, we find that Re $C = 0$.

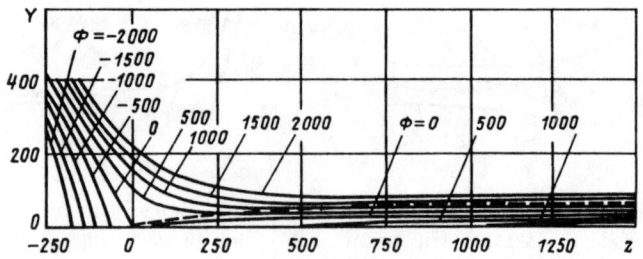

Fig. 5.9. Map equipotential lines for designing of electrodes of long-focus gun

Then, the equation of equipotential line in the plane ψ can be written down as follows:

$$u^2 = \left[2\Phi + v^2\left(1 + \frac{2}{3}v\right)\right] \bigg/ (2v+1).$$

Keeping $\Phi = \text{const} = C_1$ in this equation and varying v one can find the values of coordinate u of the points belonging to the equipotential line in question. Then coordinates Z and Y of these points in the plane ζ are calculated using (5.5).

The results of calculation of the zero equipotential line $\Phi = 0$ are given below, the values of coordinates being expressed in arbitrary units:

v	−0.49996	−0.4999	−0.45	−0.4	0	10	20	25	30
u	45	29	1.18	0.75	0	6	11.8	14.7	17.5
Z	1050	415	1.2	0.6	0	−41.6	−150	−230	−328
Y	22.6	14	0.65	0.45	0	67	250	382	540

The geometry of other equipotentials is found in a similar way.

The results of these calculations obtained in [91] are presented in Fig. 5.9. Using this diagram it is possible to determine the forms of the gun electrodes. The curves of zero potential determine the shape of cathode focusing electrodes. The anode electrode has the shape of the equipotential line with the normalized potential equal to

$$\Phi = \Phi_a = \frac{\varepsilon_0^2 \omega_c^4}{\eta j_y^2} U_a.$$

To pass from absolute values to the normalized ones (and back) the following formulae are used:

$$\Phi = \frac{10^{-12}}{2.35} \frac{B^4}{j_c^2} U; \quad Z = \frac{10^{-6}}{3.7} \frac{B^3}{j_c} z; \quad Y = \frac{10^{-6}}{3.7} \frac{B^3}{j_c} y,$$

magnetic field induction B, current density j_c and distances z and y being expressed in the units G, A/cm² and cm, respectively.

5.3 Guns for Solid Axially Symmetric Beams

5.3.1 Parallel Cylindrical Beam

The procedure of gun design for formation of a parallel cylindrical beam is the same as used for the gun with a parallel strip beam considered above. An electron beam of cylindrical form is cut out from an infinite parallel flow (Fig. 5.10). The shapes of focusing electrodes are found by computation of the equipotential lines in the region outside of the beam with the initial conditions given of the boundary of the cylinder: $U = U_a(z/d)^{4/3}$, $\partial U/\partial r = 0$. One of the possible methods of solution of this problem is described in [10]. The picture of equipotential lines is presented in Fig. 5.11. The perveance of such a gun is determined by the relation following from the 3/2 power law:

$$P = \frac{I}{U_a^{3/2}} = 2.33 \times 10^{-6} \frac{\pi r_c^2}{d^2}, \tag{5.6}$$

where r_c – cathode radius that is equal to the beam radius r_b; d – cathode–anode distance; U_a – anode voltage; I – beam current.

The anode aperture defocusing action can be approximately taken into account, if we consider it as that of a diaphragm lens with focal distance

$$f = \frac{4U_a}{(-\partial U/\partial z)_{z=d}} = -3d.$$

Then the angle of the boundary electron trajectory slope at the exit of the gun is:

$$\alpha \approx \mathrm{tg}\alpha = \frac{r_b}{|f|} = \frac{1}{3}\frac{r_b}{d}.$$

Substitution of the value of d found from (5.6) gives $\alpha \approx \tan\alpha = \sqrt{P/8.1}$, P being the gun perveance expressed in mkA/V$^{3/2}$.

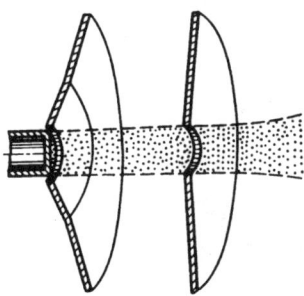

Fig. 5.10. Cylindrical beam formation

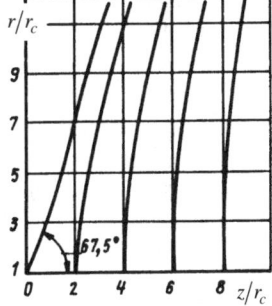

Fig. 5.11. Map of equipotential lines

5.3.2 Pierce Gun for Convergent Beam

The radial electron flow in the spherical coordinate system is used for designing the gun. This flow is characterized by the following relationships (see Sect. 4.3):

$$I = \frac{16\pi\varepsilon_0}{9}\sqrt{2\frac{|e|}{m}}\frac{U^{3/2}}{(-\alpha)^2} = 29.34 \times 10^{-6}\frac{U^{3/4}}{(-\alpha)^2};$$

$$U = U_a(-\alpha)^{4/3}/(-\alpha_a)^{4/3},$$

values of functions $(-\alpha)^2$ and $(-\alpha)^{4/3}$ being presented in Table 5.2.

For keeping the electron flow undisturbed inside the sector of angle 2θ (see Fig. 5.12), the action of the removed part of the flow is replaced by the equivalent action of focusing electrodes that should provide the following conditions at the beam boundary:

$$U = U_a\frac{(-\alpha)^{4/3}}{(-\alpha_a)^{4/3}} = \varphi(\bar{r}) \qquad \frac{\partial U}{\partial \theta} = 0.$$

The form of equipotential lines obtained in [92] by an analytical method is shown in Fig. 5.13. Here $s/\bar{r}_c$ – is the relative distance measured from the beam boundary.

The difference in the shape of equipotential lines is seen to be small enough for the two angles of beam convergence $\theta = 10°$ (hatched lines) and $\theta = 40°$

Table 5.2. Values of the function $(-\alpha)^{4/3}$ and $(-\alpha)^2$

r_c/r	$(-\alpha)^{4/3}$	$(-\alpha)^2$	r_c/r	$(-\alpha)^{4/3}$	$(-\alpha)^2$
1.00	0.000	0.0000	2.1	0.92	0.888
1.05	0.018	0.0024	2.2	1.02	1.036
1.10	0.045	0.0096	2.3	1.12	1.193
1.15	0.076	0.0213	2.4	1.21	1.258
1.20	0.110	0.0372	2.5	1.33	1.532
1.25	0.148	0.0571	2.6	1.43	1.712
1.30	0.185	0.0809	2.7	1.52	1.901
1.35	0.226	0.1084	2.8	1.63	2.098
1.40	0.268	0.1396	2.9	1.74	2.302
1.45	0.309	0.1740	3.0	1.84	2.512
1.50	0.353	0.2118	3.2	–	2.954
1.60	0.443	0.2968	3.4	–	3.421
1.70	0.535	0.3940	3.6	–	3.913
1.80	0.630	0.5020	3.8	–	4.429
1.90	0.730	0.6210	4.0	–	4.968
2.00	0.820	0.7500	4.2	–	5.528

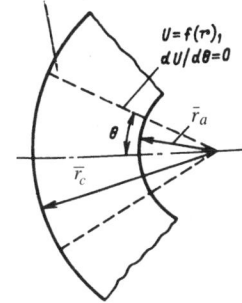
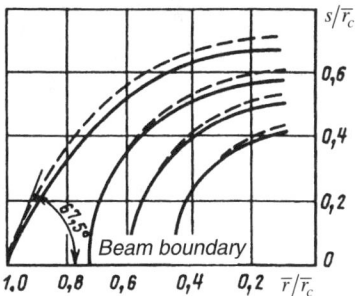

Fig. 5.12. Gun for convergent-beam formation, boundary condition

Fig. 5.13. Map of equipotential lines for conical beam design; *hatched lines – angle of beam convergence* $\theta = 10°$, *solid lines* $\theta = 40°$

(solid lines). This allows the chart of Fig. 5.13 to be used for determination of focusing electrodes of guns with intermediate values of the angle θ. The connection between the gun geometrical parameters $\bar{r}_c/\bar{r}_a$ and θ, current and anode voltage is determined by the 3/2 power law:

$$I = 29.34 \times 10^{-6} \frac{\sin^2(\theta/2)}{(-\alpha_a)^2} U_a^{3/2}, \tag{5.7}$$

where $(-\alpha_a)^2$ is the value of the tabulated function $(-\alpha)^2$, calculated for $\bar{r}_c/\bar{r} = \bar{r}_c/\bar{r}_a$.

Then the gun perveance expressed in mkA/V$^{3/2}$ is given by the formula:

$$P = \frac{I}{U_a^{3/2}} = 29.34 \frac{\sin^2(\theta/2)}{(-\alpha_a)^2}. \tag{5.8}$$

In the theory of the Pierce guns the effect of the anode aperture is taken into account only from the point of view of its defocusing action that is assumed to be equivalent to that of the diaphragm lens with the focal distance:

$$f = 4U_a/(-\partial U/\partial r)_{\bar{r}=\bar{r}_a}.$$

Substitution in this equation of $U = U_a(-\alpha)^{4/3}/(-\alpha_a)^{4/3}$ gives

$$f = -\frac{4(-\alpha_a)^{4/3}}{\frac{\partial}{\partial \bar{r}}(-\alpha)^{4/3}\big|_{\bar{r}=\bar{r}_a}}. \tag{5.9}$$

It follows from this equation that the focal distance of the anode lens depends only on the relation of cathode and anode radii of curvature. This dependence is shown in Fig. 5.14. Due to the refracting action of the anode lens the angle of electron trajectories convergence at the exit of the gun γ is less than the initial convergence angle θ (Fig. 5.15).

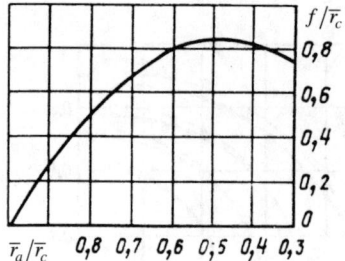

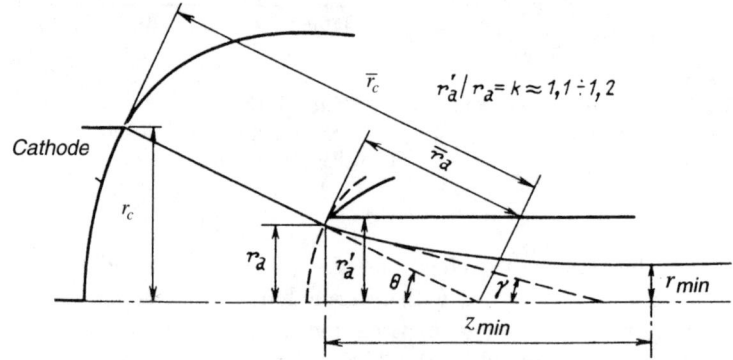

Fig. 5.14. Focusing length of the anode lens as function of electrode radii of curvature

Fig. 5.15. Scheme for designing of the gun forming convergent beam

Angles γ and θ are connected by the relation

$$\gamma = \theta - \alpha . \tag{5.10}$$

Here α – is the refraction angle of electron trajectories in the anode lens, calculated by the following formulae:

- for the gun with the aperture in the anode electrode without a grid, $\alpha \simeq \tan \alpha = r_a/f$, where r_a – beam radius at the entrance into the anode aperture, the radius of the latter being $r'_a = kr_a$ ($k \geq 1$);
- for the gun with the anode aperture covered with a grid with a square mesh $\alpha \simeq \tan \alpha = h/2f$, where h – is the grid pitch.

From comparison of the two relations it follows that the defocusing action of the anode aperture can be substantially diminished, if it is covered with a grid with a small mesh size, $h \ll r_a$. However, the presence of the grid leads to the loss of part of the beam current, henceforth guns with the grid are applied relatively rarely. Because of this the following analysis will be limited to the case of a gun without a grid. From geometrical relations (Fig. 5.15) we have $r_a = \bar{r}_a \sin \theta \simeq \bar{r}_a \theta$; then $\alpha \simeq \tan \alpha = r_a/f \simeq \bar{r}_a \theta/f$. Substituting this relation into (5.10) we obtain

$$\gamma = \theta \left(1 - \frac{\bar{r}_a/\bar{r}_c}{f/\bar{r}_c}\right) . \tag{5.11}$$

5.3 Guns for Solid Axially Symmetric Beams 137

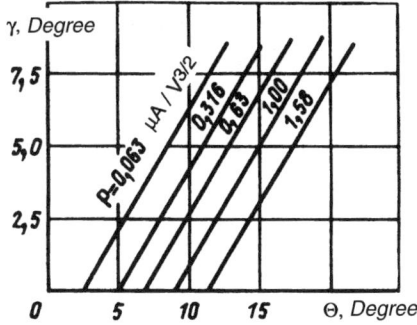

Fig. 5.16. Dependence of the angle of beam convergence γ on the angle θ and the value of perveance P

With the aid of this relation it is easy to find the angle γ, if we use the diagram in Fig. 5.14. In particular, from this diagram it follows that with $\bar{r}_a/\bar{r}_c = 0.69$, $(\bar{r}_c/\bar{r}_a = 1.45)$, $f/\bar{r}_c = 0.69$. This means that with $\bar{r}_c/\bar{r}_a = 1.45$ the angle $\gamma = 0$, i.e. at the exit of the gun, beam trajectories will be parallel to the beam axis z. When $\bar{r}_c/\bar{r}_a > 1.45$ the beam will be convergent, and when $\bar{r}_c/\bar{r}_a < 1.45$ it will be divergent.

Figure 5.16 shows the function $\gamma = f(\theta)$, with the perveance P being used as the parameter. It is obtained with the help of relations (5.8), (5.11) and the diagram of Fig. 5.14.

Electron-beam motion in the channel, located behind the gun anode, is determined by the action of Coulomb's forces and can be found with the aid of universal curves for the beam envelope and the slope angle of the beam-boundary trajectory, which are described in Sect. 4.5, or with the help of the corresponding analytical formulae.

The Scheme of Pierce Gun Design

The values of the following beam parameters are assigned:

- beam perveance P,
- minimum beam radius $r_{\min}$,
- its position on the z-axis – $z_{\min}$ (see Fig. 5.15).

At first, using the universal curves (Fig. 4.4) or equivalent analytical relations, we find the electron beam radius at the entrance of the anode aperture $r_a = r_{\min}(1 + 0.25Z^2 - 0.017Z^3)$, where $Z = 0.174\sqrt{P}\, z_{\min}/r_{\min}$ is the normalized z-coordinate, and the angle of trajectory slope $\tan\gamma = 0.174\sqrt{P\ln(r_a/r_{\min})}$.

Then, with the help of the diagram in Fig. 5.16, (5.8) and Table 5.2 we find the angle of beam convergence θ, the relation of radii of curvature $\bar{r}_c/\bar{r}_a$ $\bar{r}_a = r_a/\sin\theta$ and find values of $\bar{r}_c$ and r_c. So, the basic geometrical gun parameters have been determined.

Focusing electrodes can be found with the aid of the equipotentials map presented in Fig. 5.13. The form of the anode electrode is defined by the shape of the equipotential line, crossing the horizontal axis of the diagram at the point with the value of the relation $\bar{r}_k/\bar{r}_a$, obtained above (if the equipotential does not run through this point, it is obtained by interpolation).

Example 5.1. $P = 1\,\mu\text{A}/\text{V}^{3/2}$; $r_{\min} = 0.5$ cm; $z_{\min} = 5$ cm are the set of required beam parameters.
First, we calculate:

$$Z = 0.174\sqrt{P}\,\frac{z_{\min}}{r_{\min}} = 1.74\ ;$$

$$\bar{r}_a = r_{\min}\left(1 + 0.25Z^2 + 0.017Z^3\right)$$
$$= 0.5(1 + 0.25 \times 3.02 - 0.017 \times 5.3) = 0.85\ ;$$

$$\tan\gamma = 0.174\sqrt{P\ln(\bar{r}_a/r_{\min})} = 0.123;\quad \gamma = 7°\ .$$

Then, from the diagram (Fig. 5.16) we find $\theta = 17.5°$. For known values of θ and P using (5.8) we find $(-\alpha_a)^2 = 0.68$ and the corresponding value $\bar{r}_c/\bar{r}_a = 1.95$. Then we determine $\bar{r}_a = \bar{r}_a/\sin\theta = 0.85 : 0.3 = 2.8$ cm and $\bar{r}_c = 1.95\bar{r}_a = 5.5$ cm, $r_c = \bar{r}_c\sin\theta = 1.65$ cm. The gun electrodes are found using the equipotential chart (Fig. 5.13).

"Optimal" Gun.. In the Pierce gun theory [7] it is shown that for the set values of perveance P and initial beam radius r_a there exists such a gun geometry that provides the maximum value of the distance $z_{\min}$, that is distance from the anode electrode to the position of minimum beam cross section $r_{\min}$. This gun is sometimes called optimal. It is characterized by the following relations: $\bar{r}_c/\bar{r}_a = 2.2$; $\theta \simeq 21\sqrt{P}$; $z_{\min} = 1.05\bar{r}_c$, $r_{\min} = 0.43r_a$. Using these formulae, it is easy to calculate the gun geometry for the given beam parameters.

5.3.3 Synthesis of Guns by Ovcharov's Method

In contrast to the guns considered above, this procedure of gun design supposes obtaining a special solution of the inside problem, which should provide the required properties of the beam to be formed. The solution is carried out in the paraxial approximation using the paraxial equation, written in the special curvilinear coordinate system. The system is formed by the family of similar lines (family a) and by the family of lines orthogonal to them (family b). In the coordinate system r, z (Fig. 5.17) lines, belonging to these families, are described by equations:

$$r(z) = qR(z);\qquad \partial r/\partial z = -\frac{R(z)}{r\partial R(z)/\partial z}\ ,$$

where $R(z)$ – is some basic line from the family a; q – is the similarity parameter [93].

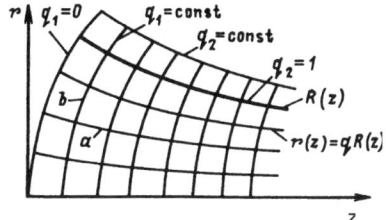

Fig. 5.17. Orthogonal curvilinear coordinate system

Such a choice of the orthogonal coordinate system corresponds to the model of a flow with similar trajectories, which is supposed to coincide with the lines belonging to family a. Parameterization of the orthogonal mesh is carried out as follows. As the parameter that determines lines from the family a, the similarity parameter $q_2 = r(z)/R(z)$ is taken. It represents the relative distance of a line of the family from symmetry axis. The parameter that determines lines from the family b, is taken to be equal to the distance at which a line of the family crosses the symmetry axis $q_1 = z$. Parameters q_1 and q_2 play a role of curvilinear coordinates, the former has dimensions of length, the latter is dimensionless.

If the base line $R(z)$ is supposed to be the beam boundary, then its paraxial equation can be written as [93, 94]:

$$R^2 \frac{d^2 U_0}{dq_1^2} + 2R \frac{dR}{dq_1} \frac{dU_0}{dq_1} + 4U_0 R \frac{d^2 R}{dq_1^2} = \frac{I}{\pi \varepsilon_0 \sqrt{2 \frac{e}{m} U_0}},$$

where U_0 is the potential on the beam axis ($q_2 = 0$); I is the beam current.

Inserting normalized variables $\bar{R} = R/R_n$, $u = U_0/U_n$ and $\bar{q}_1 = q_1/l_n$, we obtain:

$$\bar{R}^2 \frac{d^2 u}{d\bar{q}_1^2} + 2\bar{R} \frac{d\bar{R}}{d\bar{q}_1} \frac{du}{d\bar{q}_1} + 4u\bar{R} \frac{d^2 \bar{R}}{d\bar{q}_1^2} = \frac{i}{\sqrt{u}}, \tag{5.12}$$

where

$$i = \frac{I}{\pi \varepsilon_0 \sqrt{2\frac{e}{m}} U_n^{3/2} \mu^2}, \qquad \mu = \frac{R_n}{l_n},$$

R_n, U_n and l_n are normalization factors

Electron-gun calculation is carried out by the following scheme. Initially, the axial distribution of potential is set, which is to meet the following conditions:

$$u = 0, \quad \frac{du}{d\bar{q}_1} = 0, \quad \text{for} \quad \bar{q}_1 = 0 \tag{5.13}$$

$$\frac{du}{d\bar{q}_1} = \frac{d^2 u}{d\bar{q}_1^2} = 0 \quad u = 1 \text{ for } \bar{q}_1 = 1. \tag{5.14}$$

The first (5.13) corresponds to space-charge-limited cathode current flow, the latter, (5.14) are determined by the physical conditions at the transition region from the gun exit to the drift space.

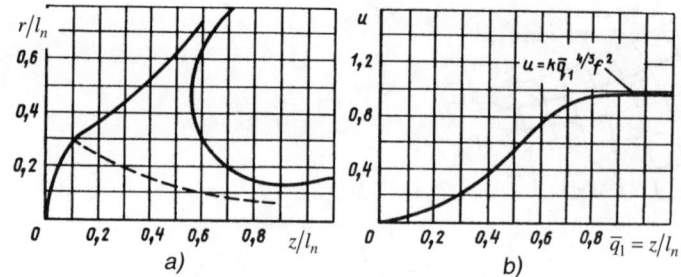

Fig. 5.18. Shape of gun electrodes (**a**) and the axial potential distribution (**b**) for a gun designed by the method of synthesis

A typical curve of potential distribution is shown in Fig. 5.18b. To set the potential distribution, the following expression can be used [94]:

$$u = k\bar{q}_1^{4/3} f^2, \qquad f = 1 + \sum_{n=1}^{5} a_n \bar{q}_1^n .$$

When setting potential in this way the conditions (5.13) are fulfilled automatically. The conditions (5.14) are realized owing to the following relations imposed on the coefficients a_3, a_4 and a_5

$$a_3 = \frac{119}{9}\frac{1}{\sqrt{k}} - 10 - 6a_1 - 3a_2 ; \quad a_4 = -\frac{187}{9}\frac{1}{\sqrt{k}} + 15 + 8a_1 + 3a_2 ;$$

$$a_5 = \frac{77}{9}\frac{1}{\sqrt{k}} - 3a_1 - a_2 - 6 .$$

To obtain the cathode of spherical form the additional condition $d^2\bar{R}/d\bar{q}_1^2 = 0$ should be fulfilled. This will be provided if coefficients k_1, a_1 and a_2 are determined by

$$k = \left(\frac{9}{4}i\right)^{2/3} ; \quad a_1 = -\frac{8}{15}\frac{d\bar{R}}{d\bar{q}_1}\bigg|_{\bar{q}_1=0} ; \quad a_2 = \frac{361}{900}\left(\frac{d\bar{R}}{d\bar{q}_1}\right)^2\bigg|_{\bar{q}_1=0} .$$

The numerical integration of (5.12) is produced, for which it is necessary to choose values of the parameter i and to set initial values $R|_{\bar{q}_1=0} = 1$ and $(d\bar{R}/d\bar{q}_1)_{\bar{q}_1=0}$. As a result of integration the electron beam form is found, which corresponds to this potential distribution.

The outside problem solution is also carried out in the curvilinear coordinate system. As is shown in [95], calculation of the potential outside of a beam can be fulfilled with the approximate formula:

$$\bar{U} = u + \mu^2 q_2^2 u \bar{R}\frac{d^2\bar{R}}{d\bar{q}_1^2} + \frac{\mu^2 i}{4\sqrt{u}}\left(1 - q_2^2 + \ln q_2^2\right) ,$$

where $\bar{U} = U/U_n$ – is normalized potential.

Setting $\bar{U} = $ const it is possible to calculate the form of the corresponding equipotential line in coordinates $\bar{q}_1$, q_2. Transform to cylindrical coordinates is carried out with the following equations:

$$r/l_n \simeq \mu q_2 \bar{R}\left[1 - \frac{1}{2}\mu^2 q_2^2 \left(\frac{d\bar{R}}{d\bar{q}_1}\right)^2\right] \quad z/l_n \simeq \bar{q}_1 - \frac{1}{2}\mu^2 q_2^2 \bar{R}\frac{d\bar{R}}{d\bar{q}_1} .$$

The typical electrodes of the gun calculated in such a way is shown in Fig. 5.18a.

5.4 Guns for Formation of Hollow Axisymmetric Beams

5.4.1 Parallel Tubular Beam

An electron gun that forms a parallel tubular beam can be designed on the basis of a gun forming a solid cylindrical beam, if we remove the core of the beam. For keeping electron motion in the tubular beam part, additional inside electrodes (Fig. 5.19) are used, which provide, along the inside boundary, the conditions $\partial U/\partial r = 0$ and $U = U_a(z/d)^{4/3}$. The form of external focusing electrodes remains unchanged as compared with that of the gun for solid beam formation.

Calculation of basic geometrical sizes of the gun can be carried out with the aid of the 3/2 power law:

$$I = 2.33 \times 10^{-6} \frac{\pi \left(r_o^2 - r_i^2\right)}{d^2} U_a^{3/2} ,$$

where r_i and r_o – are inner and outer radii of the hollow beam.

The lens action of the ring aperture in the anode electrode can be approximately made by considering its action as that of a slit lens. Then the angles of boundary trajectories slopes at the exit from the gun can be approximately determined by the formula $\alpha = \Delta y/3d$, where $\Delta y = r_o - r_i$ – is the hollow-beam thickness, external trajectories being deflected from the gun axis and internal ones to the axis.

5.4.2 Convergent Hollow Beam

An electron gun for formation of a convergent hollow beam can be designed on the basis of the gun forming a solid convergent beam. To compensate the action of the removed central part of the beam the inner electrodes are introduced, the external electrodes being unchanged (Fig. 5.20). The following boundary conditions are used in order to determine the shape of the focusing electrodes

$$U(\bar{r}) = U_a \frac{(-\alpha)^{4/3}}{(-\alpha_a)^{4/3}} , \quad \frac{\partial U}{\partial \theta} = 0 .$$

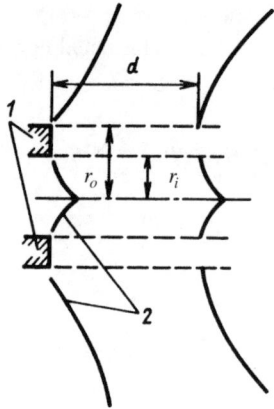
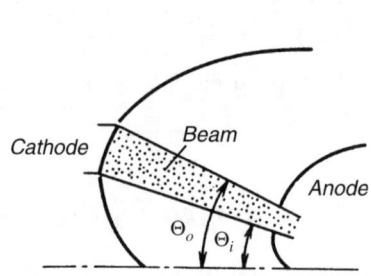

Fig. 5.19. Electron gun for formation of paralle hollow beam; 1 – cathode, 2 – focusing electrodes

Fig. 5.20. Electron gun for formation of hollow convergent beam

The general sizes of a gun can be found with an equation following from the 3/2 power law for a spherical diode:

$$I = 29.3 \times 10^{-6} \frac{\sin^2 \frac{\theta_e}{2} - \sin^2 \frac{\theta_i}{2}}{(-\alpha_a)^2} U_a^{3/2} ,$$

where θ_e and θ_i – are slope angles of external and internal beam boundaries (Fig. 5.20).

The action of the ring aperture in the anode electrode can be taken into account in the same way as has been done in previous cases. It is necessary to note that the hollow beam formed by the gun could transform into the solid one as a result of electron trajectories crossing, if some special measures are not taken.

5.4.3 Magnetron Guns for Formation of Hollow Beams

The design of magnetron guns is based on particular solutions of a system of equations that describes the flow motion in crossed fields. One of these is the solution describing flow from a flat cathode that is inclined at some angle θ with respect to field lines of a uniform magnetic field. If we arrange the coordinate system x, y, z as shown in Fig. 5.21 (coordinate x is perpendicular to the diagram plane), then a partial solution is determined by the following system of equations [96]:

$$Y = u + \frac{u^2}{6} \sin^2 \theta, \quad Z = Z_0 + \frac{u^2}{2} \sin \theta \cos \theta, \quad X = X_0 - \frac{u^2}{2} \cos \theta \tag{5.15}$$

5.4 Guns for Formation of Hollow Axisymmetric Beams

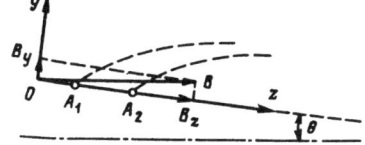

Fig. 5.21. Schematic drawing for consideration of magnetron gun design

$$\Phi = \frac{u^2}{2} + \frac{u^4}{8}\sin^2\theta, \qquad (5.16)$$

here

$$X = \frac{\varepsilon_0 \omega_c^3}{\eta j_y}x, \quad Y = \frac{\varepsilon_0 \omega_c^3}{\eta j_y}y, \quad Z = \frac{\varepsilon_0 \omega_c^3}{\eta j_y}z \quad \Phi = \frac{\varepsilon_0^2 \omega_c^4}{\eta j_y^2}U \quad \text{and} \quad u = \omega_c t$$

are normalized variables.

Equations (5.15) represent parametrical equation of the space trajectory of electrons, which leaves the cathode at the point with coordinates $Y = 0$, $Z = Z_0$, $X = X_0$. Equation (5.16) defines the potential distribution along this trajectory. This solution corresponds to the case when electrons leave the cathode with finite initial velocity $\dot{y}_0 = \frac{1}{\omega_c^2}\frac{\eta j_y}{\varepsilon_0}$. The next step in the standard procedure of beam synthesis is to cut from the infinite flow a beam of finite dimensions. In the case under consideration the cutout is the part of the flow limited with trajectories starting from cathode points A_1 and A_2 with normalized coordinates $Y = 0$, $Z = 0$ and $Y = 0$, $Z = Z_0$ the beam extent in the X direction is considered to be infinite (Fig. 5.21).

The shape of the focusing electrodes can be found as a result of solution of Cauchy's problem for the Laplace equation under initial conditions set on the beam boundaries. Results of calculation carried out by the method of analytical continuation are presented as an equipotentials chart in Fig. 5.22, an electron trajectory being shown here with a hatched line [96].

The described procedure is strictly valid for a planar magnetron gun. It is possible to adopt it for designing an axially symmetric gun if we consider the latter as the planar gun wrapped in the ring.

If the mean radius of the ring is significantly greater than the mean distance between cathode and anode, then electron trajectories in this system will only differ slightly from trajectories in the corresponding planar system. This allows us to apply for axial gun design the procedure developed for a planar gun and, particularly, to use determination of the equipotentials chart, presented in Fig. 5.22 for the electrodes form.

An axially symmetric gun designed with this procedure will have a value of perveance somewhat greater than the calculated one. The following procedure is used to remove this error. The gun is divided into elementary diodes and for each diode the correction of cathode–anode distance is carried out in order to provide the same value of the electrode field strength at the cathode surface both in planar and cylindrical diodes. This condition leads to the formula for

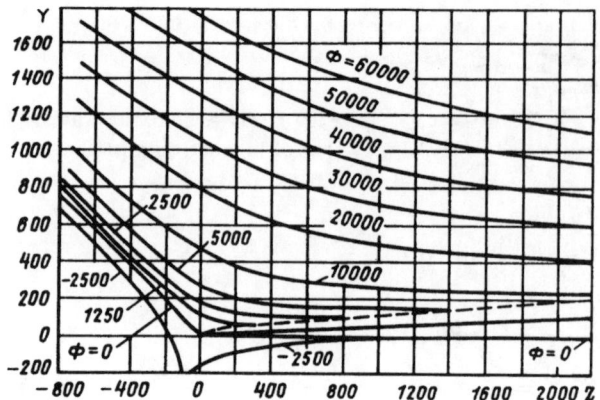

Fig. 5.22. Map of equipotential lines for magnetron gun design

correction of the cathode–anode distance:

$$d_c = d_p \left(1 - \frac{1}{2}\frac{d_c}{r_{mc}}\right)^{-1},$$

where d_p – is cathode–anode distance in an elementary planar diode, d_c – corrected cathode–anode distance of the corresponding cylindrical diode, r_{mc} – is mean radius of cathode of the given elementary diode.

Magnetron Guns of Gyrotrons. In a new class of ultrahigh-frequency tubes, known as gyrotrons, hollow axial-symmetrical beams are used in which electrons move on spiral trajectories. To transform efficiently kinetic energy of the electron beam into electromagnetic field energy it is required that the relation between the energy of electron rotation $W_\perp$ and the energy of longitudinal motion $W_\parallel$ be given by $W_\perp/W_\parallel = 2\text{--}4$.

Formation of such beams is carried out with the aid of magnetron-injection guns. Construction of such a gun is shown in Fig. 5.23 [97]. It includes cathode 1 with narrow emitting belt 2, the first anode with potential U_{a1}, the second anode with potential U_{a2} ($U_{a2} \geq U_{a1}$). The initial formation

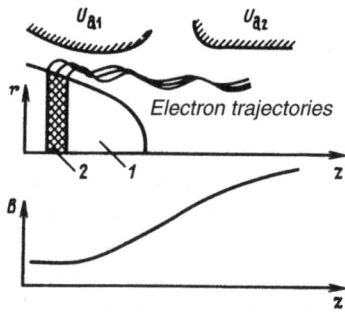

Fig. 5.23. Magnetron gun of a gyrotron. 1 – cathode, 2 – emitting belt

of the beam takes place in the cathode–first anode space in crossed electric and magnetic fields, where electrons obtain an initial rotation velocity $v_{\perp k}$. It can be shown [98] that $v_{\perp k} = E_k/B_k$, where E_k – is normal to the cathode component of electric field; B_k – is parallel to the cathode component of magnetic induction. Furthermore, electrons move in the accelerating field of the second anode and smoothly increasing magnetic field, the axial distribution of which is presented in Fig. 5.23. From the adiabatic theory of electron guns [98] it follows that this leads to an increasing rotational velocity component according to the formula $v_\perp/v_{\perp k} = (B/B_k)^{1/2}$. As a result, the relation of transverse and longitudinal electron velocities can reach the value of $v_\perp/v_{\perp k} = 1.5$–$2$, and the relation between kinetic energies of transverse and longitudinal motion, accordingly, $W_\perp/W_\| = 2$–4.

5.5 Electron Guns with Beam-Current Control

Classification and brief characteristics of this type of electron gun are presented in Sect. 1.2. The following parameters are used for description of the gun's performance:

Anode voltage U_a, control electrode voltage U_g, maximum voltage at the control electrode U_{gm}, relative value of cutoff voltage of control electrode $|U_{gc}|/U_a$, (see Fig. 5.24) triode amplification coefficient $\mu = U_a/|U_{gc}|$.

Maximum of electron beam current I_m, perveance calculated for the maximum beam current $P = I_m/U_a^{3/2}$.

The control factor, being determined as the ratio of the anode voltage to the value of control voltage that provides the change of beam current from zero to its maximum value:

$$K = \frac{U_a}{U_{gm} + |U_{gc}|} = \frac{\mu}{\mu U_{gm}/U_a + 1}. \tag{5.17}$$

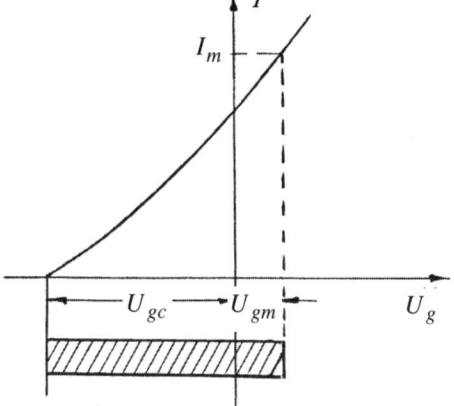

Fig. 5.24. Pulse operation of a gun with beam-current control U_{gc} – cutoff voltage of control electrode

5.5.1 Guns with Grid Control Electrodes

Guns with grid control electrodes are guns of special interest due to their high control efficiency.

A triode grid gun that provides the formation of a solid convergent beam, can be designed on the basis of a suitable diode gun, usually a Pierce-type gun. The original diode gun intended for convergent-beam formation has a cathode of spherical geometry. The equipotential surfaces spaced in the vicinity of the cathode are close to spherical ones. The control grid, of spherical shape, is placed at some small distance from the cathode. The effect of the grid on the beam formation depends on the voltage applied to the grid.

From the theory of electron tubes it follows that a grid of thin wires being placed in some initial field does not disturb it, if the grid is shaped like an equipotential surface of this field and has the applied voltage U_{gn} equal to the potential of that surface U_s. This grid voltage is known as the natural grid voltage.

Therefore, we can expect that a spherically shaped grid with applied voltage being close to the natural one will not significantly disturb the electron beam of the initial diode gun. The variation of the grid voltage with respect to the natural one leads to changing such beam parameters as the beam current and gun perveance, to the lens effect of grid meshes, which results in increasing beam emittance.

Triode tube theory can be used for predication of gridded-gun performance and design consideration.

The beam current of the diode gun is expressed by (5.7), that is

$$I_d = 29.34 \times 10^{-6} \frac{\sin^2\left(\frac{\theta}{2}\right)}{(-\alpha_a)^2} U_a^{3/2},$$

where θ is the angle of the beam convergence, $(-\alpha_a)^2$ is the function of the radii of curvature ratio $\bar{r}_c/\bar{r}_a$, which can be found from Table 5.2.

Using the results of triode tube theory one can find the following formula for the cathode current of the grid gun:

$$I_c = 29.34 \times 10^{-6} \frac{\sin^2\left(\frac{\theta}{2}\right)}{(-\alpha_g)^2} U_{geff}^{3/2}.$$

In this formula $(-\alpha_g)^2$ is the function of ratio $\bar{r}_c/\bar{r}_g$, found from Table 5.2, r_g being the radius of the grid curvature; U_{geff} is the so-called effective grid potential, which is used in electron tube theory and expressed as

$$U_{\text{geff}} = \frac{U_g + DU_a}{1 + \left(\frac{1}{\chi}\right)D},$$

where D is the grid-penetration factor, χ is a coefficient equal to the ratio of the two potentials $\chi = U_{gn}/U_a$, U_{gn} being the natural grid potential, U_a the anode potential.

5.5 Electron Guns with Beam-Current Control

For guns of spherical geometry this coefficient can be expressed as

$$\chi = (-\alpha_g)^{4/3} / (-\alpha_a)^{4/3},$$

where $(-\alpha_g)^{4/3}$ and $(-\alpha_a)^{4/3}$ are the values of the function $(-\alpha)^{4/3}$, presented in Table 5.2, and calculated for the ratios of the radii $\bar{r}_c/\bar{r}_g$ and $\bar{r}_c/\bar{r}_a$, respectively.

In the case of positive grid voltage, a part of the cathode current can be intercepted by the grid and the beam current after the grid is expressed by the formula

$$I = \delta_g I_c = \delta_g \times 29.34 \times 10^{-6} \frac{\sin^2(\theta/2)}{(-\alpha_g)^2} U_{\text{geff}}^{3/2},$$

where δ_g is coefficient of the beam-current transmission.

The following relation is obtained for the perveance of diode and triode guns:

$$\beta = \frac{I_d}{I} = \frac{P_d}{P} = \frac{1}{\delta_g}\left(\frac{U_{gn} + DU_a}{U_g + DU_a}\right)^{3/2} = \frac{1}{\delta_g}\left(\frac{\chi + D}{U_g/U_a + D}\right)^{3/2}, \tag{5.18}$$

where $U_{gn} = \chi U_a$ is the natural grid potential.

When the grid potential equals the natural one $U_g = U_{gn}$ this relation is reduced to:

$$\beta = P_d/P = 1/\delta_g = 1/(1-\sigma).$$

The beam-transmission coefficient δ, included in this formula, is equal to the geometrical grid transparency $\delta_g = 1 - \sigma$, σ is a space factor determined as the percentage of the grid area filled by the grid material.

For the grid of parallel wires the coefficient σ is equal to the ratio of wire diameter d_g to grid pitch $p\sigma = d_g/p$; for the grid with square meshes $\sigma \simeq 2d_g/p$. Usually, the coefficient $\delta_g = 0.8$–0.95, then $\beta = 1.25$–1.05.

When $U_g < U_{gn}$ or $U_g > U_{gn}$ the beam perveance in the triode gun will be accordingly less or greater than the diode gun perveance.

For the case of zero grid potential $U_g = 0$ one obtains:

$$\beta = \frac{P_d}{P} = \left(\frac{\chi + D}{D}\right)^{2/3} = (\chi\mu + 1)^{3/2}, \tag{5.19}$$

where μ – is amplification factor of triode system, $\mu = 1/D$, the coefficient of current transmission being taken equal to unity, since the grid with zero potential does not intercept the beam current.

It is seen from (5.19) that for the greater values of the amplification factor, a greater perveance of the initial diode gun is required to obtain the given perveance of the triode gun.

When the grid potential differs from the natural one the grid meshes act as the diaphragm lenses (see Sect. 3.4) and disturb the electron-beam motion. To estimate the disturbances, the expression for the focal distance of the diaphragm lens may be used

$$f = 4U_{\text{geff}}/(E_1 - E_2),$$

where U_{eff} is the effective grid potential E_1 and E_2 – electric field strength to the left and to the right of the grid plane [99]. The angle of electron trajectory refraction arising due to the lens action is determined by the formula

$$\operatorname{tg}\alpha = x/f,$$

where x is distance from the grid mesh center $0 \le x \le p/2$, p being the grid pitch.

Gun with Natural Potential on Control Grid

Let us consider in more detail designing a gun with a control grid with the natural potential at the operating point. It is assumed that there is an initial diode gun that has perveance $P_d = \beta P$ ($\beta = 1.05$–1.25) and a given coefficient of compression C_j. Designing a triode gun is reduced to the determination of the structure and position of the grid electrode to provide the required modulation parameters. To do this it is necessary to consider the dependence of modulation parameters on geometrical sizes of the triode gun.

In the first approach, calculation of the amplification factor of the triode gun with a small cathode–grid distance can be found by means of the formula for the amplification factor of a planar triode tube:

$$\mu = \frac{L'_g \left(d_{ca} - d_{cg} \right) - \Delta}{T}, \tag{5.20}$$

where L'_g – is length of grid wire per unit of its surface (for the grid with parallel wires $L'_g = 1/p$, for grid with square cells $L'_g \simeq 2/p$); Δ and T – are functions of grid space factor σ (see Table 5.3); $d_{cg} = \bar{r}_c - \bar{r}_g$ – cathode–grid distance.

As to the value d_{ca}, it is the cathode–anode distance of an equivalent spherical diode, with radius of cathode curvature and cathode current density being equal to those of the triode gun in question.

The anode curvature radius of the equivalent diode $\bar{r}_{ae}$ is found with the help of the formula for cathode current density:

$$j_c = \frac{4\varepsilon_0}{9} \sqrt{2 \frac{|e|}{m}} \frac{U_a^{3/2}}{(-\alpha_a)^2 \bar{r}_c^2},$$

where $(-\alpha_a)^2$ is the function of relation $\bar{r}_c/\bar{r}_{ae}$ tabulated in Table 5.2.

With the values j_c, $\bar{r}_c$ and U_a being known, the value $(-\alpha_a)^2$ is calculated by the formula and the radius of curvature $\bar{r}_{ae}$ is found from Tables 5.2. Then, the cathode–anode distance d_{ca} is expressed as $d_{ca} = \bar{r}_c - \bar{r}_{ae}$.

5.5 Electron Guns with Beam-Current Control

Table 5.3. Values of parameters Δ and T as function of grid space factor σ

σ	Δ	T	σ	Δ	T
0.001	–	0.9172	0.15	0.01735	0.1283
0.002	–	0.8069	0.16	0.01969	0.1192
0.003	–	0.7424	0.17	0.02217	0.1108
0.004	0.00001	0.6966	0.18	0.02479	0.10307
0.005	0.00002	0.6611	0.19	0.02751	0.09582
0.006	0.00003	0.6321	0.20	0.03042	0.08908
0.008	0.00005	0.5863	0.21	0.03342	0.08280
0.010	0.00008	0.5508	0.22	0.03656	0.07694
0.015	0.00018	0.4863	0.23	0.03952	0.07147
0.020	0.00031	0.4406	0.24	0.04319	0.06636
0.025	0.00049	0.4052	0.25	0.04664	0.06156
0.030	0.00071	0.3762	0.26	0.05030	0.05711
0.035	0.00096	0.3518	0.27	0.05402	0.05292
0.040	0.00126	0.3307	0,28	0.05785	0.04900
0.045	0.00159	0.3123	0.29	0.06178	0.04534
0.050	0.00196	0.2956	0.30	0.06581	0.04190
0.060	0.00282	0.2670	0.31	0.6995	0.3869
0.070	0.00383	0.2430	0.32	0.07418	0.03568
0.080	0.00500	0.2223	0.33	0.07850	0.03287
0.090	0.00632	0.2040	0.34	0.08291	0.03025
0.100	0.00779	0.1881	0.35	0.08741	0.02780
0.110	0.00941	0.1738	0.36	0.09199	0.02551
0.120	0.01118	0.1608	0.37	0.09664	0.02337
0.130	0.01309	0.1490	0.38	0.10137	0.02158
0.140	0.01515	0.1382	0.39	0.1062	0.01952

For systems with a high value of amplification factor μ and grid space factor $\sigma \leq 0.3$ it is possible to omit the term Δ in (5.20) and obtain:

$$\mu = \frac{L'_g (d_{ca} - d_{cg})}{T} . \tag{5.21}$$

To take into account the sphericity of gun electrodes the correction factor is introduced and the equation becomes:

$$\mu = \frac{d_{ca} - d_{cg}}{T} L'_g \frac{\bar{r}_c}{\bar{r}_{ae}} . \tag{5.22}$$

As follows from (5.21), (5.22) and Table 5.3 the amplification factor increase with increasing grid space factor σ and decreasing grid pitch p and cathode–grid distance d_{cg}. This results in reduction of the cutoff potential U_{gc}.

In the vicinity of the cathode of a spherical diode the potential distribution can be presented with the approximate formula:

$$U \approx CU_a d^{4/3}\left(1 + \frac{d}{\bar{r}_c}\right),$$

where d – is a small distance from the cathode, which is supposed to be considerably less than the cathode radius of curvature $\bar{r}_c$ ($d \ll \bar{r}_c$);
$C = 1/(-\alpha_a)^{4/3}\bar{r}_c^{4/3}$ – is a constant depending on gun geometry.

Then, the natural potential of the grid located at the distance d_{cg} is expressed as

$$U_{gn} = CU_a d_{cg}^{4/3}\left(1 + \frac{d_{cg}}{\bar{r}_c}\right). \qquad (5.23)$$

This potential is seen to be decreased with decreasing distance d_{cg}.
Substitution of (5.23) into (5.17) yields

$$K = \frac{1}{Cd_{cg}^{4/3}(1 + d_{cg}/\bar{r}_c) + 1/\mu} = \frac{\mu}{Cd_{cg}^{4/3}(1 + d_{cg}/\bar{r}_c)\mu + 1}. \qquad (5.24)$$

This equation shows that the control factor is magnified as the cathode–grid distance d_{cg} decreases and the amplification coefficient μ grows.

The above consideration and formulae can be used for designing of gridded guns with the required modulating parameters. To get the required value of the control factor three gun dimensions σ, p, d_{cg} can be varied. If two of them are chosen the third one is uniquely determined from (5.24). It should be noted that for a grid operating with positive voltage $U_{gm} > 0$, the grid space factor value σ exceeding 0.25 should not be to used since it leads to growth of the grid current interception.

Gun Operating with Zero or Negative Grid Potential

Under these operating conditions the maximum grid voltage $U_{mg} \leq 0$ and the grid does not intercept the beam current. However, the grid with zero or negative potential considerably reduces the gun perveance P as compared with the initial diode gun perveance P_d. For the case $U_{gm} = 0$ it will be:

$$P = P_d \frac{1}{(\chi\mu + 1)^{3/2}}.$$

Therefore, the initial diode gun should have much greater perveance than the nominal one of the designed gridded gun. Moreover, the grid, being at zero or negative potential, disturbs the initial potential distribution as well as the electron trajectories due to the lens action of the grid meshes. This leads to increasing of the beam cross section and beam emittance.

Tetrode Type Electron Gun with Shadow Grid

The schematic drawing of the gun is shown in Fig. 5.24. On the grid G_1, which is called shadow one, zero (cathode) potential is applied and therefore electrons moving from the cathode can not reach the grid wires and pass around them. The wires of the control grid G_2 are placed immediately behind the wires of grid G_1 in the "shadow" of the latter. Due to this the grid G_2 working with positive potential practically does not intercept the beam current. The perveance of the tetrode gun can be made close to that of the initial diode gun.

As in the forthcoming case, the zero potential grid G_1 disturbs the potential distribution and electron trajectories in the cathode vicinity leading to laminar-flow disturbance, beam cross section and emittance growth [101]. The last effect can not be eliminated by the action of focusing electrodes.

5.5.2 Electron Gun with Diaphragm Control Electrode

The schematic drawing of the gun is shown at Fig. 1.8e. The penetration factor of the control electrode (diaphragm) for the anode voltage is negligible and its cutoff voltage is practically equal to zero, $U_{gc} = 0$. Then, (5.17) for the gun control factor is reduced to:

$$K = U_a/U_{gm} \ .$$

Since the anode field practically does not penetrate the space cathode–control electrode, the systems of these two electrodes can be considered as independent beam-forming systems. Its perveance is found as:

$$P_d = PK^{3/2} \ ,$$

where P and K are the perveance and control factor of the designed triode gun.

As follows from this expression, increasing of a gun control factor is connected with the growth of the diode-system perveance P_d. This limits the values K for guns of such a type to several units. The preliminary calculation of the diode gun can be carried out with one of the methods described in Sect. 5.3.

When designing the triode gun, it is necessary to take into account the action of the electrostatic immersion lens arising in the space "control electrode – gun anode".

5.6 Principle of Computer-Aided Design of Electron Guns

Design of electron guns includes two main steps. The first one supposes determination of the initial geometry of the electron gun, which provides the

152   5 Electron Guns

required beam parameters (perveance, compression, etc.). Usually, for this purpose the method of synthesis is used, which is based on one of the simplified mathematics models of the electron gun. The second step includes electron-gun analysis with the aid of computer programs and optimization (correction) of its geometry.

5.6.1 Synthesis of Electron Gun

The program of synthesis of electron guns forming a solid axial-symmetrical beam is based on a mathematical model of an electron gun and an algorithm has been described in Sect. 5.3. The input data includes the following electron-beam parameters: I – the beam current; U'_a accelerating voltage; beam radius in its minimum cross section $r_{\min}$ and coordinate of its position $z_{\min}$. Gun electrodes geometry and boundary trajectory, being obtained as the result of the program work, are presented in graphical and tabular forms.

5.6.2 Electron-Gun Analysis

The method of successive approximation is used for computer-aided gun analysis. As a first approximation, the Laplace field is computed for the gun geometry obtained at the previous step by one of the techniques described in Chap. 2.

Then the approximate value of cathode current is found in the following way. The cathode surface is divided into a large number of small segments by some regular method. The cathode current from the k-th segment $\Delta I_k^{(2)}$ is calculated by the method of "planar virtual diode" that is created by the cathode segment and an equipotential surface of potential $U^{(1)}$ located at a small distance d_k from the cathode, the surface being considered as a virtual anode:

$$\Delta I_k^{(2)} = J_k^{(2)} \Delta S_k , \qquad J_k^{(2)} = 2.33 \times 10^{-6} \frac{\left(U^{(1)}\right)^{3/2}}{d_k^2} \alpha_r ,$$

where $J_k^{(2)}$ is the current density, ΔS_k – the area of the cathode segment, α_r is a correction factor of order of 0.3–0.5, which takes into account the fact that the potential $U^{(1)}$ is actually Laplace's potential.

This calculation is produced for all cathode segments. Then, charge particle trajectories starting from the segments are calculated, each of them is assumed to carry the corresponding value of the current. This enables the space-charge distribution to be found. Then the potential distribution in the second approximation $U^{(2)}$ is calculated, with the space-charge field being taken into account.

New values of segment currents are calculated by substituting in the above equation the potential $U^{(2)}$ and assuming $\alpha_r = 1$ and so forth.

5.6 Principle of Computer-Aided Design of Electron Guns   153

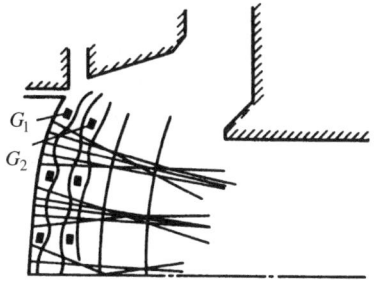

Fig. 5.25. Electron gun with shadow grid, G_1 – shadow grid, G_2 – control grid

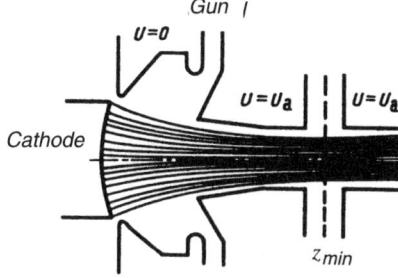

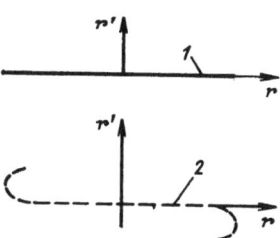

Fig. 5.26. Results of gun-trajectory analysis

Fig. 5.27. Phase curves: 1 – required (ideal) curve 2 – real curve obtained by analysis

This procedure is repeated until the potential distribution and electron trajectories in a current approximation prove to be close enough to that of the previous one.

The convergence of successive approximations can be improved if we determine the cathode current density in the n-th approximation using the formula

$$\tilde{J}_k^{(n)} = \alpha J_k^{(n)} + (1-\alpha)\, \tilde{J}_k^{(n-1)} ,$$

where $J_k^{(n)} = 2.33 \times 10^{-6} \left(U^{(n-1)}\right)^{3/2}/d_k^2$; α is the coefficient of relaxation being close to 0.7 [9].

The analysis results are data on the geometry and current of the electron beam, its distribution on the beam cross section, the value and direction of electron velocity in different points of the beam cross section.

As a standard method to represent calculation results the picture of electron trajectories is used, shown in Fig. 5.25. However, it is not informative enough to estimate a final result, namely – the quality of the beam formed.

Properties of the beam formed by the gun can be described with the aid of a phase curve, built up in transverse phase space with coordinates r and $r' = dr/dz$, r being the radial coordinate of a trajectory point and r' the trajectory slope to the gun axis. Figure 5.27 presents the phase curve of the electron

5 Electron Guns

beam of the electron gun shown in Fig. 5.26. The characteristic is obtained for the plane z_{min}, where the beam radius has its minimum value r_{min}. Such a type of phase curve shows that the beam formed by the electron gun is a nonlaminar one, its trajectories are crossed. The quality of the formed electron beam can be estimated with the aid of a parameter determining deviation of the obtained phase curve from the ideal curve, which is shown in Fig. 5.27 with a solid line:

$$\beta = \sum_{k=1}^{N} \alpha_k \Delta_k ,$$

where $\Delta_k = [a(r_k - \hat{r}_k)^2 + b(r'_k - \hat{r}'_k)^2]^{1/2}$ – is the deviation of the phase point of the k-th trajectory from the one given by values $\hat{r}_k$ and $\hat{r}'_k$; α_k – is coefficient equal the given part of the current that is carried by the given trajectory; a and b – weight factors.

In the case presented in Fig. 5.26, the ideal phase curve corresponds to the laminar beam, trajectories of which in the plane of minimum beam cross section are parallel to the system axis. Therefore, the value $\hat{r}'_k$ in the above equation for deviation Δ_k is taken equal to zero. It is evident that the less the value of parameter β the closer is the obtained phase curve to the set (ideal) one. The case $\beta = 0$ corresponds to their coincidence. So, the value β can be considered as the criteria when comparing variants of electron gun in the process of its correction (optimization).

For quantitative estimation of the electron-flow degree of order and its quality other methods and parameters can also be used, including the method of mean square emittance, which is determined by the formula [12]:

$$\bar{\varepsilon} = 4 \left(\langle r^2 \rangle \langle r'^2 \rangle - \langle rr' \rangle^2 \right)^{1/2} ,$$

where $\langle r^2 \rangle$, $\langle r'^2 \rangle$, $\langle rr' \rangle$ – are corresponding values of r^2, r'^2 and rr', averaged on the area of phase space, occupied by the beam. With regard to the electron-beam discrete model, the expression for mean square emittance can be written in the following way

$$\bar{\varepsilon} = \frac{4}{N} \left[\left(\sum_{k=1}^{N} r_k^2 \right) \left(\sum_{k=1}^{N} r'^2_k \right) - \left(\sum_{k=1}^{N} r_k r'_k \right)^2 \right]^{1/2}$$

where r_k and r'_k – are radial coordinate and slope of the k-th trajectory, summation is being made on all k trajectories.

It is not difficult to find that if the beam phase curve is a linear one, then the mean square emittance of such a beam is equal to zero. The deviation of the phase curve from the linear one gives a finite value of mean square emittance.

In [101] for quantitative estimation of transverse velocities in the minimum beam cross section the mean weighted value of electron trajectories slopes $\langle r' \rangle$ and standard (mean square) deviation σ are used:

5.6 Principle of Computer-Aided Design of Electron Guns

$$\langle r' \rangle = \left(\sum_{k=1}^{N} \Delta I_k r'_k \right) \bigg/ I$$

$$\sigma = \left[\frac{1}{I} \sum_{k=1}^{N} \Delta I_k \left(r'_k - \langle r' \rangle \right)^2 \right]^{1/2},$$

where ΔI_k – is weight, a factor that is equal to the current of the k-th current tube; I – is full beam current. The methods of electron-gun calculation applying computers, which take into account the influence of thermal velocities, were considered in [100–102].

6 Electron and Ion Sources with Field and Plasma Emitters

6.1 Some Peculiarities of Field and Plasma Emitters

Broadening of the field of application of electron and ion beams has resulted in development of their sources on the basis of field and plasma emitters.

Initially, field emitters were investigated and used as low-current point sources of electrons [103–106]. Their advantages are as follows: absence of heater, possibility to obtain high current density of order 10^5 A/cm^2 for continuous operating condition and up to 10^9 A/cm^2 for pulse operation from point sources, low energy spread of emitted electrons. This provides for forming of thin point focused beams with high energy concentration and high brightness (up to $\sim 10^{10}$ A/cm^{2}sr). On the other hand, there exist some negative properties of the sources such as low current and emission stability. They are overcome by use of multiple-tip field-emission systems and new materials of emitters [108–113].

Later, a new class of field emitters of ion and electrons was developed, known as liquid metal sources [114, 115].

Plasma is considered to be a natural source of both electrons and ions. Plasma sources can operate in continuous or pulsed operating conditions allowing narrow or wide charge particle beams to be formed. The important peculiarity of them is the possibility to operate in a low vacuum environment.

There exist a number of types of plasma charge particle sources and their numerous varieties. Series monographs are devoted to their physics, theory and technique [17, 116–119]. So, only the most typical and widely used of them will be briefly considered here.

6.2 Low-Current Field-Emission Guns

One of the first field-emission (FE) electron guns, successfully used in a scanning microscope, was developed by Crewe [106]. It consists of a single field-emission pointed cathode (1), anode (2) and accelerating electrode (4) (Fig. 6.1). An anode voltage of 20 kV applied to the anode electrode produces a field-emission current from the pointed cathode having a radius of curvature equal to 0.1 µm. The anode electrode and the accelerating electrode, operating with voltage varying from 20 to 145 kV create an electrostatic

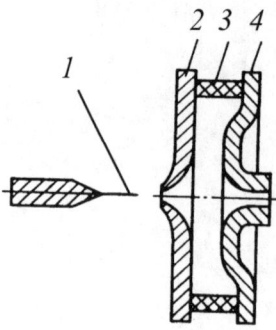 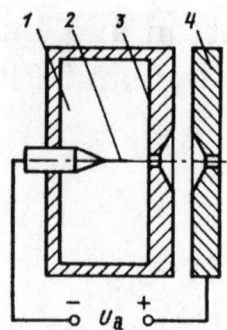

Fig. 6.1. Scheme of cold-cathode gun: 1 – cathode, 2 – first anode, 3 – insulator, 4 – second anode

Fig. 6.2. Cold-cathode microwave injector: 1 – microwave cavity, 2 – cathode, 3, 4 – first and second anodes

lens. The adjacent surfaces of these two electrodes are specially shaped to minimize lens spherical aberration. The lens is known as a Butler lens and the entire system as a Crewe–Butler optical system. The electron gun with a field-emission pointed cathode produces an electron beam with a minimum spread of electron velocity and the linear output phase curve. Its focusing by the electrostatic lens with reduced spherical aberration results in a really pointed beam focus with a very high value of brightness of 10^8A/cm^2 sr with total beam current $I = 150 \div 10^3$ µA.

Field-emission cathodes are employed in microwave injectors for linear accelerators [107]. A pointed cathode is introduced in a microwave cavity fed from a microwave source (Fig. 6.2). If the resonance condition is fulfilled an alternating voltage of a high amplitude is excited in the cavity. It produces a field-emission current in the form of short periodical pulses, with pulse duration being equal to a small part of a period of alternating voltage. In the system shown in Fig. 6.3 the pointed tungsten cathode (2) having a radius of curvature equal to 0.1 µm is introduced into a cylindrical cavity (1).

Emitted and accelerated electrons come out through the aperture in the resonator wall (3) and enter the field of a focusing electrostatic lens formed by the resonator wall and the accelerating electrode (4).

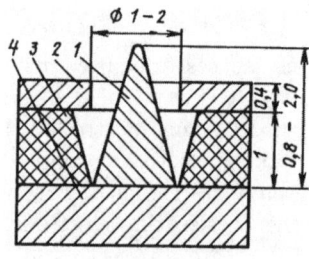

Fig. 6.3. Elementary cell of multiple-tip cathode: 1 – microcathode, 2 – control electrode, 3 – dielectric film, 4 – conducting pad; dimensions are given in µm

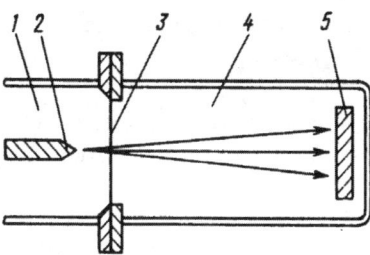

Fig. 6.4. Scheme of an electron gun with explosive emission cathode: 1 – gun region, 2 – cathode, 3 – anode, 4 – drift space, 5 – target

The parameters of this injector are as follows. The frequency and the amplitude of alternating voltage are 2.37 GHz and 1 kV, the amplitude of current pulse 100 µA, its phase duration $\sim 33°$, value of accelerating voltage $U_a = 35$ kV.

6.3 Multiple-Tip Field-Emission Cathodes

There exist multiple-tip systems of two types: cathode with randomly located tips and matrix cathodes. In the last systems microcathodes (tips) are placed in a regular and reproducible manner.

The geometry of a separate cell of a matrix cathode is presented in Fig. 6.3 [108]. A molybdenum cone microcathode (1) with radius of curvature equal to 0.05–0.06 µm enclosed by a control electrode (2). An anode electrode located at a large distance from the cathode is not shown in this picture. The matrix pitch (distance between two elementary sells) is equal to 12.5 µm. At an anode potential and control electrode potential varying in the interval 100–300 V field-emission current of the unit cell is changed from 50 µA to 150 µA. The total cathode current of a cathode containing 10^3 elementary cells comprises 10^2 mA at an average current density of 10 A/cm^2.

The geometry of cathode cells is chosen to provide the required value of beam current and zero its interception by the control electrode. The methods of numerical simulation can be used for this [109].

Recently, large-scale investigations have been conducted on application of graphite nanotubes, that is, nanometer graphite layer cylinders, for field-emission multiple-tip cathodes [110,111]. This approach is considered to be very promising for development of high-current, long lifetime and high-stability field-emission cathodes.

Field-emission multiple tip cathodes can find application in conventional vacuum devices and microwave tubes, as well as in devices of vacuum microelectronics [112, 113].

160   6 Electron and Ion Sources with Field and Plasma Emitters

6.4 High-Current Electron Sources with Explosive Emission

Explosive emission is a complicated phenomenon, which is developed when a high-voltage pulse is applied to an interelectrode gap. It is considered to be a phase of explosive gap breakdown.

A phenomenological model of the origin and evolution of explosive emission is described as follows. Initially, as the pulse voltage is applied, field electron emission appears from local centers, which are erased due to roughness and tip type inhomogeneity of real cathode surfaces. Flowing current results in self-heating of the emitting tips that in turn leads to increasing tip emission current and tip temperature, finally it involves tip matter melting and evaporation. Evaporating atoms undergo field and collision ionization. Product ions are accelerated toward the cathode and strike its surface. This leads to cathode material sputtering, and creation of additional ions and electrons. The electron current in the anode circuit is increased greatly, reaching values that are several orders greater than that of the pure field emission. Development of these processes causes formation of a dense plasma cloud, which spreads toward the anode electrode. When it reaches the anode, explosive gap breakdown occurs.

Figure 6.4 shows an electron gun for formation of a power pulse electron beam with 30 nanosecund pulse duration, beam current 5×10^4 A and energy 3×10^6 eV [104]. This consists of a pointed field emitter (1) operated in conditions of explosive emission and anode electrode of a thin metal foil. Electrons with energy of order of 10^6 eV pass through this foil and strike a target (3) to produce X-rays.

Explosive emission can be obtained not only from pointed cathodes, plane cathodes having surface microinhomogeneity can also be used as field explosive emitters. In this case the process of explosive emission is developed as was described earlier.

Cathodes with explosive emission are employed in high-power pulse microwave tubes, X-ray tubes and accelerators.

6.5 Liquid-Metal Emitters

The physics of ion field emission from a liquid-metal surface can be described in the following way [114, 115]. Suppose that a capillary tube with liquid metal is introduced into an accelerating electric field of high intensity. The liquid-metal meniscus, which exists at the end of the tube, under the action of electric forces and forces of surface tension acquires the form of a fine cone (so-called, Teilor's cone). The action of the strong field results in intensive ion emission from the tip of the cone due to the effects of field evaporation and field ionization. An example of a liquid-metal point source used for ion implantation is described in Sect. 9.3.

Field emission from an extended liquid-metal surface can also be initiated under the action of an electric field of a high strength. The field initially disturbs the smooth liquid surface, thus wave-crest-like inhomogeneities arise. They are local centers of ion field emission. Liquid field emitters can be also used as sources of electrons.

6.6 Plasma Sources of Charged Particles

Plasma sources of charged particles can be classified by the type of gas discharge producing the plasma [17]:

- Sources with hot-cathode gas discharge
- Sources with cold-cathode gas discharge
- High-frequency gas-discharge sources
- Microwave gas-discharge sources using the effect of electron cyclotron resonance (ECR)

Besides standard parameters used to characterize charge particle beam sources that is, current, voltage, perveance, emittance, some additional ones are introduced to describe performance of plasma sources: consumption of process gas; power efficiency – ratio of beam power to total source power consumption; efficiency – ratio of beam current to total source power consumption.

6.6.1 Duoplasmatrons

These sources of ions refer to hot-cathode arc-discharge type of sources with non uniform magnetic field. The design includes the directly heated cathode (7), located in the cavity of intermediate electrode (9) (this electrode can be considered as a primary anode), anode electrode (2), extracting electrode (1). Process gas is supplied through a gas inlet (4) (Fig. 6.5).

An arc discharge is initially ignited in the space cathode – intermediate electrode and then it is extended to the anode electrode. The plasma of the arc discharge is contracted, while passing through the channel in the intermediate electrode and due to the action of the nonuniform magnetic field, which is created in the channel and the gap between the intermediate electrode and the anode. The ion concentration in the vicinity of the extraction electrode can attain 10^{14}–10^{15} ions/cm^3. Under these conditions, the ion beam density can reach ~ 100 A/cm^2.

A duoplasmatron of the type shown in Fig. 6.5, operating at a gas pressure of order of 1 Pa, voltage of arc discharge equal 100 V, and extraction voltage 5 kV produces of order of 50 mA, with a power efficiency about 50%.

There exist several modifications of duoplasmatrons with better operating parameters. One of them is a duoplasmatron with oscillating electrons

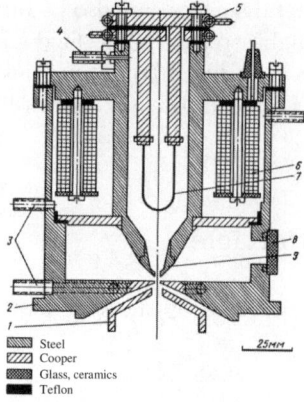 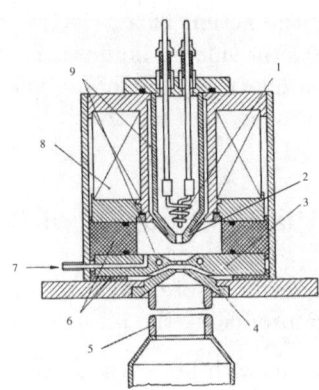

Fig. 6.5. Scheme of duoplasmatron: 1 – extractor, 2 – anode, 3 – cooling tubes, 4 – gas inlet, 5 – cathode flange, 6 – magnetic coil, 7 – cathode, 8 – window, 9 – intermediate electrode

Fig. 6.6. Duoplasmatron with oscillating electrons: 1 – cathode, 2 – intermediate electrode, 3 – anode, 4 – anti cathode, 5 – extracting electrode, 6 – insulators, 7 – gas inlet, 8 – magnetic coil, 9 – cooling system

(Fig. 6.6). Its design contains an additional electrode, which is located beyond the anode electrode.

This electrode is termed an anticathode. Its potential is the same as or more negative than that of the intermediate electrode. In this electrode system-electrons can oscillate between the intermediate electrode and the anticathode. It leads to an increase of ionization probability and, therefore, to growth of the ion current.

6.6.2 Cold-Cathode Sources

Penning-Type Ion Sources

A schematic drawing of a variant of this type sources is presented in Fig. 6.7. A penning discharge ignites in the discharge chamber containing the cold cathode (1) hollow anode (2), anticathode (3). A magnetic field directed along the chamber axis is created by a permanent magnet (6) and the magnetic circuit of mild iron.

Electrons emitted from the cathodes and those generated in the discharge volume move along complicated spiral-type trajectories. This make it possible for electrons to produce a large number of ionizations. In the source under consideration ions are extracted from the discharge chamber through an aperture in the anticathode (3) under the action of the extracting electrode (4).

The voltage of discharge ignition depends on many factors, such as gas pressure, value of magnetic field, cathode materials. A gas pressure of 0.04–

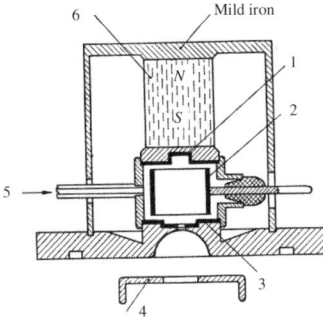

Fig. 6.7. Cold-cathode Penning-type ion source: 1 – cathode, 2 – anode, 3 – anti-cathode, 4 – extracting electrode, 5 – gas inlet, 6 – permanent magnet

3 Pa and magnetic field of 0.05–0.5 T are used in sources of this type. Aluminum, magnesium and beryllium cathodes provide ignition voltages of order of 300–400 V.

In sources with a single outlet aperture extracted ion currents vary from 1 to 300 mA depending on the source design and operating parameters.

For obtaining high-current ion beams, sources with grid or multihole outlet arrangements are used.

6.6.3 High-Frequency Ion Sources

Ion sources, using a high-frequency gas discharge, operate at a low consumption of gas and energy and provide high-intensity beams with a low spread of ion energy. They are characterized by a simple design. One of them, known as a Thonemann source, is shown in Fig. 6.8 [17]. The discharge chamber of this source, made of Pyrex glass, has two electrodes, one is a tungsten wire (1), the other is a thin dural (3) tube, which simultaneously serve as extracting electrodes. A potential difference of order of 2–5 kV is applied to them. Gas discharge is produced by the oscillatory circuit, which is supplied by high-frequency energy from a generator. Ions of the gas-discharge plasma are extracted through the channel in the extracting electrode, then, being injected in the accelerating immersion lens they are accelerated and focused.

Sources of this type produce thin ion beams with currents of the order of several milliamperes. Their efficiency can be increased by introducing, in the discharge chamber a static magnetic field oriented either normal or parallel to the axis of the chamber. Typical parameters of the sources are as follows. Operating frequency 20–180 MHz, power input 50–5000 W, magnetic induction 100–800 Gs, gas pressure 0.5–5.5 Pa, beam current 0.5–150 mA, gas economy 0.02–0.3.

6.6.4 Microwave Ion Sources

Ion sources utilizing microwave discharge and the effect of electron cyclotron resonance (ECR) can provide long-life stable ion beams for a variety of

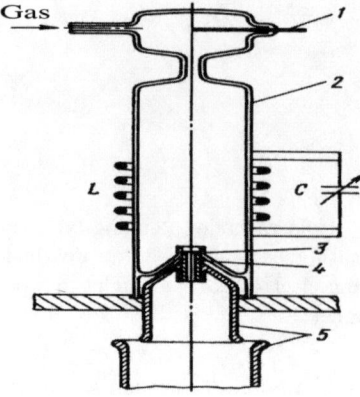

Fig. 6.8. High-frequency ion source: 1 – wire electrode, 2 – glass envelope, 3 – glass tube, 4 – extracting tubular electrode, 5 – immersion lens

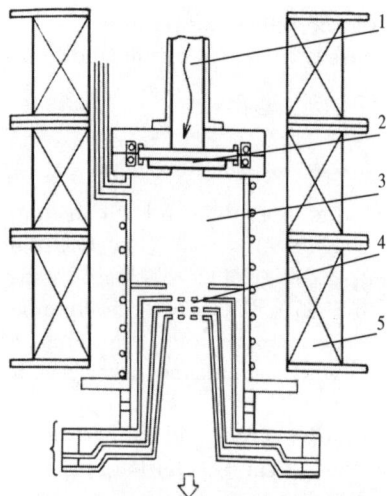

Fig. 6.9. Microwave ion source using effect of ECR: 1 – waveguide for power supply, 2 – microwave window, 3 – discharge chamber, 4 – three-grid extracting system, 5 – magnetic coil

ion species. They can produce high-current beams of single-charge ions, or multiple-charged ion beams depending on operational conditions.

A schematic drawing of a source of this type is shown in Fig. 6.9. It consists of the following components: waveguide for supplying 2.45 GHz power (1), dielectric window (2), discharge chamber (3), multihole three-grid extraction system (4), magnetic coil (5). The magnetic field intensity is 875 G, which provides the ECR at the frequency of 2.45 GHz.

The value of beam current depends on the value of microwave power. An oxygen ion current of 160 mA is obtained for 400 W power [120].

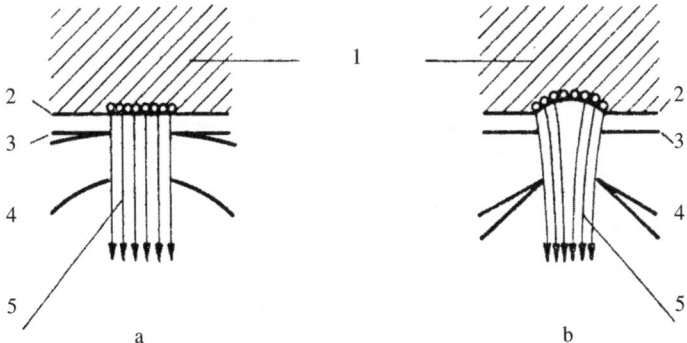

Fig. 6.10. Extraction of ions from plasma: a – planar plasma surface, b – concave plasma surface, 1 – plasma, 2 – plasma surface, 3 – focusing electrode, 4 – extractor, 5 – ion beam

6.7 Extracting of Charge Particles and Beam Forming

It is known that the wall of a discharge chamber is charged to some negative potential owing to the higher mobility of electrons as compared with that of ions. This results in formation of a layer of positive ion charges. The plasma boundary is displaced away from the wall by a distance Δ, of the order of the Debye length λ_D, $\Delta \approx 2\lambda_D$. If an aperture of a size larger than $2\lambda_D$ is made in the chamber wall for ion extraction, gas-discharge plasma will bulge out through this aperture. However, by a proper choice of geometry of the extracting electrode and the value of the negative potential applied it is possible to change the shape and the position of the plasma boundary. For instance, for the case shown at Fig. 6.10a the plasma surface coincides with the plane of the outlet aperture. If we increase the negative extracting electrode potential U_0 the plasma boundary is displaced inside the discharge chamber, its shape is changed from planar to concave (Fig. 6.10b).

6.7.1 Extraction of Ions from Planar Plasma Boundary

In the case of a planar plasma boundary in the outlet aperture (Fig. 6.10a) the physical processes in the positive-ion sheath will be the same as that in the vicinity of an undisturbed chamber wall. Then, the current density of positive ions, outgoing from the chamber aperture, will equal that going from the plasma to the chamber wall, which is determined by

$$j_i = 0.4 e n_i \left(2kT_e/m_i\right)^{1/2} = 8 \times 10^{-16} n_i \left(T_e/\mu\right)^{1/2}, \tag{6.1}$$

where n_i and m_i – density and mass of ions, e – charge of particle, k – Boltzmann constant, T_e – electron temperature, μ – molecular mass.

Using the Pierce principle it is possible to find the shapes of focusing and extractor electrodes that provide the rectilinear beam motion in the diode

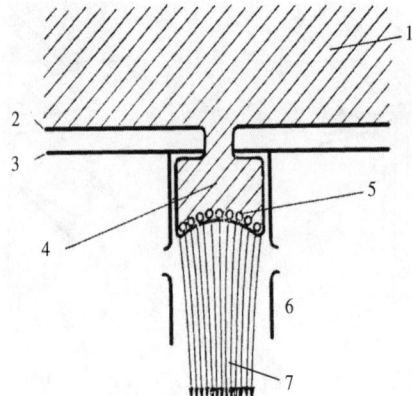

Fig. 6.11. Extractor with expansion cap: 1 – plasma, 2 – plasma surface, 3 – envelope, 4 – expansion cup, 5 – emitting plasma surface, 6 – extractor, 7 – ion beam

space created by plasma boundary and the extracting electrode. The value of space-charge-limited ion current density will obey the "3/2 power" law:

$$j_i = 4/9\varepsilon_0 \left(2e/m_i\right)^{1/2} U_0^{3/2}/d^2 \ . \tag{6.2}$$

Finding j_i from (6.1) and substituting it in (6.2) it is possible to determine the distance d, where the extracting electrode of potential U_0 should be placed.

To keep the shape and position of the plasma boundary, while U_0 is varied, a grid of high transparency can be mounted in the extraction aperture.

6.7.2 System of Extraction with Expansion Cup

A more flexible system of ion extraction is shown in Fig. 6.11. It contains an expansion cup (4). By choice of its dimensions (diameter, length) and voltage of the extracting electrode (6) it is possible to form an extended plane or concave plasma surface, which provides for forming parallel or convergent ion beams. Proper shapes of an extractor arrangement are found experimentally or by simulation with special computer codes, which take into account the effect of changing the plasma emitting surface.

Application of methods of synthesis (for instance, the Pierce method) is not so successful as in the case of guns with thermal emitters. A theoretical shape of focusing electrodes of a duoplasmatron found by the Pierce method is shown in Fig. 6.12 by dashed lines, while that found experimentally is given by solid lines [17].

6.7 Extracting of Charge Particles and Beam Forming   167

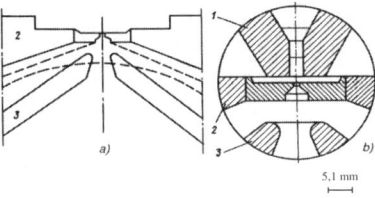

Fig. 6.12. Shape of electrodes of the extractor system of a duoplasmatron: 1 – intermediate electrode, 2 – anode, 3 – extracting electrode; theoretical shape of electrodes found by Pierce method is shown by dashed lines

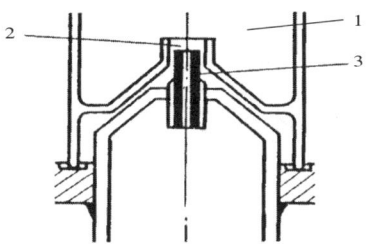

Fig. 6.13. Probe system of extraction: 1 – discharge chamber, 2 – glass tube, 3 – extracting tubular electrode

6.7.3 Probe System of Extraction

A probe-type extracting arrangement is presented in Fig. 6.13. It consists of an extracting tubular electrode (3), placed in a glass or quartz tube (2), protruding into the discharge chamber (1). The ion-emitting plasma surface is located at the upper end of this tube. The electrostatic field, which exists due to potential difference between the plasma surface and the extracting electrode, creates an immersion lens, which is used for focusing of the extracted ion beam.

Extracting systems of this type are used for forming narrow well-focused ion beams in sources with high-frequency gas discharge.

7 Magnetic Focusing Systems

7.1 Focusing Systems with Uniform Field

7.1.1 Principle of Focusing

Focusing of axially symmetric beams by means of a uniform magnetic field directed along the axis of beam propagation is frequently used in different beam applications.

Let us consider the principle of focusing using a laminar beam model and the concept of boundary trajectory. The magnetic field should produce the focusing force to balance the space-charge force and thus restrict the transverse beam expansion.

The resulting radial magnetic force in uniform magnetic field, as shown in Sect. 3.2, is expressed as:

$$F_r = -\frac{1}{4}\frac{e^2}{m}B^2 r \left(1 - \frac{\psi_k^2}{\pi^2 r^4 B^2}\right), \qquad (7.1)$$

where B – induction of uniform field, r – current radial coordinate of a particle, ψ_k – magnetic flux passing through cathode surface (Fig. 7.1).

Equation (7.1) shows that, with other things being equal, the magnetic force depends on the cathode magnetic flux. As condition $\psi^2/\pi r^4 B^2 < 1$ is fulfilled the magnetic force is directed toward the axis of symmetry and its action is focusing.

Introducing the value of the beam radius at the entrance to the uniform beam region r_0 and magnetic flux ψ_0 enclosed by the circle $2\pi r_0$ we can consider the three types of magnetic focusing:

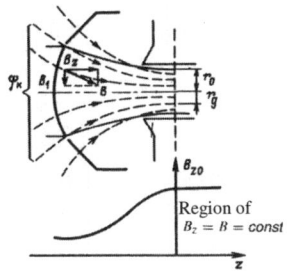

Fig. 7.1. Magnetic fluxes in the gun region

$\psi_k = 0$ – completely shielded cathode
$\psi_k < \psi_0$ – partially shielded cathode
$\psi_k = \psi_0$ – unshielded cathode

The degree of cathode shielding can be characterized by a coefficient of shielding G determined as:

$$G = \left(\frac{\psi_k}{\psi_0}\right)^2 = \frac{\psi_k^2}{(\pi r_0^2 B)^2} . \tag{7.2}$$

Sometimes, for this purpose a parameter r_g is used:

$$r_g^2 = \psi_k/\pi B . \tag{7.3}$$

If the magnetic flux passing through the cathode surface is directed in the same direction as that in the uniform field region, then r_g has the physical meaning of the radius of a magnetic flux tube $\psi_k = \text{const}$ at the entrance a uniform field (Fig. 7.1). In the opposite case such an interpretation is not valid. Parameters G and r_g are connected by the ratio

$$r_g = r_0 \sqrt[4]{G} . \tag{7.4}$$

For a completely shielded cathode $G = r_g = 0$ and for an unshielded cathode $G = 1$, $r_g = r_0$.

7.1.2 Balanced (Brillouin) Axially Symmetric Beam

The magnetic field is assumed to be uniform and the cathode is completely shielded from the magnetic field ($G = 0$). In this case, motion of a charge particle beam in the hydrodynamic approximation is described by the following system of equations:

$$\ddot{r} = -\frac{e}{m}\frac{\partial U}{\partial r} - \frac{1}{4}\left(\frac{e}{m}\right)^2 B^2 r , \quad \ddot{z} = -\frac{e}{m}\frac{\partial U}{\partial z} , \quad \dot{\theta} = -\frac{1}{2}\frac{e}{m}B ,$$

$$\frac{1}{r}\frac{\partial}{\partial r}\left(r\frac{\partial U}{\partial r}\right) + \frac{\partial^2 U}{\partial z^2} = -\frac{\rho}{\varepsilon_0} , \quad \text{div}\,\rho v = \frac{\partial}{\partial r}(\rho \dot{r}) + \frac{\partial}{\partial z}(\rho \dot{z}) = 0 .$$
$$\tag{7.5}$$

It is easy to check that this system of equations has the following partial self-consistent solution: $r = \text{const}$ ($\dot{r} = \ddot{r} = 0$), $\dot{z} = \text{const}$, $\rho = \text{const}$, $\partial U/\partial z = 0$, $\partial U/\partial r = -\rho r/2\varepsilon_0$. The necessary condition of such a solution is the balance of electric and magnetic forces:

$$\frac{e}{m}\frac{\partial U}{\partial r} = -\frac{1}{4}\left(\frac{e}{m}\right)^2 B^2 r .$$

This determines the connection between space-charge density ρ and magnetic induction B. Substitution in it of $\partial U/\partial r = -\rho r/2\varepsilon_0$ yields:

$$\frac{\rho}{\varepsilon_0} = \frac{1}{2}\frac{e}{m}B^2 .$$

7.1 Focusing Systems with Uniform Field

For the case of electron beams, introducing plasma and Larmor's frequencies $\omega_p^2 = \frac{e}{m}\frac{\rho}{\varepsilon_0}$ and $\omega_L^2 = \frac{1}{4}\left(\frac{e}{m}\right)^2 B^2$, it is possible to present this condition in the form:

$$\omega_L^2 = \frac{1}{2}\omega_p^2. \tag{7.6}$$

A particle beam being injected in the uniform field region with zero radial velocities and uniform space-charge density distribution ($\rho = $ const) will proceed with balanced motion ($\dot{r} = 0$, $r = $ const) if the above conditions are fulfilled. The beam will be rotated with constant angular velocity $\dot{\theta} = -\frac{e}{m}B$ as a rigid rod. The longitudinal velocity $\dot{z}$ of beam particles can be found from the law of energy conservation:

$$\dot{z}_0^2 = 2\frac{|eU_0|}{m} - \left(r\dot{\theta}\right)^2.$$

At the axis of symmetry ($r = 0$) it is reduced to:

$$\dot{z}_0^2 = 2\frac{|eU_0|}{m},$$

where U_0 – potential at the axis of symmetry.

The potential at an arbitrary point of the beam cross section can be expressed in terms of U_0 and space-charge density ρ from Poisson's equation:

$$U = -\left(\rho/4\varepsilon_0\right)r^2 + U_0.$$

Substitution of this equation in the law of energy conservation yields:

$$\dot{z}^2 = \frac{e\rho}{2\varepsilon_0 m}r^2 + \frac{2|eU_0|}{m} - \left(r\dot{\theta}\right)^2.$$

Since

$$(r\dot{\theta})^2 = \frac{1}{4}\left(\frac{e}{m}\right)^2 B^2 r^2 = \frac{e}{2m}\frac{\rho}{\varepsilon_0}r^2,$$

then:

$$\dot{z}^2 = \frac{e\rho}{2m\varepsilon_0}r^2 + \frac{2|eU_0|}{m} - \frac{e\rho}{2m\varepsilon_0}r^2 = \frac{2|eU_0|}{m}.$$

It thus follows that the value of the longitudinal component of velocity does not depend on the particle radial position and equals the velocity of particles moving along the symmetry axis.

Taking account of conditions, $\rho = $ const, $\dot{z} = $ const one finds that current density $j = \rho\dot{z}$ is also constant over the beam cross section, $j = \rho\dot{z} = const$. The results obtained above allow the value of magnetic induction that provides the equilibrium beam motion to be found:

$$B = \left(\frac{2m}{|e|\varepsilon_0}\frac{I}{\pi r^2 \dot{z}}\right)^{1/2} = \left(\frac{2m^{3/2}}{|e|\varepsilon_0}\frac{I}{\pi r^2 \sqrt{2(eU_0)}}\right)^{1/2}.$$

The value of U_0 can be expressed as $U_0 = U_a + \Delta U$, where U_a is the potential of a metallic electrode (tube), ΔU – potential difference due to space charge ($|\Delta U| < 0$ for negative space charge, $|\Delta U| > 0$ for positive one). Usually $|\Delta U| \ll |U_a|$, then:

$$B = B_b = \left(\frac{2m^{3/2}}{|e|\,\varepsilon_0}\,\frac{I}{\pi r^2\sqrt{2|eU_a|}}\right)^{1/2}. \tag{7.7}$$

For an electron beam it is written as:

$$B = B_b = \frac{830}{r}\,\frac{I^{1/2}}{U_a^{1/4}}.$$

Parameters B, r, I, U included in this equation are expressed in Gs, cm, A, V, respectively.

The radius of a balanced beam, which corresponds to a given value of B is found from the above equations and expressed as:

$$r = r_b = \frac{1}{B}\left(\frac{2m^{3/2}I}{|e|\,\varepsilon_0\pi\sqrt{2|eU_a|}}\right)^{1/2}. \tag{7.8}$$

Parameters B_b and r_b are usually called Brillouin induction and Brillouin radius.

7.1.3 Unbalanced Beam in Uniform Field

A strictly balanced beam is difficult to realize in practice since several conditions need to be fulfilled for this, such as zero radial component of velocity at the entrance in the uniform field region, exact value of beam radius, uniform space-charge density distribution. Real beams are usually unbalanced ones.

Let us consider unbalanced beam motion assuming a laminar beam structure and using the concept of beam-boundary trajectory.

Approximately, assuming $\partial U/\partial z = 0$ and $\dot{z} = \text{const}$, one finds the value of electric self-field strength acting on a beam-boundary particle:

$$-\frac{\partial U}{\partial r} = \frac{q_1}{2\pi\varepsilon_0 r} = \pm\frac{I}{2\pi\varepsilon_0 r\dot{z}}, \tag{7.9}$$

where q_1 – value of charge per unit beam length, I – beam current, r – beam radius, $\dot{z}$ – longitudinal charge particle velocity, assumed to be constant over beam length and cross section.

This equation follows from the Gauss theorem as shown in Sect. 4.4. Its substitution in the first equation of system (7.5) and excluding of time t gives the following trajectory equation:

$$\dot{z}^2\frac{d^2r}{dz^2} - \frac{|e|I}{2\pi\varepsilon_0 mr\dot{z}} + \frac{1}{4}\left(\frac{e}{m}\right)^2 B^2 r = 0. \tag{7.10}$$

7.1 Focusing Systems with Uniform Field

After introduction of the normalized radius $R = r/r_b$ it can be transformed to the form:

$$\frac{\dot{z}^2}{\omega_L^2}\frac{d^2R}{dz^2} - \frac{1}{R} + R = 0 , \qquad (7.11)$$

where $\omega_L = \frac{1}{2}\frac{|e|}{m}B$ – Larmor frequency.

The last equation describes nonlinear oscillation of a boundary particle around the equilibrium radius $R_e = 1$ ($r_e = r_b$). With a small amplitude of oscillation being assumed it can be linearized. Let $R = R_e(1+\delta) = 1+\delta$, where $\delta \ll 1$, then $1/R \approx 1 - \delta$, $d^2R/dz^2 = d^2\delta/d^2z$ and the linearized equation is written as:

$$\frac{d^2\delta}{dz^2} + 2\frac{\omega_L^2}{\dot{z}^2}\delta = 0 .$$

Its solution is

$$\delta = C_1 \cos\sqrt{2}\omega_L\frac{z}{\dot{z}} + C_2 \sin\sqrt{2}\omega_L\frac{z}{\dot{z}} .$$

Arbitrary constants C_1 and C_2 are found from initial conditions at the entrance in the uniform-field region ($z = 0$):

$$\delta_0 = R_0 - 1 = \frac{r_0}{r_b} - 1 = \frac{(r_0 - r_b)}{r_b} ,$$

$$\delta_0' = \left(\frac{d\delta}{dz}\right)_0 = \left(\frac{dR}{dz}\right)_0 = \frac{r_0'}{r_0} ,$$

where r_0 and r_0' – initial radial coordinate and initial trajectory slope of the boundary trajectory.

Determination of C_1 and C_2 results:

$$\delta = \delta_0 \cos\sqrt{2}\omega_L\frac{z}{\dot{z}} + \frac{\delta_0'\dot{z}}{\sqrt{2}\omega_L} \sin\sqrt{2}\omega_L\frac{z}{\dot{z}} .$$

Since $r = r_b(1+\delta)$, then the following equation is obtained for the boundary trajectory:

$$r = r_b + (r_0 - r_b)\cos\sqrt{2}\frac{\omega_L}{\dot{z}}z + \frac{r_0'\dot{z}}{\sqrt{2}\omega_L}\sin\sqrt{2}\frac{\omega_L}{\dot{z}}z . \qquad (7.12)$$

Therefore, a boundary particle periodically oscillates around the equilibrium radius $r_e = r_b$ when moving along the z-axis (Fig. 7.2). The amplitude Δr_m and wavelength λ of oscillations are determined by equations:

$$\Delta r_m = \left[(r_0 - r_b)^2 + \frac{(r_0'\dot{z})^2}{2\omega_L^2}\right]^{1/2} , \qquad (7.13)$$

$$\lambda = \frac{2\pi\dot{z}}{2\omega_L} \qquad (7.14)$$

As can be seen $\Delta r_m \to 0$, when $r_0 \to r_b$ and $r_0' \to 0$.

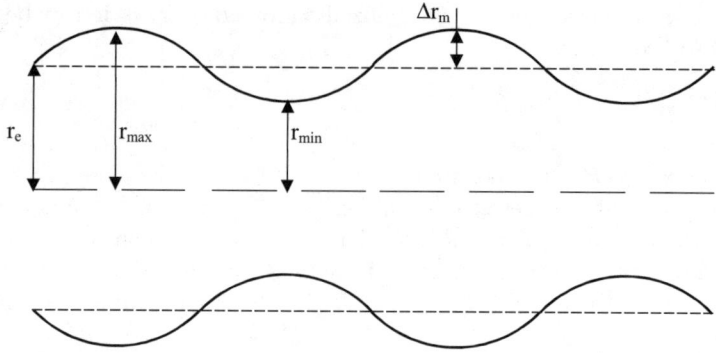

Fig. 7.2. Oscillation of a boundary particle

It is interesting to note that in the case of electron beams, small beam-boundary oscillations have the wavelength that approaches that of plasma oscillations. Indeed, taking account of (7.6) we can rewrite (7.14) thus:

$$\lambda = \frac{2\pi \dot{z}}{\sqrt{2}\omega_L} = \frac{2\pi \dot{z}}{\omega_p} = \lambda_p \ .$$

7.1.4 Beam Formed with Partially Shielded Gun

In this case, taking into account (3.40), (7.3) and (7.9), it is possible to derive the following equation for the beam-boundary trajectory:

$$\frac{d^2 r}{dt^2} - \frac{|e| I}{2\pi \varepsilon_0 m r \dot{z}} + \frac{1}{4}\left(\frac{e}{m}\right)^2 B^2 r \left[1 - \frac{r_g^4}{r^4}\right] = 0 \ ,$$

or, introducing normalized radius $R = r/r_b$ and excluding time t,

$$\frac{\dot{z}^2}{\omega_L^2}\frac{d^2 R}{dz^2} - \frac{1}{R} + R\left[1 - \frac{R_g^4}{R^4}\right] = 0 \ , \qquad (7.15)$$

where $R_g = r_g/r_b$.

Assuming $d^2 R/dz^2 = 0$ it is possible to determine the equilibrium radius for the boundary trajectory:

$$R_e^2 = \left(1 - \frac{R_g^4}{R_e^4}\right)^{-1},$$

or

$$R_e^2 = \frac{1}{2} + \frac{1}{2}\sqrt{1 + 4R_g^4} \ . \qquad (7.16)$$

Equation (7.15) describes nonlinear oscillation of a boundary particle around the equilibrium radius R_e. Assuming a small value of oscillation it is possible to make its linearization and obtain:

7.1 Focusing Systems with Uniform Field

$$\frac{d^2\delta}{dz^2} + \frac{2\omega_L^2}{\dot{z}^2}\left(1 + \frac{R_g^4}{R_e^4}\right)\delta = 0, \qquad (7.17)$$

where $\delta = (R - R_e)/R_e \ll 1$.

Its solution is:

$$\delta = C_1 \cos\sqrt{2}\alpha\omega_L \frac{z}{\dot{z}} + C_2 \sin\sqrt{2}\alpha\omega_L \frac{z}{\dot{z}},$$

where $\alpha = \left(1 + R_g^4/R_e^4\right)^{1/2}$, C_1 and C_2 are arbitrary constants found from the initial conditions:

$$z = 0 \begin{cases} \delta_0 = \dfrac{(R_0 - R_e)}{R_e} = \dfrac{r_0 - r_e}{r_e}; \\ \left(\dfrac{d\delta}{dz}\right)_0 = \dfrac{r_0'}{r_e}. \end{cases}$$

After determination of C_1 and C_2 the solution of (7.17) takes the form:

$$\delta = \frac{r_0 - r_e}{r_e}\cos\sqrt{2}\alpha\omega_L\frac{z}{\dot{z}} + \frac{r_0'\dot{z}}{\sqrt{2}\alpha r_e\omega_L}\sin\sqrt{2}\alpha\omega_L\frac{z}{\dot{z}},$$

where $r_e = R_e r_b$.

Since $r = r_e(1+\delta)$ the final form of trajectory equation is written as:

$$r = r_e + (r_0 - r_e)\cos\sqrt{2}\alpha\omega_L\frac{z}{\dot{z}} + \frac{r_0'\dot{z}}{\sqrt{2}\alpha\omega_L}\sin\sqrt{2}\alpha\omega_L\frac{z}{\dot{z}}.$$

The amplitude and wavelength of boundary trajectory oscillations are determined by the expressions:

$$\Delta r_m = \left[(r_0 - r_e)^2 + \frac{(r_0'\dot{z})^2}{2\omega_L^2\alpha^2}\right]^{1/2}, \qquad (7.18)$$

$$\lambda = \frac{2\pi\dot{z}}{\sqrt{2}\alpha\omega_L}.$$

As (7.18) shows, for obtaining a completely smooth beam without any oscillation the two conditions need to be fulfilled: $r_0' = 0$, $r_e = r_0$. They are attained by matching of the electron gun and the magnetic focusing system. The following equations are used for determination of the required value of magnetic induction:

$$B = \frac{1}{r_0(1 - r_g^4/r_0^4)^{1/2}}\left(\frac{2m^{3/2}I}{|e|\varepsilon_0\pi\sqrt{2|eU_a|}}\right)^{1/2},$$

and

$$B = \frac{1}{r_0(1 - G)^{1/2}}\left(\frac{2m^{3/2}I}{|e|\varepsilon_0\pi\sqrt{2|eU_a|}}\right)^{1/2}.$$

It is seen that the value of B depends on the rate of cathode shielding. It reaches a minimum for a completely shielded cathode ($G = r_g = 0$).

For electron beams, after substitution of corresponding numerical constants, one obtains:

$$B = \frac{1}{r_0(1-G)^{1/2}} \frac{830 I^{1/2}}{U_a^{1/4}},$$

with the following dimensions of variables being used: Gs, cm, A, V. It can be also written as:

$$B = B_b/(1-G)^{1/2},$$

where B_b – Brillouin magnetic induction (7.7).

There exists an important peculiarity of beam focusing with a partially shielded gun. As is seen from (7.18) the amplitude of the beam-boundary oscillation caused by the initial radial velocities ($r_0' \neq 0$) can be reduced by increasing the magnetic induction B (since $\omega_L = |e|/mB$). At the same time the difference ($r_0 - r_e$), which determines the amplitude of the first oscillating term can be diminished by proper choice of the coefficient of gun shielding G, at $G = 1 - (B_b/B)^2$ it becomes zero.

7.1.5 Balanced Nonlaminar Beam

This is a uniform axially symmetric beam, which trajectories are crossed but their envelope is a smooth cylindrical surface. It is known as a Kaptchisky–Vladimirsky beam (K–V beam). Let us consider the conditions of its realization, following a scheme described in [68].

The magnetic field is assumed to be uniform and directed along the z-axis of a cylindrical system of coordinates. For the case under consideration a system of equations of charge particle motion can be obtained from the system of equations (3.29)–(3.31) if we put in them $B_r = 0$, $E_z = 0$ and $B_z = B$:

$$m(\ddot{r} - r\dot{\theta}^2) = eE_r + er\dot{\theta}B, \qquad (7.19)$$

$$m\frac{1}{r}\frac{d}{dt}\left(r^2\dot{\theta}\right) = -e\dot{r}B, \qquad (7.20)$$

$$m\ddot{z} = 0. \qquad (7.21)$$

In these equations: B – external uniform magnetic field, E_r – radial component of self-field, its z-component E_z is assumed to equal zero on account of the axial beam uniformity ($E_z = -\partial U/\partial z = 0$).

From (7.20) one finds:

$$\frac{d}{dt}\left(mr^2\dot{\theta} + \frac{1}{2}eBr^2\right) = 0.$$

7.1 Focusing Systems with Uniform Field

The parameter in brackets is nothing less than the generalized azimuthal momentum:

$$p_\theta = mr^2\dot{\theta} + \frac{1}{2}eBr^2 . \tag{7.22}$$

It is conserved during particle motion $p_\theta = \text{const}$. Expressing $\dot{\theta}$ from (7.22), $\dot{\theta} = p_\theta/mr^2 - (e/2m)B$ and substituting it in (7.19) one derives the following equation for radial particle motion:

$$\frac{d^2r}{dt^2} + \omega_L^2 r - \frac{p_\theta^2}{m^2 r^3} - \frac{e}{m}E_r = 0 . \tag{7.23}$$

Particles move in the axial direction with a constant velocity $\dot{z} = \text{const}$, as follows from (7.21).

Multiplication of (7.23) by $2\frac{dr}{dt}$ and some algebra transforms yields:

$$\frac{d}{dt}\left[\left(\frac{dr}{dt}\right)^2 + \frac{p_\theta^2}{m^2 r^2} + \omega_L^2 r^2 + \frac{2e}{m}U(r)\right] = 0 .$$

Therefore, the expression in square brackets is an integral of motion:

$$J = \left(\frac{dr}{dt}\right)^2 + \frac{p_\theta^2}{m^2 r^2} + \omega_L^2 r^2 + \frac{2e}{m}U(r) = \text{const} .$$

Keeping in mind that the Hamilton function for the case under consideration is expressed as (see, for instance, Sect. 3.6):

$$H = \frac{1}{2m}\left[p_r^2 + \frac{1}{r^2}\left(p_\theta - \frac{er^2 B}{2}\right)^2 + eU(r)\right] ,$$

it is possible to show that this integral of motion is a linear combination of two independent integrals of motion H and p_θ.

For the further analysis the term $p_\theta^2/m^2 r^2$ is transformed in the following manner. From (7.22) we find $p_\theta/mr = r\dot{\theta} + (e/2m)Br$ and $p_\theta^2/m^2 r^2 = \left(r\dot{\theta} + (e/2m)Br\right)^2$.

The last equation can be presented in the form $p_\theta^2/m^2 r^2 = (v_\theta + \bar{v}_\theta)^2$, where $v_\theta = r\dot{\theta}$ is the linear azimuthal velocity, $\bar{v}_\theta = (e/2m)Br = \omega_L r$ - average azimuthal velocity caused by magnetic field action.

With account of the above results the integral of motion J can be written as:

$$J = v_\perp^2 + \omega_L^2 r^2 + 2\frac{e}{m}U(r) , \tag{7.24}$$

where $v_\perp^2 = (dr/dt)^2 + (v_\theta + \bar{v}_\theta)^2$ is the square of transverse velocity $v_\perp$.

As shown in Chap. 4, the solution of a self-consistent problem for non-laminar beam is reduced to the common solution of the Poisson equation for the electrostatic field (4.1), the Vlasov equation for the function of phase-space density (4.5) and (4.8) and (4.9) expressing space charge and current densities through phase-space density f. For the Vlasov equation, its partial solution can be an arbitrary function of an integral of motion. In the case under consideration this is a function of an integral of motion (7.24).

Referring to [68] the following function for phase space density is to be used to provide a self consistent nonlaminar balanced flow:

$$f = f_0 \delta(J - J_0),$$

where $\delta(J - J_0)$ – delta function of argument $(J - J_0)$, with J_0 being some constant.

Such a choice of the function means that motion of a set of charge particles with the same value of integral of motion J is considered.

Let us determine the space-charge density distribution, which corresponds to the given phase-space density, using (4.8):

$$\rho(x, y, z) = e \int f dv,$$

where integration is produced in the velocity space.

As applied to the motion in transverse plane xy the last equation is written as:

$$\rho(x, y) = e \int_{-\infty}^{+\infty} \int_{-\infty}^{+\infty} f dv_x dv_y.$$

Introducing in the velocity space a polar coordinate system we obtain:

$$v_x = v_\perp \cos\alpha, \qquad v_y = v_\perp \sin\alpha,$$

where $v_\perp$ – module of transverse velocity, α – polar angle.

Then,

$$\rho(x, y) = e \int_0^{2\pi} \int_0^\infty f v_\perp dv_\perp d\alpha.$$

As f does not depend on α we obtain:

$$\rho(x, y) = 2\pi e \int_0^\infty f v_\perp dv_\perp.$$

Taking account of the axial beam symmetry allows us to express $\rho(x, y)$ in the polar coordinate system r, θ in the following way:

$$\rho(r) = 2\pi e \int_0^\infty f_0 \delta(v_\perp^2 + \omega_L^2 r^2 + 2\frac{e}{m}U(r) - J_0) v_\perp dv_\perp.$$

If we introduce a new variable $u = v_\perp^2$, $du = 2v_\perp dv_\perp$, then it is possible to write:

$$\rho(r) = \pi e \int_0^\infty f_0 \delta(u - u_0) du , \qquad (7.25)$$

where

$$u_0 = J_0 - \omega_L^2 r^2 - 2\frac{e}{m} U(r) . \qquad (7.26)$$

The value of the integral $\int_0^\infty \delta(u - u_0) du = 0$ depends on the sign of u_0. If $u_0 < 0$, then everywhere the argument of the δ-function $u - u_0 > 0$ and $\int_0^\infty \delta(u - u_0) du = 0$ owing to a property of the δ-function. In the opposite case, when $u_0 > 0$, the argument $u - u_0$ becomes zero at the value of the variable $u = u_0$, which gives $\int_0^\infty \delta(u - u_0) du = 1$. The value and sign of u_0 depend on the radial coordinate r. Denoting as R the radial coordinate, at which u_0 becomes zero and changing its sign, we can write:

$$\rho(r) = \pi e f_0 = \text{const at } r \leq R, \qquad \rho(r) = 0 \text{ at } r > R .$$

Therefore, the beam under consideration has a sharply defined boundary of constant radius R and uniform space-charge density expressed as:

$$\rho = I/\pi R v_z .$$

The potential distribution inside the beam at $r \leq R$ is found from the Poisson equation:

$$\frac{1}{r}\frac{\partial}{\partial r}\left(r\frac{\partial U}{\partial r}\right) = -\frac{\rho}{\varepsilon_0} .$$

Its integration gives:

$$U(r) = -\frac{1}{4}r^2 \frac{\rho}{\varepsilon_0} + C .$$

Without loss of generality, it is possible to assume the potential at the axis of symmetry ($r = 0$) equal to zero, then $C = 0$ and

$$U(r) = -\frac{1}{4}r^2 \frac{\rho}{\varepsilon_0} = -\frac{Ir^2}{4\pi\varepsilon_0 R^2 v_z} .$$

Substitution of the potential found in the integral of motion and some algebraic transforms lead to the following expression for the integral of motion J:

$$J = \left(\frac{dr}{dt}\right)^2 + \frac{p_\theta^2}{m^2 r^2} + \omega^2 r^2 = J_0 = \text{const} , \qquad (7.27)$$

where:

$$\omega_r^2 = \omega_L^2 - \frac{\omega_p^2}{2} , \qquad \omega_p^2 = \frac{\rho e}{m\varepsilon_0} = \frac{I|e|}{\pi\varepsilon_0 R^2 v_z m} .$$

180   7 Magnetic Focusing Systems

It follows from this equation that particles with $p_\theta = 0$ reach the maximum distance from the axis of symmetry and determine the radius R of the balanced beam:

$$\omega_r^2 R^2 = J_0 \ .$$

7.1.6 Magnetic Systems for Uniform Field

Solid or sectional solenoids with magnetic shields made of soft-magnetic materials (Fig. 7.3) as well as shielded permanent magnets (Fig. 7.4) can be used for creation of a uniform magnetic field. Magnetic shields are intended to concentrate a magnetic field in the region of beam propagation and to form

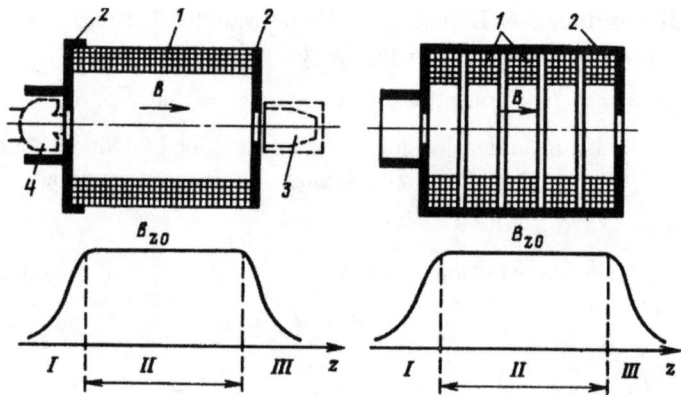

Fig. 7.3. Solenoid-type magnetic systems producing a uniform field

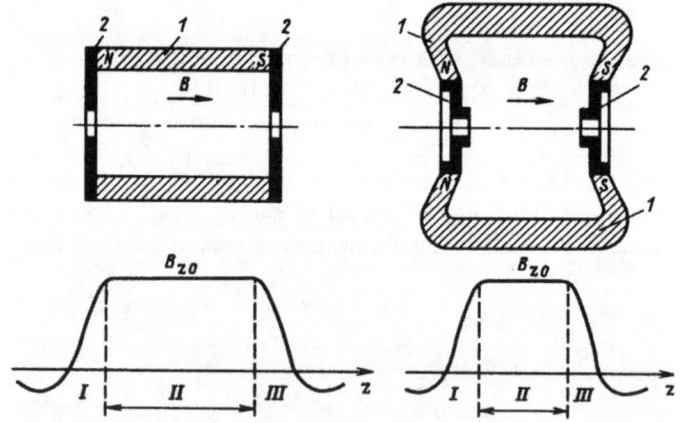

Fig. 7.4. Permanent-magnet systems producing a uniform field

the required field distribution in gun and collector regions. Typical curves of axial magnetic field distribution (B-curves) are shown in Figs. 7.3 and 7.4.

The main disadvantage of focusing systems with a uniform field is low efficiency of magnetic-field utilization. In order to provide a uniform magnetic field throughout a length l the transverse dimension of the magnetic pole pieces should be of the same order as l. So, the total volume V occupied by the magnetic field is of order l^3, while the useful volume is $V_p = \pi r_a^2 l$, r_a being radius of the channel, where the charge particle beam propagates.

It is possible to introduce magnetic field utilization factor as the ratio of magnetic field energy concentrated in these two volumes:

$$K_B = \int_{V_p} \frac{B^2}{2\mu_0} dV \Big/ \int_V \frac{B^2}{2\mu_0} dV. \tag{7.28}$$

For the case of uniform field it is expressed as:

$$K_B \approx \pi r_a^2 l / l^3 \approx \pi r_a^2 / l^2 .$$

As a rule $r_a \ll l$, and it results in very low magnetic-field utilization in focusing systems with uniform field. It can be increased by using magnetic-field reversals.

7.2 Systems of Reversed- and Periodic-Field Focusing

7.2.1 Principle of Reversed-Field Focusing

As follows from (7.1) the radial magnetic focusing force is proportional to the square of the magnetic induction. This allows focusing a charge particle beam with a field that reverses at intervals along the beam axis [121].

The reversals in magnetic field considerably reduce the weight of magnetic material required to focus a beam with given parameters. One or several reversals are used in practice. Figures 7.5 and 7.6 illustrate the curves of magnetic-field distribution (B-curves) for two single-reversed systems: an "ideal" one and a real one. In the "ideal" system the direction of magnetic field is changed instantly. The magnetic focusing force does not change its value when a charge particle passes the reversal.

In a real permanent-magnet reversed system the reversal zone has a finite length L_r where the magnetic force is decreased. A charge particle beam being balanced in the first zone of a uniform field undergoes some disturbances when passing the reversal zone. This leads to beam ripples in the second zone of a uniform field (Fig. 7.7).

To get a minimum beam disturbance the beam reversal should be as sharp as possible. This means, in practice, that the hole in the reversal pole piece should be as small as practicable.

Compensating peaks, located before and after the reversal zone (Fig. 7.8) can significantly compensate the beam disturbance in a reversed system.

182   7 Magnetic Focusing Systems

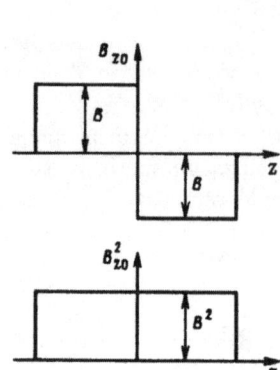

Fig. 7.5. Axial induction distribution for an ideal reversal

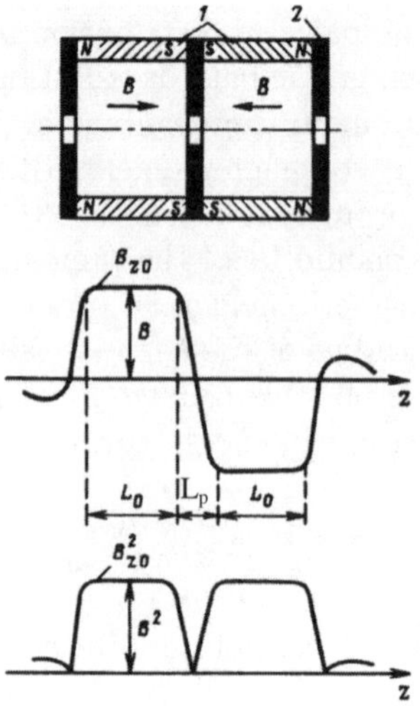

Fig. 7.6. Axial induction distribution for a real reversal: 1 – permanent magnets, 2 – pole pieces

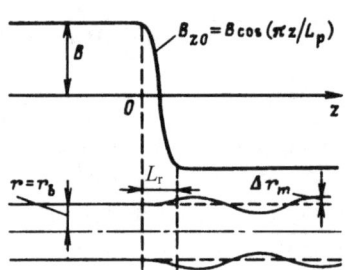

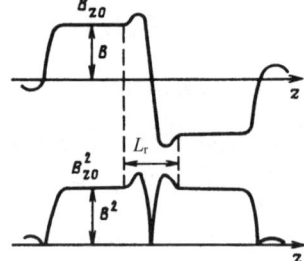

Fig. 7.7. Disturbances introduced in charge particle beam by field reversal

Fig. 7.8. Distribution of magnetic induction in a reversed system with compensating picks

Charged particles passing the field ejection obtain an additional radial impulse directed toward the beam axis, which compensate the effect of decreasing the magnetic force in the reversal.

The value of compensating peaks is matched to provide equality of the mean-square magnetic induction in the reversal zone and induction of a uniform field [122,123]:

$$\bar{B}_p^2 = \frac{1}{L_p} \int_{L_p} B_{z0}^2 dz = B^2 \ . \tag{7.29}$$

Application of reversal allows reduction of stray magnetic fluxes and increases the coefficient of magnetic-field utilization. For instance, in a system with single-stage reversal in which the length of the uniform zone $L_0 \approx l/2$ (l – total length of the system) this coefficient is increased approximately by four times $K_{B1} \approx r_a^2/L_0^2 = 4r_a^2/l^2$. For a system with N reversals the coefficient of utilization increases as $K_{BN} \approx r_a^2/L_0^2 = (N+1)^2 r_a^2/l^2$. Increasing of magnetic-field utilization results in considerable reduction of the weight of the magnetic system approximately by $1/(N+1)^2$ times. With increasing number of reversals the length of the uniform field zone decreases and the reversed magnetic system is transformed into a periodic system with continuously varying magnetic field.

7.2.2 Estimation of Beam Disturbances

For estimation of beam disturbances produced by a field reversal we use the paraxial boundary trajectory equation:

$$\dot{z}^2 \frac{d^2 r}{dz^2} - \frac{|e| I}{2\pi \varepsilon_0 m r \dot{z}_0} + \frac{1}{4} \left(\frac{e}{m}\right)^2 B_{z0}^2 r = 0 \,. \tag{7.30}$$

Assuming balanced beam motion in the uniform field before the reversal we can write the following initial conditions for computation of the boundary trajectory in the reversal zone $r = r_b$, $dr/dz = 0$, with r_b being the balanced beam radius in the uniform field region. Introducing normalized variables $R = r/r_b$ and $b = b(z) = B_{z0}/B$ (B being the induction of the uniform-field zone) we can rewrite (7.30) as:

$$\frac{\dot{z}^2}{\omega_L^2} \frac{d^2 R}{dz^2} - \frac{1}{R} + b^2(z) R = 0 \,, \tag{7.31}$$

or, denoting $\omega_L^2/\dot{z}^2$ as $2a$,

$$\frac{d^2 R}{dz^2} = 2a \left[\frac{1}{R} - b^2(z) R\right] \,. \tag{7.32}$$

An approximate solution of this equation can be found if we assume that the reversal extension is sufficiently small and the radial coordinate of a boundary particle practically does not change when the particle traverses the reversal zone $r \approx r_b = \text{const}$ and $R = r/r_b = 1$. Then (7.32) is reduced to:

$$\frac{dR}{dz} = 2a \int_{L_p} [1 - b^2(z)] dz \,. \tag{7.33}$$

For analytical estimation of the trajectory slope after the reversal, the distribution of magnetic induction $b(z)$ is approximated by an analytical function [123]:

$$b(z) = \cos(\pi z/L_p) \,. \tag{7.34}$$

Then,

$$\frac{dR}{dz} 2a \int_0^{L_r} \left(1 - \cos^2 \frac{\pi z}{L_r}\right) dz = 2a \int_0^{L_r} \sin^2 \frac{\pi z}{L_r} dz = aL_r,$$

and

$$\frac{dr}{dz} = \frac{1}{2} \frac{r_b \omega_L^2}{\dot{z}^2} L_p.$$

It is seen that the value of trajectory slope dr/dz is reduced as extension of the reversal decreases. Using $r = r_b$ and dr/dz, determined by (7.34), as entrance conditions for the uniform-field region after the reversal and taking account of the results of Sect. 7.1 we find that the boundary trajectory will oscillate with amplitude:

$$\Delta r_m = \frac{r_b \omega_L}{2\sqrt{2}\dot{z}} L_p.$$

7.2.3 Principle of Periodic-Field Focusing

It is possible to overcome the diverging effect of beam space charge and provide a confined beam cross section using periodical magnetic fields, which vary periodically along the beam axis. One of a simplest systems of this kind is shown in Fig. 7.9a. It consists of a number of equally spaced magnetic shielded coils.

The focusing effect of such fields can be explained in the following way. Suppose that in the region of strong magnetic field (for the system mentioned above it is the region near the middle plane of a coil) the magnetic focusing force exceeds the spreading force of space charge. The charge particle will be deflected here toward the axis. In the region of a weak magnetic field they will be spread owing to space-charge repulsion. When particles enter the strong-field region they again undergo the focusing action and so on. Therefore, a periodical field can balance, on average, the diverging effect of space charge. Evidently beam pulsations are unavoidable for focusing of this type, though they can be made sufficiently small.

The great advantages of periodical focusing appear when permanent magnets are used to produce magnetic field. A permanent-magnet focusing system consists of a number of ring magnets separated by pole pieces, which form a near-sinusoidal varying field creating a series of reversed magnetic lenses (Fig. 7.9d). This system provides a high concentration of magnetic field inside the region of beam propagation and has minimum stray magnetic fluxes. It practically means that a system of this type produces an intense focusing field with low energy stored in the magnetic field, that is to say, the integral of B^2 over the volume occupied by the field. Since, for properly designed permanent magnets, their weight is proportional to this energy, periodical systems of this type can be designed for low weight and small dimensions as compared with focusing systems of a uniform field.

7.2 Systems of Reversed- and Periodic-Field Focusing

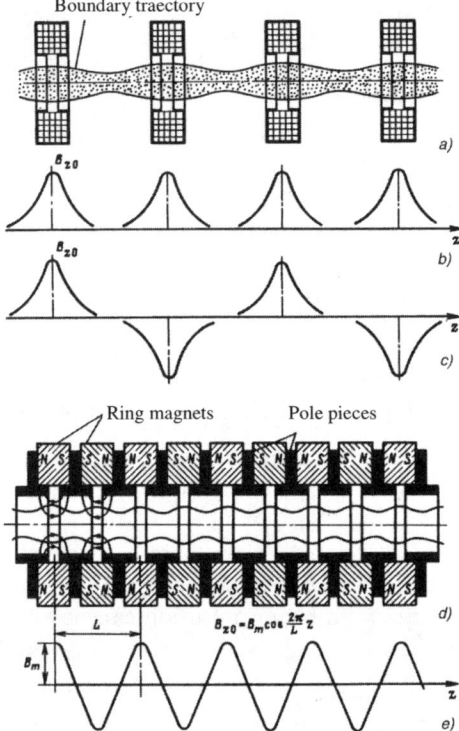

Fig. 7.9. Periiodic magnetic systems and curves of magnetic-field distribution: **a, b, c** – systems with solenoids, **d, e** – with permanent magnets

7.2.4 Analysis of Periodic Focusing

Cosinusoidal variation of magnetic induction along the beam axis is assumed:

$$B_{z0} = B_m \cos \frac{2\pi}{L} z ,$$

where L is the period of the system.

The paraxial equation of the boundary trajectory in this field is written as:

$$\dot{z}^2 \frac{d^2 r}{dz^2} - \frac{|e| I}{2\pi\varepsilon_0 mr\dot{z}} + \frac{1}{4}\left(\frac{e}{m}\right)^2 B_{z0}^2 r \left[1 - \left(\frac{r_k^2 B_{z0k}}{r^2 B_{z0}}\right)\right] = 0 .$$

It can be transformed to the canonical form, which is usually used in periodic focusing analysis. The normalized variables are introduced for that: $R = r/r_0$, $Z = 2\pi z/L$, $b(z) = B_{z_0}/B_m$, with the following normalization factor being used; r_0 – initial beam radius at the entrance to the regular part of focusing system, L – period of focusing system, B_m – amplitude of magnetic induction.

After some algebra, one obtains the paraxial trajectory equation in its canonical form:

$$\frac{d^2 R}{dZ^2} + 2\alpha b^2(Z)R - \frac{2\alpha G}{R^3} - \frac{\beta}{R} = 0 .$$

In this equation: $\alpha = \frac{1}{32\pi^2}\left(\frac{e}{m}\right)^2 \frac{B_m^2 L^2}{\dot{z}^2}$ – parameter of magnetic field, $\beta = \frac{1}{8\pi^3 \varepsilon_0} \frac{|e|}{m} \frac{IL^2}{r_0^2 \dot{z}^3}$ – parameter of space charge, $G = \frac{r_k^4 B_{z0k}^2}{r_0^4 B_m^2}$ – coefficient of cathode shielding, $b_z = \frac{B_{z0}}{B_m} = \cos Z$ – normalized magnetic field.

In the case of cosinusoidal variation of magnetic field $b(z) = \cos Z$ the previous trajectory equation can be transformed to:

$$\frac{d^2 R}{dz^2} + \alpha(1 + \cos 2Z)R - \frac{2\alpha G}{R^3} - \frac{\beta}{R} = 0 . \tag{7.35}$$

This nonlinear differential equation has no analytical solution and is investigated by computer-aided analysis. The main results of such an analysis obtained in [124] can be summarized in the following way:

- A necessary condition of minimum beam pulsation is a zero radial velocity of charged particles at the entrance in the regular part of a periodic focusing system.
- There exist optimal relations of beam parameters α, β, G, which provide a minimum beam pulsation. For a small value of space-charge parameter β the condition of minimum beam pulsation can be expressed in analytical form: $\alpha \approx \beta/(1 - 2G)$

For a system with a completely shielded cathode it is reduced to $\alpha \approx \beta$.

Using these equations it is possible to find the expressions for the optimum value of the amplitude of the magnetic induction B_m for given beam parameters I, U_a, r_0:

$$B_m = \frac{1}{r_0 (1 - 2G)^{1/2}} \left(\frac{4m^{3/2} I}{|e|\varepsilon_0 \pi \sqrt{2|eU_a|}}\right)^{1/2} ,$$

where B_m – amplitude of magnetic induction, r_0 – initial beam radius, I – beam current, U_a – accelerating voltage.

For the case of electron beams, after substitution of numerical values of constants, this equation is written as:

$$B_m = \sqrt{2} \frac{830}{r_0 (1 - 2G)^{1/2}} \frac{I^{1/2}}{U^{1/4}} ,$$

where B, I, U_a, r_0 are expressed, respectively, in Gs, A, V and cm.

Comparing this equation with that for Brillouin induction for uniform magnetic field one finds that they differ only by a factor of $\sqrt{2}$. Practically, this means that the root-mean-square value of magnetic induction of the periodic magnetic field, which provides the balance, on average, of space-charge repulsion, is equal to the Brillouin field of a uniform-field focusing system.

In general, relations of parameters α, β, G that deliver a minimum beam pulsations are presented in graphical form in Fig. 7.10 [124]. Minimum beam

7.2 Systems of Reversed- and Periodic-Field Focusing

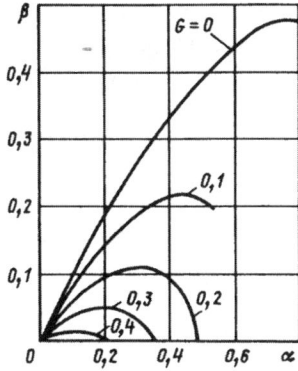

Fig. 7.10. Curves for determination of optimal parameters that deliver minimum beam pulsation

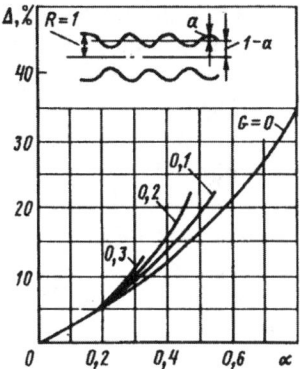

Fig. 7.11. Curves for calculation of the amplitude of pulsation

pulsations are usually described by a simple periodical function of the longitudinal coordinate with period equal to half of the period of the magnetic system L. For instance, they can be approximated as [124]:

$$R = (1-a) + a\cos 2z ,$$

where a – normalized amplitude of beam pulsation, $(1-a)$ – average normalized beam radius.

The relative amplitude of beam-boundary pulsation $a/(1-a)$ plotted against magnetic-field parameter α is shown in Fig. 7.11.

7.2.5 Stability of Periodic Focusing

Results presented in Fig. 7.11 indicate that the amplitude of beam pulsation is increased with increasing α, but they do not indicate that beginning from a certain value of α instead of periodical pulsation growing ones are erased, with amplitude quickly increasing with z-coordinate.

This can be predicted theoretically because trajectory equation (7.35) is a differential equation of Mathieu type:

$$\frac{d^2 R}{dZ^2} + \alpha(1+\cos 2Z)R = \frac{2\alpha G}{R^3} + \frac{\beta}{R} .$$

If we omit the terms of its right part, it is reduced to the uniform Mathieu equation, of which the canonical form is:

$$\frac{dR}{dZ} + (a + 2q\cos 2Z) R = 0 .$$

For the case under consideration coefficients a and q are equal $a = \alpha$, $2q = \alpha$, so $2q = a$.

It is known that this equation may have confined or growing solutions depending on the values of a and q. As far as periodical focusing is concerned these two types of solutions correspond accordingly to zones of stable and unstable focusing.

It allows Mathieu diagrams for stable and unstable solutions to be used for prediction of the beam-focusing stability. For instance, application of these diagrams yields the following interval for the first zone of stable focusing $0 < \alpha < 0.66$ and for the second one $1.75 < \alpha < 3.7$. Focusing in the first zone is usually used in practice.

These results are considered to be approximate ones, since the total trajectory equation is nonuniform and a nonlinear equation of Mathieu type. A more precise analysis shows that the first stable zone lies within the limits $0 < \alpha < 0.4/(1 + 1.6G)$ [10].

Periodic Lenses System

This focusing system is created by a set magnetic lenses located along the beam axis. Shielded solenoids (Fig. 7.9a) or ring-shaped permanent magnets can be used as focusing magnetic lenses. Computer-aided design of such a system is considered in Sect. 7.5.

7.3 Focusing of Hollow Beams in Uniform Field

There exists a peculiarity of hollow-beam focusing in a uniform magnetic field. It is caused by specific properties of the space-charge field distribution in hollow beams [125–127].

Applying the Gauss theorem to a cylindrical volume confined by the inner beam boundary one can easily find that the radial component of the field is equal to zero at this boundary (the beam is supposed to be infinitely long).

Therefore, for obtaining balanced motion of the inner charge particles, the magnetic focusing force is to be zero at the inner boundary or it should be balanced by some other radial forces such as the external electrostatic force or centrifugal force. In the opposite case the inner charge particles will be deflected toward the beam axis and hollow beam will break down.

For this reason for obtaining an equilibrium hollow-beam, a magnetic system with partially shielded cathodes are used. As follows from the general equations of motion (3.37) and (3.39) and the equation for the focusing force (7.1), the radial magnetic force becomes zero at the radial distance that is determined by equalities $\psi^2 = \psi_k^2$ or, $\pi^2 r^4 B^2 = \psi_k^2$, where B – induction of uniform field; $\psi = \pi r^2 B$ – magnetic flux confined by the circle of radius r, with r being current radial position of a charge particle; ψ_k – cathode magnetic flux through a surface confined by the circle of radius r_k, which is the radial coordinate of the particle at the cathode surface, as shown in Fig. 7.1.

7.3.1 Equilibrium Hollow Beam

The following system of equations describes charge particle beam motion in a uniform field when the hydrodynamic beam model is being used:

$$\ddot{r} = -\frac{e}{m}\frac{\partial U}{\partial r} - \frac{1}{4}\left(\frac{e}{m}\right)^2 B^2 r \left(1 - \frac{\psi_k^2}{\pi^2 r^4 B^2}\right), \qquad (7.36)$$

$$\ddot{z} = -\frac{e}{m}\frac{\partial U}{\partial z}, \qquad (7.37)$$

$$\dot{\theta} = -\frac{1}{2\pi}\frac{e}{m}\frac{\pi r^2 B - \psi_k}{r^2}, \qquad (7.38)$$

$$\frac{1}{r}\frac{\partial}{\partial r}\left(r\frac{\partial U}{\partial z}\right) + \frac{\partial^2 U}{\partial z^2} = -\frac{\rho}{\varepsilon_0}, \qquad (7.39)$$

$$\mathrm{div}\,\rho\vec{v} = \frac{\partial}{\partial r}(\rho\dot{r}) + \frac{\partial}{\partial z}(\rho\dot{z}) = 0. \qquad (7.40)$$

Let us find a partial self-consistent solution of this system of equations, which provides equilibrium beam motion ($\ddot{r} = \ddot{z} = 0$, $\dot{r} = 0$). Substitution of $\ddot{r} = 0$ in (7.36) yields:

$$\frac{\partial U}{\partial r} = -\frac{1}{4}\frac{e}{m}rB^2\left(1 - \frac{\psi_k^2}{\pi^2 r^4 B^2}\right). \qquad (7.41)$$

This is the equation of balance of the radial space charge and magnetic focusing forces. With $\dot{r} = 0$ (7.40) is reduced to $\partial(\rho\dot{z})/\partial z = 0$. This equation is satisfied if we put $\rho(z) = \mathrm{const}$ and $\dot{z}(z) = \mathrm{const}$. The last equality involves $\ddot{z} = 0$ and $\partial U/\partial z = 0$. As a result of this the Poisson equation (7.39) is simplified to:

$$\frac{1}{r}\frac{\partial}{\partial r}\left(r\frac{\partial U}{\partial r}\right) = -\frac{\rho}{\varepsilon_0}, \qquad (7.42)$$

which is compatible with the condition $\rho(z) = \mathrm{const}$.

Integration of this equation and substitution of the result in (7.41) yields:

$$\frac{4m}{\varepsilon_0 e}\int_{r_{\mathrm{int}}}^{r} r\rho(r)\,dr = r^2 B^2\left(1 - \frac{\psi_k^2}{\pi^2 r^4 B^2}\right), \qquad (7.43)$$

where r_{int} – internal beam radius.

For particles at the internal beam boundary it is reduced to:

$$\psi_{k\,\mathrm{int}}^2 = \pi^2 r_{\mathrm{int}}^4 B^2. \qquad (7.44)$$

Let us consider particular conditions that provide the balance equation and corresponding schemes of balanced-beam focusing:

a) $\psi_k = \mathrm{const}$. In this case the cathode magnetic flux ψ_k is supposed to be constant for all particles of beam cross section $\psi_k = \psi_{k\,\mathrm{int}} = \pi r_{\mathrm{int}}^2 B = \mathrm{const}$.

Substitution of this equation in (7.43) and its solution concerning $\rho(r)$ results in:

$$\rho(r) = \frac{1}{2}\varepsilon_0 \frac{e}{m} B^2 \left(1 + \frac{r_{int}^4}{r^4}\right). \tag{7.45}$$

Therefore, a nonuniform space-charge density distribution over the beam cross section is required to get the balanced beam motion.

b) $\psi_k = Cr$. In this case, ψ_k is a linear function of radial position. The constant C is found from the condition (7.44) $\psi_{k\,int}^2 = (Cr_{k\,int})^2 = \pi^2 r_{int}^2 B^2$, then $C = \pi r_{int} B$ and

$$\psi_k = Cr = \pi r_{int} Br. \tag{7.46}$$

Substitution of (7.46) in (7.43), after some algebra, gives:

$$\rho = \frac{1}{2}\varepsilon_0 \frac{e}{m} B^2 = \text{const}.$$

So, equilibrium beam motion theoretically can be realized with $\rho = \text{const}$ over the beam cross section.

c) $\psi_k = \text{const}$ – quasibalanced beam. In this case the exact balance of radial forces is fulfilled only at internal and external beam boundaries. As to the inner layers of the beam their motions are performed under the action of unbalanced forces. This results in the radial displacement of particles. But if a beam has a finite extension the inner layers do not succeed to cross the beam boundaries and the beam can conserve its tubular form.

As in previous cases equilibrium motion of internal charge particles is obtained if condition $\psi_{k\,int}^2 = \pi^2 r_{int}^4 B^2$ is fulfilled. The condition of balanced motion of charge particles at the external boundary ($r = r_{ext}$) is found if we substitute in (7.43) $\psi_k = \psi_{k\,int} = \pi r_{int}^2 B$ and $r = r_{ext}$:

$$\frac{2m}{\varepsilon_0 e\pi} \int_{r_{int}}^{r_{ext}} 2\pi r\rho(r)dr = r_{ext}^2 B^2 \left(1 - \frac{r_{int}^4}{r_{ext}^4}\right). \tag{7.47}$$

The integral in this equation is equal to the value of space charge per unit beam length:

$$\int_{r_{int}}^{r_{ext}} 2\pi r\rho(r)dr = q_1 \approx \frac{I}{\dot{z}}.$$

Then, substituting the last equation in (7.43), one finds the value of magnetic induction that provides balanced motion of the external boundary trajectory:

$$B^2 = \frac{2I}{\varepsilon_0 \left(|e|/m\right) \pi r_{ext}^2 (1 - r_{int}^4/r_{ext}^4)\dot{z}},$$

or, admitting approximately $\dot{z} \approx \sqrt{2|eU_a|/m}$, with U_a being the potential of an electrode enclosing the beam:

7.3 Focusing of Hollow Beams in Uniform Field

$$B^2 = \frac{2I}{\varepsilon_0 \left(|e|/m\right) \pi r_{\text{ext}}^2 (1 - r_{\text{int}}^4/r_{\text{ext}}^4) \sqrt{2|eU_a|/m}}.$$

For electron beams, after substitution of numerical constants, it is written as:

$$B = \frac{830}{r_{\text{ext}} \left(1 - r_{\text{int}}^4/r_{\text{ext}}^4\right)^{1/2}} \frac{I^{1/2}}{U_a^{1/4}}, \tag{7.48}$$

where magnetic induction B, beam current I, electrode potential U_a and linear dimensions are expressed, respectively, in Gs, A, V, cm.

7.3.2 Injection of Hollow Beams in Uniform Field

An important point of hollow-beam focusing is injection of beams in the regular region of the focusing field. Several possible systems of injection are shown in Fig. 7.12. In systems a and b the magnetic flux ψ_k passing through the cathode is equal to the magnetic flux passing through the inner rod of a magnetic circuit. It is evident that in this case the value of ψ_k is the same for all particles of the beam.

There exists the following difference between these two systems. In system a particles creating the inner beam boundary are injected in the uniform field without crossing of magnetic force lines. As a result of this their azimuthal velocity in the uniform field appear to be equal to zero, they are not affected by the magnetic focusing force and propagate farther along a straight line, which creates the inner beam boundary. The remaining particles of the beam on their way to the regular region of uniform field cross magnetic field lines and acquire azimuthal velocities, which produce radial focusing forces when particles are moving in the uniform field.

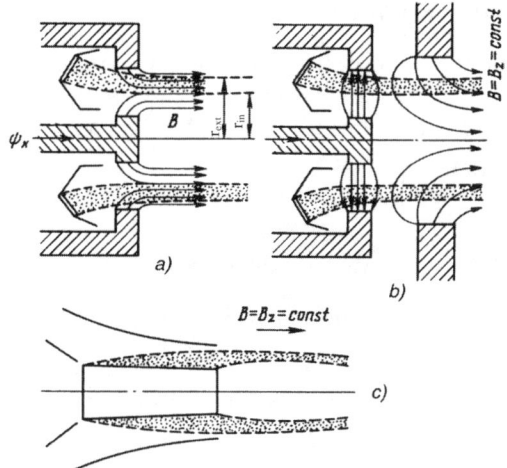

Fig. 7.12. Systems of forming and injection of hollow beams

In the system b, shown in Fig. 7.12, all particles of the beam passing through the ring gap in the magnetic circuit acquire azimuthal velocities $\dot{\theta}$, which are determined by the relation $\dot{\theta} = \frac{e}{2\pi m}\frac{\psi_k}{r^2}$. Subsequently, when moving in the transient-field region, which proceeds to the regular uniform field, they change the azimuthal velocity due to interaction with the radial component of the magnetic field of the transient region. The resulting azimuthal velocities at the entrance to the uniform field will be determined by, following from Bush's theorem:

$$\dot{\theta} = -\frac{e}{2\pi m}\frac{\psi - \psi_k}{r^2}.$$

When ψ and ψ_k have the same directions (same signs) the azimuthal velocity becomes zero in the uniform-field region at a radial distance that is determined from the equation $r = (\psi_k/\pi B)^{1/2}$.

With ψ and ψ_k being of opposite signs the azimuthal velocity of all beam particles will be different from zero. Therefore, all of them are affected by the magnetic focusing force. However, equilibrium motion of inner particles can also be achieved. It is based on the balance of magnetic focusing and centrifugal forces. Assuming $\psi > 0$ and consequently $\psi_k < 0$ it is possible to show that the balance of these forces in the uniform field will take place at a radial distance $r = (\psi_k/\pi B)^{1/2}$. But, in contrast to the previous case, balanced motion in the radial direction will be accompanied by rotational motion.

It should be noted that the condition of $\psi_k = $ const is approximately fulfilled for a magnetron injection gun (Fig. 7.12c) with a small angle of a cathode cone, being immersed in a uniform magnetic field.

As has been shown above obtaining a balanced hollow beam with constant space-charge density $\rho = $ const requires a special magnetic flux distribution in the vicinity of the particle emitter.

7.3.3 Boundary Trajectory Pulsation

In the case of improper injection in a uniform field the boundary trajectories of a hollow beam will oscillate. The equation of the external boundary trajectory will be the same as that for a solid beam-boundary trajectory (7.15):

$$\frac{\dot{z}^2}{\omega_L^2}\frac{d^2R}{dz^2} - \frac{1}{R} + R\left(1 - \frac{R_{g\,\text{ext}}^4}{R^4}\right) = 0, \qquad (7.49)$$

where $R = r/r_b$ – normalized radius with normalization factor r_b being equal $r = r_b = \frac{1}{B}\left(\frac{2m^{3/2}I}{|e|\varepsilon_0\pi\sqrt{2|eU_a|}}\right)^{1/2}$, $R_{g\,\text{ext}}$ – normalized radius defining value of magnetic flux ψ_k for external trajectory $\psi_{k\,ext} = R_{g\,\text{ext}}^2 r_b^2 \pi B$.

Equation (7.49) describes nonlinear oscillations of the external trajectory around the equilibrium radius $R_{e\,\text{ext}} = \left(1 - R_{e\,\text{ext}}^4/R_{\text{ext}}^4\right)^{-1/2}$. Its linearization gives:

$$\frac{d^2\delta}{dz^2} + \frac{2\omega_L^2}{\dot{z}^2}\left(1 + \frac{R_{e\,\text{ext}}^4}{R_{e\,\text{ext}}^4}\right)\delta = 0 ,$$

where $\delta = (R - R_{e\,\text{ext}})/R_{e\,\text{ext}}$.

The following initial conditions are used for its solution:

$$\delta = \delta_0 = \frac{R_0 - R_{e\,\text{ext}}}{R_{e\,\text{ext}}}, \quad \frac{d\delta}{dz} = \left(\frac{d\delta}{dz}\right)_0 = \frac{1}{R_{e\,\text{ext}}}\left(\frac{dR}{dz}\right)_0 .$$

The result of this solution is written as:

$$\delta = \delta_0 \cos\sqrt{2}\alpha\omega_L \frac{z}{\dot{z}} + \frac{\dot{z}}{\sqrt{2}\alpha\omega_L}\left(\frac{\partial\delta}{dz}\right)_0 \sin\sqrt{2}\alpha\omega_L \frac{z}{\dot{z}} ,$$

where $\alpha = \left(1 + R_{e\,\text{ext}}^4/R_{e\,\text{ext}}^4\right)^{1/2}$.

It is seen that the external trajectory oscillates with frequency equal to $\sqrt{2}\alpha\omega_L$, with amplitude depending on an initial condition.

The equation for the inner boundary trajectory is obtained from (7.49), if we omit in it the term $1/R$ and replace $R_{e\,\text{ext}}$ for $R_{g\,\text{int}}$:

$$\frac{\dot{z}^2}{\omega_L^2}\frac{d^2R}{dz^2} + R\left(1 - \frac{R_{g\,\text{int}}^4}{R^4}\right) = 0 . \tag{7.50}$$

It follows from this equation that the equilibrium radius of internal trajectory $R_{e\,\text{int}}$ is equal to $R_{e\,\text{int}} = R_{g\,\text{int}}$, with $R_{g\,\text{int}}$ being a function of magnetic flux $\psi_{k\,\text{int}}$, $R_{g\,\text{int}} = \left(\psi_{k\,\text{int}}/r_b^2\pi B\right)^{1/2}$. Its linearized version is:

$$\frac{d^2\delta}{dz^2} + 4\frac{\omega_L^2}{\dot{z}^2}\delta = 0 .$$

The solution of this equation

$$\delta = \delta_0 \cos 2\omega_L \frac{z}{\dot{z}} + \frac{\dot{z}}{2\omega_L}\left(\frac{d\delta}{dz}\right)_0 \sin 2\omega_L \frac{z}{\dot{z}}$$

describes the oscillation of the inner trajectory around R_{int} with a frequency equal to $2\omega_L = \omega_C$, that is, the cyclotron frequency.

Reversed and Periodic Focusing

These types of focusing theoretically can be used for hollow beams. As in the case of uniform-field focusing there exists the problem of equilibrium of the inner-beam trajectory. The ways of its solution seem to be somewhat more complicated due to reversals of magnetic field. At present, systems of this type have not found practical application.

7.4 Focusing of Strip Beams

7.4.1 Uniform-Field Focusing

For analysis we assume that the charge particle beam is a laminar one and its width (dimension along the x-coordinate) is much greater than its thick-

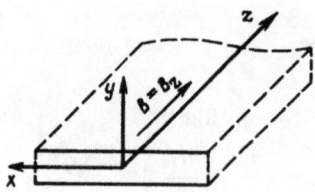

Fig. 7.13. Strip beam in uniform magnetic field

ness (Fig. 7.13). Taking account of the results of Sect. 3.3 we can write the following system of equations for particle beam motion:

$$\ddot{y} = \frac{e}{m}E_y + \frac{e}{m}\dot{x}B ,\qquad(7.51)$$

$$\dot{x} = -\frac{e}{m}(\psi - \psi_k) ,\qquad(7.52)$$

$$\ddot{z} = 0 \left(\dot{z} \approx \sqrt{2\frac{|eU_a|}{m}} \right) ,\qquad(7.53)$$

where B – magnetic induction of uniform magnetic field directed along the z-axis; U_a – accelerating potential; ψ – magnetic flux passing through a surface of unit width and height equal to the y-coordinate of a charged particle at the point under consideration (see Fig. 3.2) $\psi = By$; ψ_0 – magnetic flux passing through the surface of unit width and height equal to the y-coordinate at the point of the particle start, $y = y_0$; E_y – single component of electric field, which is produced by the beam space charge.

By using the Gauss theorem this can be expressed in terms of beam parameters:

$$E_y = \pm \frac{I_1}{2\varepsilon_0 \dot{z}} ,\qquad(7.54)$$

where I_1 – beam current per unit beam width, the field components E_z and E_x are assumed to equal zero.

The differential equation of the beam-boundary trajectory resulting from equations of motions (7.51)–(7.53) has the following form:

$$\frac{d^2y}{dz^2} + \left(\frac{e}{m}\right)^2 \frac{B^2}{\dot{z}^2}\left(y - \frac{B_k}{B}y_k\right) - \frac{|e|I_1}{2\varepsilon_0 m \dot{z}^3} = 0 .\qquad(7.55)$$

Assuming $d^2y/dz^2 = 0$ we determine the equilibrium coordinate y_e:

$$y_e = y_b + \frac{B_k}{B}y_k ,\qquad(7.56)$$

where $y_b = mI_1/2\varepsilon_0 |e| \dot{z}B^2$ – equilibrium coordinate for a charge particle source completely shielded from the magnetic field (Brillouin equilibrium coordinate).

Solution of (7.55) describing the boundary trajectory in the plane yz has the form:

$$y = y_e + \delta, \tag{7.57}$$

where

$$\delta = (y_0 - y_e)\cos\omega_c \frac{z}{\dot z} + \left(\frac{dy}{dz}\right)\frac{\dot z}{\omega_c}\sin\omega_c \frac{z}{\dot z}. \tag{7.58}$$

In general, the boundary trajectory oscillates around the equilibrium coordinate y_e. For the optimal condition of beam injection ($y_0 = y_e$ and $dy/dz = 0$) oscillations vanish. The value of magnetic induction that provides the equilibrium beam motion is determined by:

$$B^2 = \frac{mI_1}{(1 - G^{1/2})2y_e\varepsilon_0 |e| \dot z},$$

where $G = \left(\frac{B_k y_k}{B y_e}\right)^2$ $G = (\psi_k/By_e)^2$ – coefficient of shielding of charge particle source.

Motion of Particles in xz Plane

Along with particle motion in the yz plane there exist their displacement in xz plane. It is caused by the x-component of velocity, which, being found from (7.52) and (7.56)–(7.58), is expressed as:

$$\dot x = -\frac{e}{m}By_b - \frac{e}{m}B\delta,$$

where δ is determined by (7.58).

It is seen that the x-component of velocity consists of two parts. One of them ($-eBy_b/m$) does not depend on the z-coordinate, the other periodically changes with z-coordinate. The constant component of $\dot x$ leads to permanent particle displacement in the x-direction.

More detailed analysis shows that particles of the internal beam layers also undergo displacement in the x-direction, which is proportional to their y-coordinate. For a negative value of y the displacement of particles occurs in the opposite direction. The effect of displacement becomes important for beams of a finite width, as it leads to deformation of the beam cross section.

7.4.2 Periodic Magnetic Focusing

A periodic system with cosinusoidal variation of magnetic induction $B_z = B_m \cos 2\pi(z/L)$ has been investigated in [128]. Conditions of obtaining the minimum beam pulsation and beam stability were found. In particular, the condition of focusing-system operation in the first zone of stability, which is expressed as (for electron beams):

$$B_m L / U_a^{1/2} < 24.5,$$

here B_m – amplitude of magnetic induction (Gs), L – period of system (cm), U_a – accelerating voltage (V).

It should be noted that this effect of particle displacement is not essential for this type of focusing since reversal of magnetic fields involves the change of direction of displacement.

7.5 Computer-Aided Design of Magnetic Systems

Computation and design of magnetic focusing systems usually is performed in two stages. The first one is performed with simple computer codes, which are based on simplified models of magnetic systems and charged-particle beams.

For instance, for computation of solenoids and permanent magnets different analytical models can be used. Laminar beam models and conception of beam boundary trajectory are usually used for computation of beam motion. These codes are known as "express codes". They allow quick analysis of many variants of a system in order to find the one that largely meets the requirements. The obtained results are considered to be approximate ones.

In the second stage of the design procedure the system found is analyzed and corrected with the help of high-level computer codes, which are based on numerical models of beams and focusing systems, which take into account the real properties of magnetic materials and the charge particle beam nature.

Let us consider, as an example, the correction procedure of a solenoid-type magnetic lens. This lens is intended for a multiple lens focusing system (like that shown in Fig. 7.9a), which transports the electron beam for a long distance with the following parameters: beam perveance $P = 0.8$ µA/V$^{3/2}$, accelerating voltage $U_a = 6.5$ kV, period of system $L = 9.6$ cm, minimum beam radius $r_{\min} = 0.6$ cm.

In accordance with the lens application its main function is the symmetric transform of the beam from input to output of the lens, when the output beam parameters are the same as the input ones.

The initial variant of the lens was found with the aid of express computer codes. Results of its trajectory analysis are shown in Fig. 7.14. They are obtained for lens excitation equal to 556 Aw (Ampere turns). It is seen that the electron beam is not symmetrical for the given value of lens excitation and correction of excitation is required.

Results of lens correction are presented in Fig. 7.15, where output phase curves and the value of parameter β are given as a function of lens excitation,

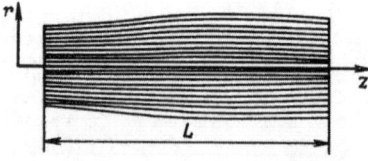

Fig. 7.14. Results of trajectory analysis of the initial variant of the lens: lens excitation 556 Aw

7.5 Computer-Aided Design of Magnetic Systems   197

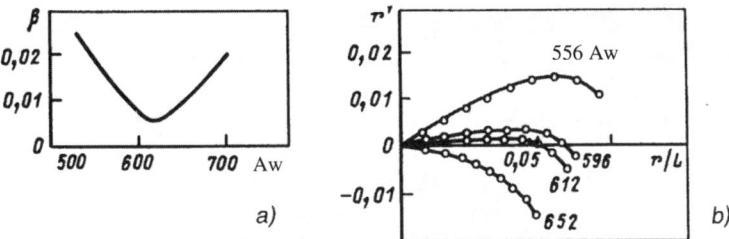

Fig. 7.15. Results of lens correction: parameter β (**a**) and output phase curves (**b**) as function of lens excitation

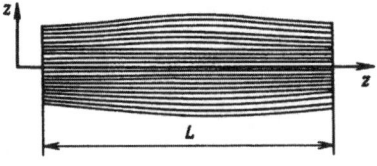

Fig. 7.16. Electron trajectories in the lens for the optimal value of lens excitation – 612 Aw

with β being a measure of the difference between input and output beam phase curves:

$$\beta = \sum_{k=1}^{N} \alpha_k \left[\left(\frac{r_k - \hat{r}_k}{L} \right)^2 + (r'_k - \hat{r}'_k)^2 \right]^{1/2},$$

where r_k and $\hat{r}_k$ – radius of the k-th trajectory at the input and output of the lens, r'_k and $\hat{r}'_k$ are the trajectory slopes.

In the case under consideration the beam is supposed to enter the lens with zero trajectory slopes, $r'_k = 0$ the above expression is reduced to:

$$\beta = \sum_{k=1}^{N} \alpha_k \left[\left(\frac{r_k - \hat{r}_k}{L} \right)^2 + (\hat{r}'_k)^2 \right]^{1/2}.$$

From the results shown in Fig. 7.15 it follows that optimal excitation of a magnetic lens, which delivers a minimum of β and provides the greatest degree of beam symmetry, is equal to 612 Aw. Results of trajectory tracing for this optimal value of excitation are shown in Fig. 7.16. After determination of the required parameter of a single lens, trajectory analysis of the focusing system as a whole is performed.

8 Electrostatic Focusing Systems

8.1 Focusing System for Solid Axially Symmetrical Beams

8.1.1 Equilibrium Cylindrical Beam

An electrode system for focusing of a cylindrical equilibrium beam can be designed on the basis of one-dimensional particle flow in a planar diode formed by electrodes A_1 and A_2 with equal potential $U_1 = U_2 = U_a$ (Fig. 8.1). Charge particles are assumed to be injected into the diode space with uniform current density and uniform velocity normal to the plane of electrode A_1. The rectilinear motion of charge particles will take place in the diode, with potential being changed only in the z-direction in accordance with:

$$\frac{z}{d} = G^{1/2} \left[\left(\frac{U}{U_m} \right)^{1/2} - 1 \right]^{1/2} \left[\left(\frac{U}{U_m} \right)^{1/2} + 2 \right] . \tag{8.1}$$

In this equation:

$$G = \frac{4}{9} \varepsilon_0 \sqrt{2 \frac{|e|}{m}} \frac{(U_m/U_a)^{3/2}}{jd^2/|U_a|^{3/2}} ,$$

where U_a is potential of the electrodes of the diode; U_m is the value of the potential in the middle plane of the diode; j is the current density; d is the length of the diode space.

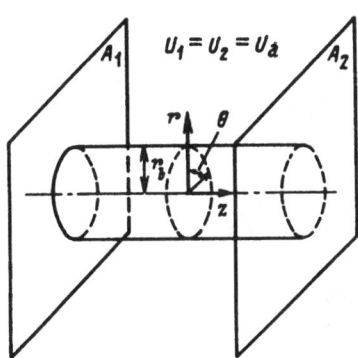

Fig. 8.1. Parallel equilibrium beam

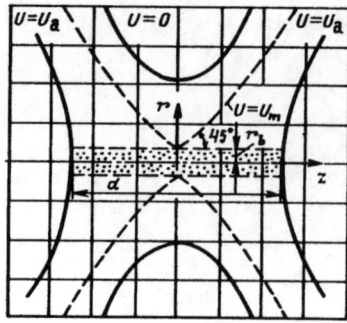

Fig. 8.2. Geometry of electrodes for parallel-beam formation

The value of U_m depends on the parameter $jd^2/|U_a|^{3/2}$ and can be found from the formula

$$\frac{16}{9}\varepsilon_0\sqrt{2\frac{|e|}{m}}\left[1-\left(\frac{U_m}{U_a}\right)^{3/2}\right]\left[1+2\left(\frac{U_m}{U_a}\right)^{1/2}\right] = \frac{jd^2}{|U|^{3/2}}. \quad (8.2)$$

In the diode cross section the potential remains constant, hence

$$\partial U/\partial r = 0, \quad \partial U/\partial \theta = 0. \quad (8.3)$$

In accordance with the Pierce principle one cuts from this flow a cylindrical beam of radius r_b and replaces the effect of the discarded part of the flow by the equivalent action of the focusing electrodes. The forms of such electrodes are found by solution of the Cauchy problem for Laplace's equation with boundary conditions, determined by ratios (8.1) and (8.2). A simple analytical solution of this problem may be obtained when $U_m/U_a > 0.7$. For this case the distribution of potential in the diode space may be described with good accuracy (approximately 0.5%) by the approximate expression:

$$U = U_m + Cz^2, \quad C = \frac{4}{d^2}(U_a - U_m).$$

Solution of the Cauchy problem for the Laplace equation

$$\frac{1}{r}\frac{\partial}{\partial r}\left(r\frac{\partial U}{\partial r}\right) + \frac{\partial^2 U}{\partial r^2} = 0,$$

with boundary conditions: $U = U_m + Cz^2$, $\partial U/\partial r = 0$ at $r = r_b$ gives [129]:

$$\left(\frac{z}{d}\right)^2 = \left(\frac{r_b}{d}\right)^2\left[\frac{(r/r_b)^2-1}{2} - \ln\frac{r}{r_b}\right] + \frac{U-U_m}{4(U_a-U_m)}.$$

Using this expression one can calculate the form of the equipotential lines in exterior region. Figure 8.2 shows the form of the equipotential lines $U = 0, U = U_a, U = U_m$ for a particular case of a beam with ratio $U_m/U_a = 0.73$. It is interesting to note that the slope angle of the central electrode to the beam boundary is 45° whereas the characteristic value in the Pierce-type guns is 67.5°.

8.1.2 Focusing System with Periodic Potential Distribution

This is formed by a sequence of high-voltage 1 and low-voltage 2 electrodes (Fig. 8.3). The potential $U_f < U_a$, and in some cases $U_f = 0$. If the beam is matched with the focusing system, its radius periodically changes along the system with periodicity L.

Figure 8.4 shows the axial distribution of potential within the one period of the system determined by:

$$U_0 = U_{00} + U_m \cos 2\pi (z/L) ,$$

where U_{00} is the average value of potential, U_m is the amplitude of potential variation.

The figure also demonstrates the boundary trajectories for different values of U_m ($U_{m3} > U_{m2} > U_{m1}$). It is seen that with other things being equal, there exists an optimum value of U_m, which provides the beam symmetry at the entrance and exit of the focusing cell of the periodic system, in the case shown in Fig. 8.4 the optimum value is U_{m2}.

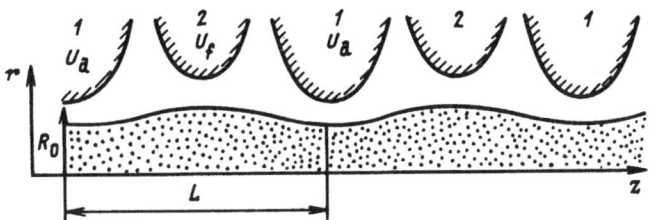

Fig. 8.3. Periodic electrostatic focusing systems. L – period of the system, R_0 – initial beam radius

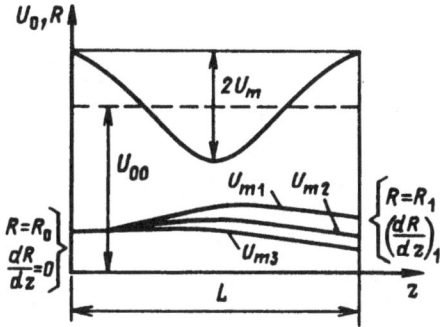

Fig. 8.4. Boundary trajectories for different values of U_m ($U_{m3} > U_{m2} > U_{m1}$)

Synthesis of Electrodes for the Periodic Focusing System

The following parameters are the input data for the synthesis: beam current I, the value of average potential U_{00}, initial beam radius R_o. For the synthesis, the paraxial equation of the boundary trajectory is used (see Sect. 4.3):

$$\frac{d^2R}{dz^2} + \frac{1}{2U_0}\frac{dU_0}{dz}\frac{dR}{dz} + \frac{1}{4U_0}\frac{d^2U_0}{dz^2}R = \frac{I}{4\pi\varepsilon_0 R\sqrt{2|e|/m}\,U^{3/2}}.$$

With the value of potential U_m chosen, numerical integration of this equation is made for one period of the system with the following initial conditions: $z = 0$, $R = R_0$, $(dR/dz)_0 = 0$.

The obtained results are estimated with respect to the symmetry of beam geometry and a new value of U_m is chosen. This procedure is repeated until the required degree of symmetry is achieved.

Finding the optimum value U_m can be done automatically with the help of one of the standard procedures of finding extremum of a criterial function, which expresses the deviation of output trajectory parameters R_1 and $(dR/dz)_1$ from input ones: R_0 and $(dR/dz)_0 = 0$:

$$\beta = \left[\left(\frac{R_0 - R_1}{L}\right)^2 + \left(\frac{dR}{dz}\right)_1^2\right]^{1/2}.$$

Full symmetry of the beam input and output parameters corresponds to a zero value of the criterial function ($\beta = 0$).

The form of electrodes of the system is found as a result of the solution of the exterior problem of synthesis. An approximate solution of the problem yields the following expression for the potential outside of the beam:

$$U = U_{00}\left[1 + \frac{U_m}{U_{00}}I_0\left(\frac{2\pi r}{L}\right)\cos\frac{2\pi z}{L} + \frac{0.0152}{\sqrt{U_0/U_{00}}}P\left(1 + 2\ln\frac{r}{R}\right)\right],$$

where $P = I/U_{00}^{3/2}$ – is the perveance of electron beam in $\mu A/V^{3/2}$, being calculated for average potential U_{00}.

This formula allows calculation of the family of equipotential lines in the exterior region ($r > R$) and determination of the form of electrodes for the periodic system. A typical form of electrodes found in this way is demonstrated in Fig. 8.3. Regular periodic motion with period L, shown in this figure, will be obtained when the parameters of the beam and conditions of its injection into the focusing system are strictly in accordance with the calculated ones. Otherwise, the character of the beam motion will be more complicated and beam ripple with a far longer period will be imposed on the regular beam motion (Fig. 8.5) [130].

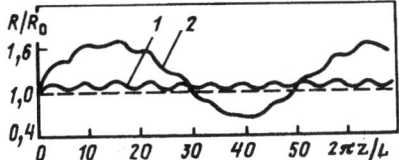

Fig. 8.5. Effect of initial conditions on beam-boundary trajectories: 1 – matched beam $(dR/dz)_0 = 0$; 2 – beam with positive slope of the boundary trajectory $(dR/dz)_0 = 0.05\,(2\pi R_0/L)$.

8.1.3 Periodic System of Apertured Disks

Periodical electrostatic focusing can be carried out by a system of apertured disks with alternating potentials (Fig. 8.6) [131]. The axial potential distribution in this system is analytically presented as a superposition of a number of space harmonics, with the period of the first harmonic being equal to that of the focusing system. In the system formed by disks of equal thickness the amplitude of the first harmonic considerably exceeds the amplitudes of higher ones. This allows us to neglect them as beam motion is being analyzed [132]. A detailed analysis of beam motion in the systems of this type including the problem of beam-focusing stability have been considered in [131, 132].

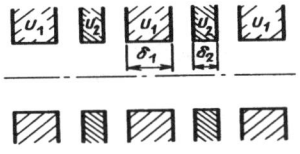

Fig. 8.6. Periodic system of apertured disks

8.1.4 System of Unipotential Electrostatic Lenses

A focusing system formed by a set of unipotential lenses and its equivalent scheme is presented in Fig. 8.7. Moving through the lens field, beam particles obtain the radial impulse directed toward the system axis. In between the lenses beam motion is determined by Coulomb repulsion. As a result of this, the initially convergent beam is compressed to a certain minimum radius $r_{\min}$ and then starts to spread until entering the field of the next lens. By adequately selecting the refracting lens power it is possible to provide the periodic beam motion and its transport for a long distance.

Lens parameters, including refracting power, depend on lens geometry and the potential being applied to the focusing electrode U_f. Lenses are usually designed to operate with zero potential of the focusing electrode $U_f = 0$, since in this case there is no need for a source of focusing voltage. With $U_f = 0$ the lens parameters (focal distance, coefficient of spherical aberration) are completely determined by lens geometry and can be found from the diagrams of Figs. 8.8 and 8.9.

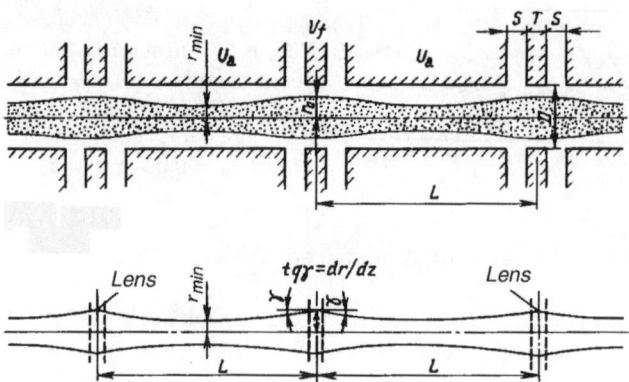

Fig. 8.7. System of unipotential electrostatic lenses and its equivalent scheme

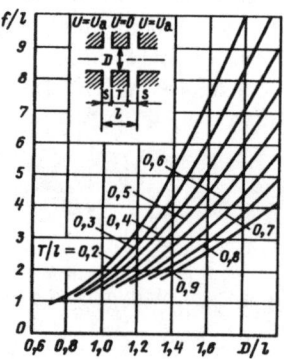 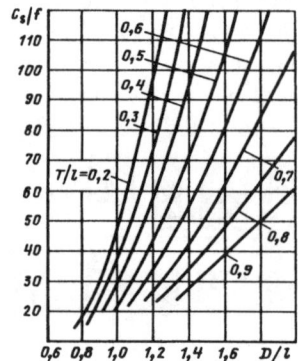

Fig. 8.8. Focal length as function of lens geometry for $U_f = 0$

Fig. 8.9. Coefficient of spherical aberration as function of lens geometry for $U_f = 0$

Designing of the lens focusing systems includes preliminary determination of lens geometry and the following beam-trajectory analysis and correction of lens dimensions. The basic relation connecting parameters of a particle beam with those of focusing lenses can be found, if we assume that lenses are thin and beam motion in between them is determined by Coulomb repulsion only.

When the beam perveance P, the beam radius in minimum beam cross section $r_{\min}$ and the period of lens system L are given, it is possible to find the beam radius and the slope of the boundary trajectory at the middle plane of the lens. The universal curve of beam spreading, considered in Sect. 4.5, can be used for this calculation as well as the following analytical expressions:

$$r_0 \approx r_{\min}\left(1 + 0,25Z^2 - 0,017Z^3\right)$$

8.1 Focusing System for Solid Axially Symmetrical Beams 205

$$Z = \left(\frac{P}{2\pi\varepsilon_0\sqrt{2|e|/m}}\right)^{1/2}\frac{L}{2r_{\min}}, \tan\gamma \left(\frac{P}{2\pi\varepsilon_0\sqrt{2|e|/m}}\right)^{1/2}\sqrt{\ln\frac{r_0}{r_{\min}}}.$$

For electron beams $\left(\frac{P}{2\pi\varepsilon_0\sqrt{2|e|/m}}\right)^{1/2} = 0.174\sqrt{P}$, where perveance $P = I/U_a^{3/2}$ is expressed in $\mu A/V^{3/2}$.

The angle of trajectory refraction, with spherical aberration being taken into account, is determined by (3.65)

$$\alpha' = \alpha + (C_s/f)\,\alpha^3\cos^2\alpha\,,$$

where f and C_s are the lens focal length and coefficient of spherical aberration $\alpha = r_0/f$ is the paraxial angle of refraction.

The required angle of boundary trajectory refraction is equal to 2γ. Its substitution in the previous equation yields:

$$2\gamma = \alpha + (C_s/f)\,\alpha^3\cos^2\alpha\,. \tag{8.4}$$

Using this equation and the diagrams of Figs. 8.8 and 8.9 it is possible to find the lens geometry, which provides regular periodical motion of a beam with prescribed parameters.

The calculation is performed in the following sequence. The angle of trajectory refraction 2γ and beam radius at the middle plane of the lens r_0 are calculated. Values of two of the lens dimensions, for instance T and l, are prescribed, the third one D is determined with the help of the diagrams of Figs. 8.8 and 8.9 to satisfy (8.4), connecting beam and lens parameters.

Example 8.1. The following beam parameters are given: beam current I = 0.5 A, beam voltage $U_a = 10^4$ V, focusing electrode voltage $U_f = 0$, minimum beam radius $r_{\min} = 4\,mm$, period of focusing system $L = 63$ mm. Values to be calculated:
Beam perveance $P = I/U_a^{3/2} = 0.5/\left(10^4\right)^{3/2} = 0.5\,\mu A/V^{3/2}$,

$$Z = 0.174\sqrt{P}L/2r_{min} = 0.174\sqrt{0.5}\times 63/(2\times 4) = 0.98\,,$$

$$r_0 = r_{min}\left(1 + 0.25\times 0.96 - 0.017\times 0.94\right) = r_{min}\times 1.224 = 4.9\text{ mm},$$

$$tg\gamma = 0.174\sqrt{P\ln\left(r_0/r_{\min}\right)} = 0.174\sqrt{0.5\ln 1.224} = 0.055$$

$$2\gamma \approx 0.11\,.$$

Lens dimensions S and T are chosen on the basis of design considerations, taking into account the value of field strength at the gap S: $S = 6.6$ mm $T = 2.7$ mm. Then, $l = 2 + T = 15.9$ mm, $T/l = 0.164$, $r_0/l = 0.308$.

The unknown value of the diameter of the aperture in the central lens electrode is determined by solution of (8.4). For convenience, it is expedient to rewrite it as:

$$\gamma = \frac{r_0/l}{f/l} + \frac{C_s}{f} \left(\frac{r_0/l}{f/l}\right)^3 \cos\left(\frac{r_0/l}{f/l}\right).$$

The left part of this equation is equal to $2\gamma = 0.11$, while the right part is a function of D/l. The solution of this equation consists of finding the value of D/l, which turns it into equality. It can be calculated by the trial and error method using the diagrams presented in Figs. 8.8 and 8.9. In the given case, D/l is found to be equal to $D/l = 1.25$ and, consequently, $D = 20$ mm.

The method described above allows us to find the initial geometry of the lenses. Since the method is an approximate one, the trajectory analysis is used to check the results obtained and make the necessary correction of the lens geometry to get the required symmetry of beam parameters at the entrance and the output of the lens.

Trajectory Analysis and Correction of Lens Geometry

The trajectory analysis involves solution of the self-consistent problem for intense charged particle beams as has been formulated in Chap. 4. It includes joint solution of equations of motion and field equations. The algorithm of this solution is described in Sect. 4.9. Results of the analysis are the map of electron trajectories and the phase curve of the beam at the exit of the lens. They are used for correction of lens geometry to get the symmetry input and output beam parameters. For quantitative estimation of the beam symmetry a criteria function β, characterizing the deviation of the output characteristic from the input one, is introduced:

$$\beta = \sum_k \alpha_k \left[\left(\frac{r_{k1} - r_{k0}}{L}\right)^2 + (r'_{k1} - r'_{k0})^2\right]^{1/2},$$

where r_{k0}, r'_{k0}, r_{k1}, r'_{k1} – radii and slopes of the k-th trajectory at the entrance and the exit of the lens, L is the period of the lens system, α_k is a weighting coefficient equal to a part of the total beam current I that carries the k-th trajectory $\alpha_k = I_k/I$.

Since, in the case in question the input phase curve is supposed to be the ideal one, that is $r'_{k0} = f(r_{k0}) = 0$, it is necessary to put in the above equation $r'_{k0} = 0$. When input and output beam parameters coincide the parameter β becomes zero.

Motion of electrons in the lens depends on the focusing electrode potential and geometry of lens electrodes. Therefore, β is an implicit function of these parameters and changes as they are varied. Figure 8.10 shows the variation of the output phase curve and parameter with the changing aperture D of the central electrode of the lens presented in Fig. 8.11. The nonlinearity of the output phase characteristics is explained by the strong effect of the spherical aberration of the lens. This effect does not allow the beam to be obtained with strictly symmetric input and output parameters. The minimum value of β and the best degree of the beam symmetry are obtained for the relative diameter

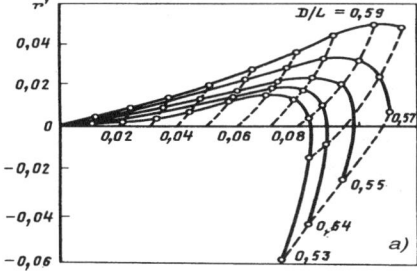
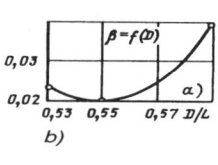

Fig. 8.10. Output beam phase curves obtained for different values of aperture in central lens electrode D/L (**a**) and corresponding variation of the criterial function β (**b**)

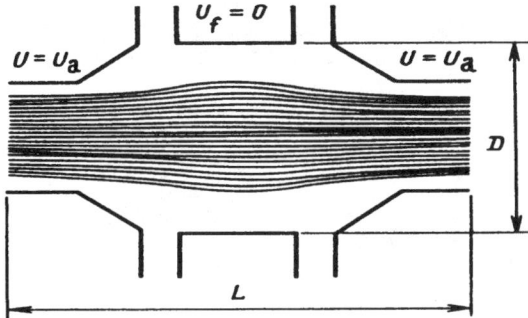

Fig. 8.11. Results of trajectory analysis of a single lens having the optimal value of $D/L = 0.55$

of the central electrode $D/L = 0.55$. Results of trajectory calculation for this case indicates a sufficient degree of beam symmetry (Fig. 8.11).

The process of lens-geometry correction is reduced to a search for the extreme value of β when the dimensions of the lens and the potential of its central electrode are changed.

After finding the geometry of a single lens of the trajectories analysis of the lens system is performed (Fig. 8.12). It is seen that while moving in the lens systems, the beam loses its initial laminar structure. This result is explained by the effect of spherical lens aberration.

Synthesis of Electrostatic Lenses for Periodic Beam Focusing

The following parameters are used as input data for lens synthesis: period of lens system L, initial beam radius R_0 and beam perveance P, estimated on the axial potential at the lens entrance U_{00} (Fig. 8.13). In the considered case the process of synthesis is an iterative one and performed as follows: an axial potential distribution in the lens is set up and the boundary trajectory of the beam is computed. The results of computation are estimated from the

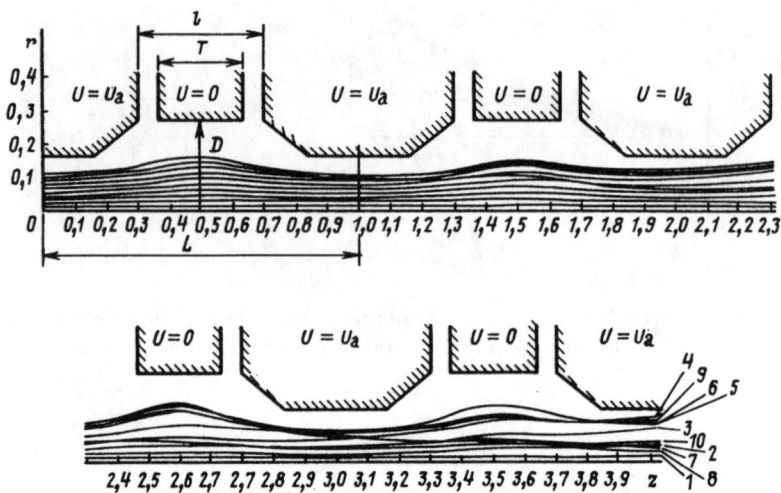

Fig. 8.12. Results of trajectory analysis of the periodic lens system

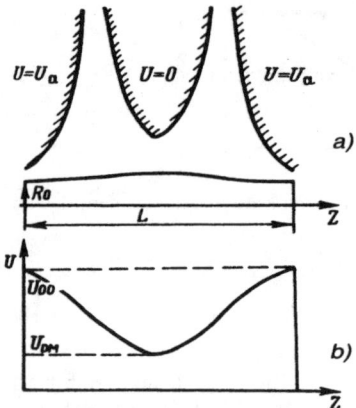

Fig. 8.13. Lens electrodes found by method of synthesis and axial potential distribution

point of view of beam symmetry and corresponding changes in the axial the potential distribution are made. This procedure is repeated until the required symmetry of the beam at the entrance and output is achieved.

The paraxial equation of the beam-boundary trajectory is used for the lens synthesis (see Sect. 4.3) When the potentials are normalized to U_{00} this equation is written as:

$$\frac{d^2R}{dz^2} + \frac{1}{2U}\frac{dU}{dz}\frac{dR}{dz} + \frac{1}{4U}\frac{d^2U}{dz^2}R = \frac{P}{4\pi\varepsilon_0\sqrt{2|e|/m}U^{3/2}R},$$

where U is the normalized potential, P is beam perveance calculated on the potential U_{00}.

8.1 Focusing System for Solid Axially Symmetrical Beams

To set the axial potential distribution of one may use its representation by splines [5]. For a unipotential lens the axial distribution is described by two cubic splines:

$$U = U_{00} + Az^2 + Bz^3 \quad 0 \le z \le L/2,$$

$$U = U_{0m} - A(z - L/2)^2 - B(z - L/2)^3 \quad L/2 \le z \le L.$$

In this equation coefficients A and B are expressed through values of the axial potential at the entrance and exit of the lens U_{00} and potential at the middle of the lens U_{0m}:

$$A = 12(U_{0m} - U_{00})/L^2, \quad B = 16(U_{00} - U_{0m})/L^3.$$

At the center of the lens, splines are coupled correctly to the second derivative.

With the value of U_{00} fixed, changing of the axial potential distribution is obtained by variation of U_{0m}.

The procedure of lens synthesis includes setting the axial potential distribution, numerical integration of the boundary trajectory equation and determination of the optimal value of potential U_{0m} that provides the best beam symmetry. The search for the optimal value of U_{0m} can be done automatically by searching for the extreme of a criterion function using some standard procedure. For the given case the criterial function is formulated as below:

$$\beta = \left[\left(\frac{R_0 - R_1}{L}\right)^2 + \left(\frac{dR}{dz}\right)_1^2\right]^{1/2}.$$

It shows a deviation of output parameters of the boundary trajectory R_1 and $(dR/dz)_1$ from input ones R_0 and $(dR/dz)_0 = 0$.

The geometry of the lens electrodes is found as a result of solution of the exterior problem of synthesis, which is formulated as the Cauchy problem for the Poisson equation, a method of its solution being considered in Sect. 2.3.

In the considered case, the solution has the following form:

$$U(r,z) = U - \frac{1}{4}\frac{d^2U}{dz^2}r^2 + \frac{0.0152}{\sqrt{U}}P\left(1 + 2\ln\frac{r}{R}\right),$$

where U is normalized potential, P is the beam perveance.

Figure 8.13 demonstrates an example of the typical geometry of lens electrodes, obtained by the method of synthesis with the spline interpolation of the axial potential distribution (beam perveance is $P = 0.5 \times 10^{-6} A/V^{3/2}$, ratio $R_0/L = 0.1$).

Stability of Beam Focusing by Lens Systems

The stability of a single particle motion in a system of thin ideal lenses, separated by drift spaces free from any fields including the space-charge field, has been investigated by Pierce who obtained the following condition of stability of motion [7]

$$f > L/4,\qquad(8.5)$$

where f is the paraxial focal length of the lenses, L is the period of the system.

When the condition (8.5) is fulfilled, the radial coordinate of the particle moving through the focusing system remains a limited (nonincreasing) value.

It is possible to obtain a more general condition of stability, if we use a more sophisticated model of the system "beam – channel – lens". In particular, the lenses are considered to be thick ones and the effect of space charge is taken into account in the linear approximation. For this a cloud of the space charge with constant density, which is equal to the average density of space charge in a real beam, is introduced in the space between the lenses. Analysis of stability, based on this model and the matrix formalism gives the condition of focusing stability [133]:

$$f > \frac{h\left(1+\mathrm{th}^2\varphi l\right) + \frac{l}{2}\left(1+h^2\varphi^2\right)\frac{\mathrm{th}\varphi l}{\varphi l}}{1 + h\varphi\,\mathrm{th}\varphi l},\qquad(8.6)$$

where h is distance, which determines the position of the lens principal plains (see Sect. 3.5); L is the period of the system, $2l = L - 2h$ is the space between lenses filled with a cloud of the space charge of constant density $\rho = \mathrm{const}$, φ is a parameter of the space charge determined by the relation:

$$\varphi^2 = \frac{I}{4\pi r_{ev}\varepsilon_0\sqrt{2\,|e|/m}\,U_a^{3/2}}\qquad r_{ev} = (r_{min}+r_0)/2,$$

where $r_{ev} = (r_{min}+r_0)/2$ is the assumed average radius of the beam. The condition of stability (8.6) is reduced to (8.5) for the case $h = 0$ and $\varphi = 0$.

8.2 Systems for Focusing of Strip Beams

8.2.1 Equilibrium Strip Beam

Electrostatic systems for focusing of an equilibrium strip beam can be designed by analogy with the focusing system for an axially symmetrical beam described in Sect. 8.1.1. The strip beam is cut from an infinite rectilinear space-charge flow in a planar diode.

Assuming the width of the strip beam is much more than its thickness, it is possible to consider the problem as a two-dimensional one. In this case, determination of the form of focusing electrodes is reduced to solution of the Cauchy problem for the two-dimensional Laplace equation $\partial^2 U/\partial y^2 + \partial^2 U/\partial z^2 = 0$ in the exterior region.

Then, using the approximate expression for potential distribution in diode space:

$$U = U_m + Cz^2 \qquad C = \frac{4}{d^2}(U_a - U_m),$$

and the additional condition at the beam boundary $\partial U/\partial y = 0$, one obtains the following result [134]:

$$U = U_m + \frac{2}{9}\frac{U_m}{C}\left(\frac{z}{d}\right)^2 (1 + \tan^2\gamma)\cos 2\gamma,$$

where $\tan\gamma = y/z$, U_m is the potential minimum in the middle plane ($z = 0$) of the diode space.

This equation allows us to calculate the equipotential surface and find the form of the focusing electrodes.

8.2.2 Periodic Electrostatic System

This system is analogous to the system used for the axially symmetric beam (Sect. 8.1.2) and is designed with the same algorithm. It includes determination of the amplitude of the axial potential variation and calculation of the potential distribution in the exterior region with the following approximate formula:

$$U = U_{00}\left[1 + \frac{U_m}{U_{00}}\cosh\frac{2\pi}{L}y\cos\frac{2\pi}{L}z \right.$$
$$\left. + \frac{P_1}{4\varepsilon_0\sqrt{2\frac{e}{m}U_0/U_m}}(Y + 2(y - Y))\right]$$

where U_{00} – average value of the axial potential U_m – amplitude of potential variation, P_1 – beam perveance per unit beam width calculated on the average potential U_{00}, L – period of the system, Y – coordinate of the boundary trajectory.

8.3 Focusing Systems for Hollow Beams

8.3.1 Periodic Electrostatic Focusing

There is a peculiarity of hollow-beam focusing. Spatial measures are to be used in order to provide the balanced motion of the particles belonging to the inner beam boundary, since there exist no outward forces to counteract the periodic electrostatic focusing forces.

Several methods are used to get the balanced motion of the inner charged particles: injection in the focusing system of a rotating hollow beam

212   8 Electrostatic Focusing Systems

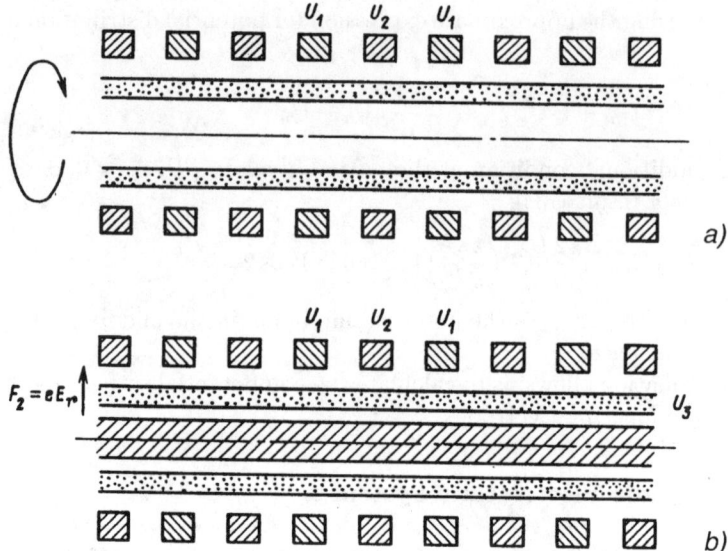

Fig. 8.14a,b. System of periodic focusing of hollow beams

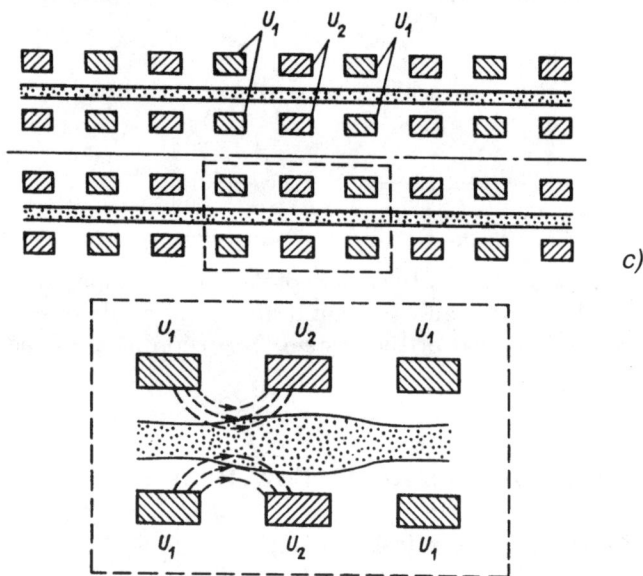

Fig. 8.15. System of bilateral periodic focusing of hollow beams

(Fig. 8.14a), creation of an additional radial electrostatic field (Fig. 8.14b), and use of a bilateral periodic focusing system (Fig. 8.15) [136–138].

In the first case, steady motion of inner electrons is obtained due to a balance (on average) between the periodic focusing and centrifugal forces. In the second case an additional radial defocusing force is created to counteract the periodic focusing force. The process of hollow-beam focusing in the bilateral periodic focusing system corresponds to that of periodic strip beam focusing.

8.3.2 Centrifugal Electrostatic Focusing

Centrifugal focusing of a rotating hollow beam is performed in the field of a cylindrical diode (Fig. 8.16) [30, 31]. The initial beam rotation is created by special guns. The focusing is a result of a balance of the centrifugal force and electrostatic field force:

$$m\left(r\dot\theta\right)^2/r = |eE_r|.$$

Here, E_r is the radial electrostatic field of the diode $|E_r| = U_f/r \ln \frac{b}{a}$, U_f is applied focusing voltage, a and b are radii of the inner and outer electrodes.

For a particle moving in the field of a cylindrical diode its angular momentum $p_\theta = mr^2\dot\theta$ is conserved $p_\theta = $ const. Then, the above equation of the balance of the forces can be written as:

$$\frac{p_\theta^2}{(mr^3)} = |e|\,U_f/r \ln \frac{b}{a}.$$

When the initial value of $p_\theta = p_{\theta 0}$ is known and the beam radius is given this equation allows us to determine the required value of the focusing potential U_f.

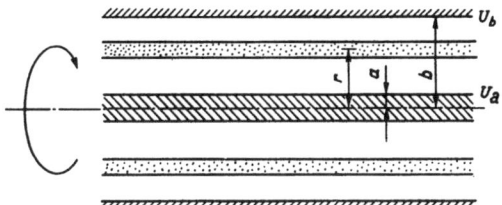

Fig. 8.16. Centrifugal hollow-beam focusing

9 Optical Systems of Technological Installations

9.1 Electron Probe of Welding Installation

A charge particle probe is considered to be a narrow charge particle beam of high brightness, high current or power density in its focus spot. Electron probes with power density in the range $(0.5\text{--}10) \times 10^6$ W/cm^2 are used for beam welding.

A typical electron-beam welding system is shown in Fig. 9.1. Electron gun (1) provides the initial electron-beam formation and acceleration. Magnetic lens (2) focuses the beam in a small spot to provide beam power density in the range $5\text{--}10^2$ kW/mm^2. Deflecting system (3) is used to change the beam spot position. Typical values of accelerating voltage and beam current lay in the following intervals: $U = 20\text{--}150$ kV, $I = 0.1\text{--}1$ A.

The following main factors determine the minimal size of focal spot in electron-beam welding installations:

- initial (thermal) electron velocities;
- aberrations of optical system;
- self-fields;
- scattering of electrons by metal vapor arising in the welding process.

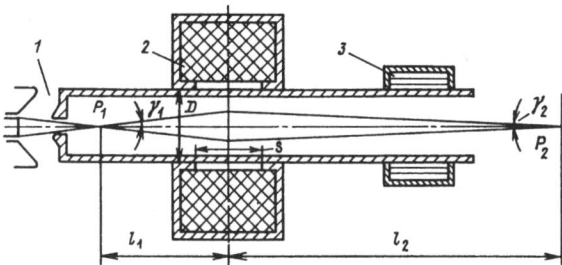

Fig. 9.1. Scheme of an electron-beam welding installation: 1 – electron gun, 2 – magnetic focusing lens, 3 – deflecting system

9.1.1 Effect of Thermal Velocities

The minimal radius of a focus spot resulting from thermal velocities can be estimated with the help of the equation for the electron-beam emittance. Emittance of an electron gun with a thermal cathode is expressed by the equation (see Appendix):

$$\varepsilon_p = C_p r_k \left(kT_k/|e|U_a\right)^{1/2},$$

where r_k – cathode radius, T_k – cathode temperature, U_a – accelerating voltage, C_p – coefficient equal to 1 or $\sqrt{2}$, depending on conditions of emittance determination (see Appendix).

Taking into account the law of emittance invariance one obtains the following equation for radius of the focus spot resulting from the effect of thermal velocities:

$$r_T = \varepsilon/\gamma_2 = C_p r_k \left(\frac{kT_k}{|e|U_a}\right)^{1/2} \frac{1}{\gamma_2}, \tag{9.1}$$

or

$$r_t = C_p r_k \left(\frac{U_T}{U_a}\right)^{1/2} \frac{1}{\gamma_2}, \tag{9.2}$$

where $U_T = kT_k/|e| = T_k/11600$, γ_2 – angle of beam convergence.

It is evident that the value of r_T is decreased with increasing U_a and γ_2.

9.1.2 Aberrations of Optical System

It is considered that the main contribution in increasing the size of the focal spot is introduced by spherical aberration of the magnetic lens. Along with this, as analysis shows, nonlinearly of the gun phase curve can additionally increase the radius of the focal spot.

In the case of an aberration-free lens all rays leaving a point A are focused in the image point B (Fig. 9.2).

If the lens has spherical aberration the peripheral rays are refracted more strongly. The additional angle of refraction is determined by the equation:

$$\Delta\gamma = r_0^3 C_\alpha/f, \tag{9.3}$$

where f – paraxial focal distance, C_α – angle coefficient of spherical aberration.

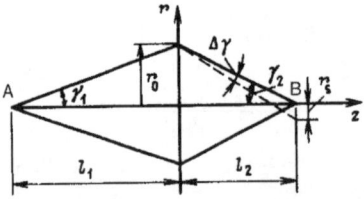

Fig. 9.2. Effect of lens spherical aberration

9.1 Electron Probe of Welding Installation

Typical values of C_α for shielded magnetic solenoids lay in the range $C_\alpha = 0.01$–0.05 cm^{-2}.

The radius of a focusing spot in the plane of paraxial focus is found as

$$r_s = \Delta \gamma l_2 / \cos^2 \gamma_2 \approx \Delta \gamma l_2 \ .$$

In the plane of best focus the radius of the focus spot reaches its minimum:

$$r_s = \frac{1}{4}\Delta \gamma l_2 = \frac{1}{4} r_0^3 \frac{C_\alpha l_2}{f} \ .$$

Taking into account that $1/f = 1/l_1 + 1/l_2$ and $r_0 \approx l_2 \gamma_2$ one finds:

$$r_s = \frac{1}{4} l_2^3 \gamma_2^3 C_\alpha \left(1 + \frac{l_2}{l_1}\right) \ . \tag{9.4}$$

Assuming, approximately, that the total radius focus spot is determined as $r = (r_T^2 + r_s^2)^{1/2}$ one obtains:

$$r = \left\{ \left[r_k \left(\frac{U_T}{U_a}\right)^{1/2} \frac{1}{\gamma_2} \right]^2 + \frac{1}{4} \left[l_2^3 \gamma_2^3 C_\alpha \left(1 + \frac{l_2}{l_1}\right) \right]^2 \right\}^{1/2} \ .$$

As is seen, the value of r is a complicated function of angle γ_2. It is possible to find its optimal value that yields a minimum of r. Omitting intermediate calculations we obtain:

$$r_{\min} = \left(\frac{4}{3}\right)^{1/2} r_k \left(\frac{U_T}{U_a}\right)^{1/2} \frac{1}{\gamma_{2\,\mathrm{opt}}} \ ,$$

where $\quad \gamma_{2\,\mathrm{opt}} = \left(\frac{1}{3}\right)^{1/2} \left(\frac{4}{3} r_k \sqrt{\frac{U_T}{U_a}} \frac{1}{K}\right)^{1/4} , \quad K = \left(1 + \frac{l_2}{l_1}\right) C_\alpha l_2^3 \ .$

Aberrations of an electron gun result in a nonlinear phase curve of the beam at the exit plane of the gun. It is known that a beam having a nonlinear phase curve can not be focused at a point spot. In this case, the radius of the focus spot can be found by using the concept of transverse phase space and the method of a phase parallelogram [143].

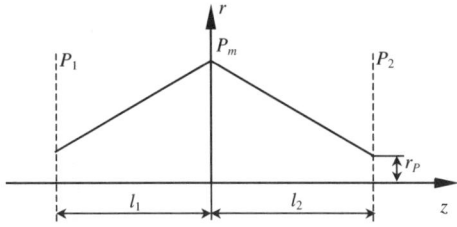

Fig. 9.3. For calculation of the focus spot resulting from nonlinearly of phase curve

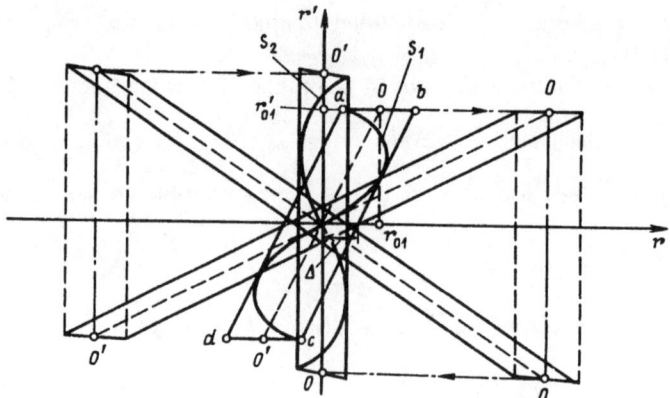

Fig. 9.4. Method of phase parallelogram

Let us suppose that the beam phase curve at the exist of a gun at plane P_1 (Fig. 9.3) has the form S_1 (Fig. 9.4). Such a shape of the phase curve corresponds to a nonlaminar diverging beam. Let us consider how this curve will be transformed by a converging thin lens for plane P_2 located at distance l_2 from the lens.

In accordance with the method of the phase parallelogram phase curve S_1 is enclosed in parallelogram $abcd$, with its upper and lower sides being lines of $r' = $ const. Lateral sides are drawn in such a way as to provide the minimum of parallelogram area as is shown at Fig. 9.4. At any linear transformation the phase curve will remain inside this parallelogram and the problem is reduced to analysis of its transformation.

At the interval between plane P_1 and the middle plane of the lens P_m electrons move in the field-free region, their trajectories being stright lines and phase coordinates r' of parallelogram boundary points remain constant. As a result of this parallelogram sides ab and cd shift parallel with the r-axis, the point of intersection of lateral parallelogram sides ad and bc with the r-axis do not change their position. The initial phase parallelogram is transformed into a new one, as shown in Fig. 9.4, with its area being equal to that of the initial parallelogram $A_1 = |r'_{01}| 4\Delta$.

At the middle plane P_m, due to lens action, the r'-coordinates of phase points are changed by a jump in the value r/f, where f is the lens focusing distance, whereas the r-coordinates remain unchanged as the lens is assumed to be thin. In the space after the lens the phase parallelogram is transformed in such a way that the r'-coordinates of its vertices remain invariable. The radius of the beam spot at the exit plane P_2 will be minimum, if the large axis of the parallelogram coincides with the vertical coordinate axis r' and lateral sides ad and bc are parallel to it.

When the input phase curve S_1 is given as well as the base distances l_1 and l_2, it is possible to find the lens power $1/f$ that provides a minimum radius of beam spot and to determine its value [143]:

$$\frac{1}{f} = \frac{1}{l_2} + \frac{C}{1+Cl_1}, \quad r_p = \Delta \frac{Cl_2}{1+Cl_1},$$

where C is a constant, which determines the slope of the large axis OO' of the initial parallelogram, $C = r'_{01}/r_{01}$, Δ – its half-width.

Computation of spot sizes of different electron-beam welding systems shows that the value of spot size resulting from nonlinearly of the gun phase curve is the same order as that arising due to spherical aberration of magnetic lenses.

The resulting beam spot size r is accepted for calculating by the following ate equation:

$$r = \left(r_T^2 + r_s^2 + r_p^2\right)^{1/2}.$$

There is no strict proof of this equation, but it gives results reasonably close to practice. The criteria and principles of optimization of the concentrating optical systems can also be found in [139]

9.1.3 Effects of Self-Fields and Scattering of Electrons

As shown in [140] the electron space charge is neutralized by ions created as a result of ionization of atoms of residual gas and metal vapor by beam electrons. Therefore, the total electrostatic self-field is absent practically everywhere (with the exception of the gun region) and does not affect beam focusing. As for a self-magnetic field, it produces an additional focusing action, especially at high accelerating voltage. It can be taken into account by using the results of Sect. 4.6.

The effect of electron-beam scattering by atoms of metal vapor arising in the welding process appears only in the vicinity of the welded joint. Its estimation is given in [32].

9.2 Optical System for Electron-Beam Lithography

The following parameters of electron beams are used in these systems: accelerating voltage $U_a = 15 - 25$ kV, beam currents $I = 0.1$–1 μA radius of focusing spot $r = 0.05$–0.25 μm. Such electron beam probes are produced by complicated optical systems, which include electron guns with thermal or cold field emitters, a set of magnetic lenses and beam-deflecting systems.

9.2.1 System with Thermal Cathode

The gun incorporates a lanthanum hexaboride emitter (1) heater (2), focusing electrode (3) and anode (4) (Fig. 9.5). The beam current is emitted from the tip of the emitter, with the effective radius r_k being equal to 5–10 μm.

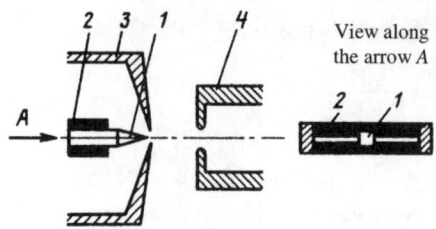

Fig. 9.5. Electron gun with thermal cathode: 1 – cone cathode, 2 – heater, 3 – focusing electrode, 4 – anode

The radius of the beam spot after the first magnetic lens in plane P_2 (Fig. 9.6), can be found with the help of (9.2), with C_p being equal to unity, $C_p = 1$:

$$r_2 = r_k \left(\frac{U_T}{U_a}\right)^{1/2} \frac{1}{\gamma_2},$$

where γ_2 – angle of beam convergence after the first magnetic lens.

Diaphragm (3) cuts out a part of the beam, its angle of divergence is decreased and becomes equal to γ_2^*. Then, after the second magnetic lens in plane P_3 the radius of the beam spot will be:

$$r_3 = r_2 \gamma_2^* / \gamma_3.$$

Objective lens (6) focuses the beam to the object plane P. The total radius of the beam spot at this plane is determined as:

$$r = \left(r_T^2 + r_S\right)^{1/2},$$

where r_T – spot radius resulting from the initial thermal velocities, $r_T = r_3 \gamma_3^* / \gamma_4$, r_S – spot radius arising due to spherical aberration of the objective lens.

The last equation does not contain parameter r_P, which takes into account the nonlinearity of the electron-gun phase curve. This is explained in the following way. Diaphragms (3) and (5) used in the optical system cutoff the peripheral part of the electron beam and the final focal spot is formed only by its central part, where the phase curve is practically a linear one.

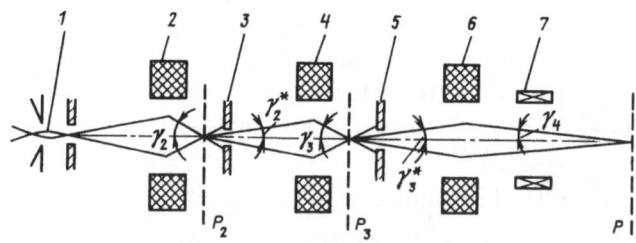

Fig. 9.6. Electron-optical system for electron-beam lithography: 1 – electron gun, 2, 4, 6 – magnetic lenses, 3, 5 – cutting out diaphragms, 7 – deflecting system

The value of current of the formed electron probe can be expressed as:

$$I_3 = \pi^2 r_T^2 \gamma_4^2 B_0 ,$$

where B_0 is brightness of the electron source (electron gun). For guns with point lanthanum hexaboride emitters it is equal to $(1-5)\,10^6$ A/cm^{2}sr.

9.2.2 System with Cold Field Emitter

The system consists of an electron gun, magnetic focusing lens and deflecting system (Fig. 9.7). The electron gun includes a cold field emitter (1) and two anodes (2), (3), which have different potential U_1 and U_2 (usually $U_1 \approx 5$ kV, $U_2 = 20\text{--}25$ kV). This version of the gun is known as a Crewe–Butler gun [145, 146]. It provides the initial formation and acceleration of the electron beam. Magnetic lens (4) focuses it to a final spot. In this gun, electrons are emitted from a fine tip, with the apparent radius of emitter being $\sim 10^{-3}$ µm. Practically, this permits the emitter to be considered as a point one.

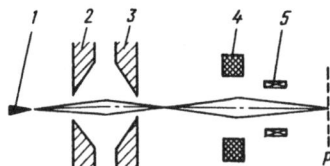

Fig. 9.7. Optical system with cold field emitter: 1 – field emitter, 2, 3 – first and second anodes, 4 – magnetic lens, 5 – deflecting system

An electrostatic immersion lens created by the anodes 2 and 3 has considerable spherical aberration. This results in a nonlinear phase curve of the beam formed by the gun. The final size of the focus spot will be determined by two factors: nonlinearly of the beam phase curve and spherical aberration of the magnetic lens.

It can be expressed as:

$$r = \left(r_s^2 + r_p^2\right)^{1/2} .$$

This equation does not include radius r_T as the emitter is considered to be a point one. Systems with a cold field emitter allow us to form electron probes with radius of order of 0.1 µ/A and higher values of currents as compared with systems using thermal cathodes [145].

9.3 Ion Probes

Liquid-metal ion sources are used in ion-probe-forming systems to obtain probes of high quality with a probe diameter of order of (0.1–1) µm and currents lying in the interval (0.1–10) nA [145, 147].

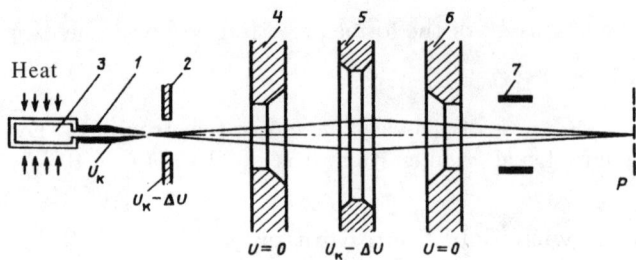

Fig. 9.8. Scheme of optical system for ion-probe formation: 1 – liquid-metal emitter, 2 – extractor, 3 – chamber with molten metal, 4 – accelerating electrode, 5, 6 – electrodes of electrostatic lens, 7 – deflecting system

An ion-probe-forming system with a liquid-metal emitter is shown in Fig. 9.8. It consists of a triode ion gun, electrostatic unipotential lens and deflecting system. The ion gun includes the liquid-metal emitter (1), extractor (2) and accelerating electrode (4). Approximate values of gun electrode potentials are $U_k \approx 10^5$ V, $\Delta U \approx 5 \times 10^3$ V.

Molten metal (for instance, gallium) exists in the externally heated chamber (3). From this chamber it proceeds along a tungsten capillary tube (1), a liquid-metal meniscus being created at its end. Under the action of electrostatic forces and forces of surface tension the meniscus acquires the form of a fine cone. Ions of the metal are emitted from its tip under the action of the field of extractor electrode (2) and accelerated by electrode (4), then they are focused by the unipotential electrostatic lens.

Parameters of the ion beam formed by the ion gun are as follows: current density of order of 10^5 A/cm^2, brightness 10^6 A/cm^2 sr, total current (0.01–10) μA. The dimension of the focus spot is mainly determined by chromatic aberration of the optical system, with its value being equal to (0.1–1) μm [147].

10 Intense Relativistic Charged-Particle Beams

10.1 Relativistic Equations of Motion

One of the effects of the special theory of relativity is the dependence of mass of a moving particle on particle velocity, which is expressed as:

$$m = m(v) = \frac{m_0}{\sqrt{1 - v^2/c^2}} = \gamma m_0 , \qquad (10.1)$$

where m_0 – is rest mass of the particle, v – its velocity, $\gamma = 1/\sqrt{1 - v^2/c^2}$ – relativistic factor.

In this case, the relativistic equation of particle motion is to be written as:

$$\frac{d}{dt}(m\vec{v}) = e\vec{E} + e(\vec{v} \times \vec{B}) . \qquad (10.2)$$

The left part of this equation can be rewritten as:

$$\frac{d}{dt}(m\vec{v}) = m\frac{d\vec{v}}{dt} + \vec{v}\frac{dm}{dt} . \qquad (10.3)$$

As

$$\frac{dm}{dt} = m_0 \frac{d}{dt}(1 - v^2/c^2)^{-1/2} = \frac{m_0 v}{c^2(1 - v^2/c^2)^{3/2}} \frac{dv}{dt} . \qquad (10.4)$$

Then (10.3) can be presented in the following form:

$$\frac{d}{dt}(m\vec{v}) = \frac{m_0}{(1 - v^2/c^2)^{1/2}} \frac{d\vec{v}}{dt} + \vec{v} \frac{m_0 v}{c^2(1 - v^2/c^2)^{3/2}} \frac{dv}{dt} . \qquad (10.5)$$

Substitution of (10.5) in (10.2) yields:

$$\frac{m_0}{(1 - v^2/c^2)^{1/2}} \frac{d\vec{v}}{dt} + \vec{v} \frac{m_0 v}{c^2(1 - v^2/c_2)^{3/2}} \frac{dv}{dt} = e\vec{E} + e(\vec{v} \times \vec{B}) . \qquad (10.6)$$

Multiplying the left and right parts of this equation by vector $\vec{v}$ one finds:

$$\frac{m_0 v}{c^2(1 - v^2/c^2)^{3/2}} \frac{dv}{dt} = \frac{e}{c^2}(\vec{v} \cdot \vec{E}) . \qquad (10.7)$$

Substitution of the last equation in (10.6) gives:

$$m_0 \gamma \frac{d\vec{v}}{dt} = e\vec{E} + e(\vec{v} \times \vec{B}) - \frac{e\vec{v}}{c^2}(\vec{v} \cdot \vec{E}) . \qquad (10.8)$$

This equation is considered to be the relativistic equation of motion written in Newton's form [59].

10.1.1 Law of Energy Conservation

For particle motion in static electric and magnetic fields there exists the law of total energy conservation.

Multiplying (10.8) by vector $\vec{v}$, after some algebra one obtains:

$$\frac{m_0 v}{(1-v^2/c^2)^{1/2}} \frac{dv}{dt} = -e\left(1 - \frac{v^2}{c^2}\right)\frac{dU}{dt},$$

or, taking into account (10.4),

$$\frac{d}{dt}(mc^2) = -e\frac{dU}{dt}.$$

This equation yields the following law of total energy conservation:

$$E = m_0 \gamma c^2 + eU = \text{const}. \tag{10.9}$$

Assuming the potential of a particle emitter and initial particle velocity equal zero one finds that this constant is equal to $m_0 c^2$. Then (10.9) is reduced to:

$$m_0 c^2 (\gamma - 1) = -eU, \tag{10.10}$$

where $-eU > 0$.

The left part of this equation is the kinetic energy of the particle $T = m_0 c^2 (\gamma - 1)$.

Using the above equations it is possible to deduce the following equation for particle velocity:

$$v = \sqrt{2\frac{|eU|}{m_0}}\left(\frac{\gamma+1}{2}\right)^{1/2}\frac{1}{\gamma}, \tag{10.11}$$

where $\gamma = 1 + |eU|/m_0 c^2$.

For electrons, these are written as:

$$v = 5.95 \times 10^7 \sqrt{U}\left(\frac{\gamma+1}{2}\right)^{1/2}\frac{1}{\gamma}, \qquad \gamma = 1 + 1.96 \times 10^{-6} U,$$

where v and U are expressed in cm/s and V.

10.1.2 Equations of Motion in Axially Symmetry Fields

These equations can be derived with the help of the Lagrange equations (3.8)

$$\frac{d}{dt}\left(\frac{\partial L}{\partial \dot{q}_i}\right) - \left(\frac{\partial L}{\partial q_i}\right) = 0 \qquad (i = 1, 2, 3),$$

and the relativistic Lagrange function written in cylindrical coordinates r, θ, z [2,60]:

$$L = -m_0 c^2 \left[1 - \frac{1}{c^2}\left(\dot{r}^2 + r^2\dot{\theta}^2 + \dot{z}^2\right)\right]^{1/2}$$
$$-eU + e\left(\dot{r}A_r + r\dot{\theta}A_\theta + \dot{z}A_z\right).$$

Omitting intermediate transforms, it is possible to write the following system of equations:

$$\frac{d^2r}{dr^2} = -\frac{e}{\gamma m_0}\left[\frac{\partial U}{\partial r} - B_z r\dot{\theta} - \frac{\gamma m_0}{e} r\dot{\theta}^2 + B_\theta \dot{z} - \left(\dot{r}\frac{\partial U}{\partial r} + \dot{z}\frac{\partial U}{\partial z}\right)\frac{\dot{r}}{c^2}\right], \quad (10.12)$$

$$\frac{d^2z}{dt^2} = \frac{e}{\gamma m_0}\left[\frac{\partial U}{\partial z} + B_r r\dot{\theta} - B_0 \dot{r} - \left(\dot{r}\frac{\partial U}{\partial r} + \dot{z}\frac{\partial U}{\partial x}\right)\frac{\dot{z}}{c^2}\right], \quad (10.13)$$

$$r^2\dot{\theta} - r_0^2 \dot{\theta}_0 = -\frac{e}{\gamma m_0}(rA_\theta - r_0 A_{\theta 0}). \quad (10.14)$$

In these equations $r_\theta^2 \dot{\theta}_0$ – initial angular momentum, $A_{\theta 0}$ – azimuthal component of vector magnetic potential in an initial point having initial radial coordinate r_0. Scalar electrostatic and vector magnetic potentials were assumed to be independent of the azimuthal coordinate $\frac{\partial A}{\partial \theta} = \frac{\partial U}{\partial \theta} = 0$ while deriving these equations.

10.2 Intense Relativistic Beams in Vacuum Channels

10.2.1 Electron Beam Spread under Self-Fields Action

Let us consider divergence of an axially symmetric laminar electron beam moving in a vacuum channel free from external fields. In this case, beam motion will be determined only by beam self-fields, that is, the radial component of the electrostatic field $E_{\rho r}$ created by the space charge and azimuthal magnetic self-field $B_{s\theta}$ produced by the beam current.

Allowing for these assumptions the relativistic equations of electron motion are written as:

$$m_0\gamma\frac{dv_r}{dt} = eE_{\rho r} - ev_z B_{S\theta} - \frac{ev_r^2 E_{\rho r}}{c^2} \quad (10.15)$$

$$m_0\gamma\frac{dv_z}{dt} = ev_r B_\theta - \frac{ev_r v_z E_{\rho r}}{c^2} \quad (10.16)$$

$$r^2\dot{\theta} = 0. \quad (10.17)$$

Limiting consideration to laminar beams we can analyze the motion of beam-boundary electrons. The parameters $E_{\rho r}$ and B_θ can be expressed in terms of beam parameters in the following way:

$$E_{\rho r} = -\frac{I}{2\pi\varepsilon_0 r v_z}, \quad B_\theta = -\frac{\mu_0 I}{2\pi r},$$

where I – beam current, r – beam radius.

Substitution of these equations in (10.15) and (10.16) yields:

$$m_0 \gamma \frac{dv_r}{dt} = \frac{|e| I}{2\pi\varepsilon_0 r v_z \gamma^2} \tag{10.18}$$

$$\frac{dv_z}{dt} = 0. \tag{10.19}$$

Equation (10.18) shows that taking account of the self-magnetic field leads to decreasing the defocusing force by γ^2 times.

Expressing the longitudinal velocity through the accelerating potential U_a

$$v_z \approx v = \sqrt{2\frac{|e| U_a}{m}} \left(\frac{\gamma+1}{2}\right)^{1/2} \frac{1}{\gamma},$$

and excluding time in (10.18) one derives the equation of boundary electrons:

$$\frac{d^2 r}{dz^2} = \frac{K}{r \left(\frac{\gamma+1}{2}\right)^{3/2}}, \tag{10.20}$$

where $K = \frac{1}{4\pi\varepsilon_0 \sqrt{2|e|/m}} \frac{I}{U_a^{3/2}} = 1.5 \times 10^{-2} P$, P being the microperveance of the electron beam expressed in $\mu A/V^{3/2}$.

Comparing (10.20) with the nonrelativistic trajectory equation (4.30) one finds that they differ only by a factor $1/\left(\frac{\gamma+1}{2}\right)^{3/2}$. This allows us to apply some results of the analysis of (4.30) for the case under consideration. In particular, introducing the effective beam perveance $\overset{*}{P} = P \left(\frac{2}{\gamma+1}\right)^{3/2}$ and normalized variable $Z = 0.174 \sqrt{\overset{*}{P}} \frac{z}{r_{min}} = 0.174 \sqrt{P \left(\frac{2}{\gamma+1}\right)^{3/2}} \frac{z}{r_{min}}$, we can use universal curves, presented in Fig. 4.4, for computation of the relativistic beam spread. Results of such a computation are shown in Fig. 10.1. It is seen that the spread of a beam of constant perveance $P = 1 \ \mu A/V^{3/2}$ depends on the accelerating voltage, decreasing with its growth. This effect is explained by the focusing action of the self-magnetic field.

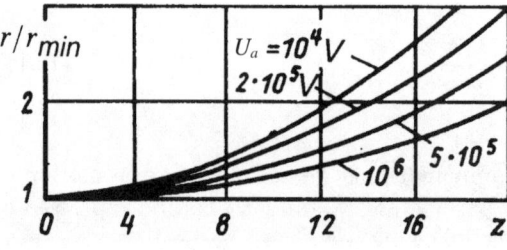

Fig. 10.1. Beam-boundary trajectories of a beam of perveance $P = 1 \ \mu A/V^{3/2}$ for different accelerating voltages U_a

10.2.2 Beam Current Limitation

There exists a limit of electron beam current, transmitted through a cylindrical channel. This limit is caused by depression of the potential due to beam space charge and the creation of a "virtual" cathode (see also Chap. 4).

For relativistic electron beams the value of critical current is determined by the following semiempirical equation, which is known as the Bogdankevich–Rukhadze equation [13, 148]:

$$I_{cr}^{BR} = \frac{17 \times 10^3 \left(\gamma_a^{2/3} - 1\right)^{3/2}}{G}, \qquad (10.21)$$

where I_{cr}^{BR} – critical beam current, $G = 1 + 2\ln\frac{a}{b}$, b – beam radius, a – channel radius, γ_a – relativistic factor, expressed through the accelerating voltage $\gamma_a = 1 + (1.96 \times 10^{-6})U_a$.

A more exact equation, obtained by Dgenny and Procktor is written as [13]:

$$I_{cr}^{DP} = I_{cr}^{BR}\gamma_a^{2/3}\left\{\left[\left(\gamma_a^{2/3} + G\right)^2 - \gamma_a^{2/3}\right]^{1/2} - G\right\}^{-1}. \qquad (10.22)$$

Results of computation of critical perveance $P_{cr} = I_{cr}/U_a^{3/2}$ are shown in Fig. 10.2. Curve 1 is computed with (10.21), curve 2 with (10.22).

As is seen, the results of the two computations differ insignificantly, both curves showing decreasing critical perveance with accelerating voltage growth. This effect can be explained by the increasing relativistic mass of electrons.

It is important to note that (10.21) and (10.22) are valid for cylinders of infinite length. For cylinders of a finite length the value of the critical current is increased compared to that given by (10.21) and (10.22). However, when condition $l/a \geq 2.58\,(b/a)^{0.133}$ is fulfilled this excess is not more than 10% [149].

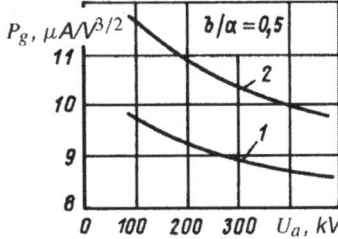

Fig. 10.2. Results of computation of the critical perveance

Dynamic Model of Virtual Cathode

The virtual cathode is considered to be an unsteady dynamic state of the electron beam [150]. A description of this state can not be given within the hydrodynamic theory of electron beams. Discrete models of charge particle beams (sheet charge model, ring charge model, etc.) are used for analysis of this state [150–154].

Computations produced on the basis of discrete models demonstrate the following picture of formation of the virtual cathode when an electron beam is injected in a cylindrical cavity [153, 154]. If the value of beam current is less than the critical one the potential inside the cavity decreases gradually as discrete charges fill it. The steady state electron motion is established during a time span equal approximately to two times the transit time of electrons through the cavity. A potential minimum is created at the central plane of the cavity. Its depth depends on the value of the injected current (Fig. 10.3 curves I_1 and I_2 The value of the potential minimum is decreased with increasing injected current.

For values of beam current greater than the critical one the picture is entirely different. As a beam is injected in the cylindrical cavity the potential inside it drops to zero or even becomes negative in the vicinity of the injecting plane (Fig. 10.3, curve I_3). The value of the potential minimum is periodically changed in time with a frequency of the order of the plasma frequency of the beam. A dynamic state of the injected beam is established, when a part of the beam current is rejected from the potential minimum whereas the other part proceeds to the end of the cavity.

Oscillations of the virtual cathode can be used for microwave generation [155–158]. Theoretical and experimental results indicate that there exist two types of microwave generation, one is produced by the virtual-cathode oscillation, the another is caused by electron oscillation in the gap "cathode–anode–virtual cathode" (oscillation of Bargauzen–Kurtz type) [159].

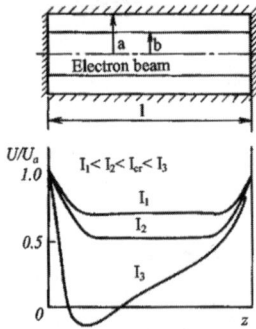

Fig. 10.3. Injection of an electron beam in a cylindrical channel: curves of potential distribution for different beam currents.

10.2.3 Brillouin Relativistic Electron Beam

The conditions of obtaining a nonrelativistic equilibrium beam in a uniform magnetic field have been considered in Sect. 7.1. For a relativistic beam it is necessary to take into account the action of the self-magnetic field.

The vector of magnetic induction of the beam of a self-field is expressed by the Maxwell equation

$$\operatorname{rot} \vec{B}_s = \mu_0 \vec{j},$$

where $\vec{j}$ – vector of beam current density.

Projection of this equation on cylindrical coordinates r, θ, z yields:

$$\frac{1}{r}\frac{\partial B_{sz}}{\partial \theta} - \frac{\partial B_{s\theta}}{\partial z} = \mu_0 j_r \tag{10.23}$$

$$\frac{\partial B_{sr}}{\partial z} - \frac{\partial B_{sz}}{\partial r} = \mu_0 j_\theta \tag{10.24}$$

$$\frac{1}{r}\frac{\partial}{\partial r}(rB_{s\theta}) - \frac{\partial B_{sr}}{\partial \theta} = \mu_0 j_z \tag{10.25}$$

In the case of an equilibrium cylindrical beam $dr/dt = 0$, $r = $ const, $\partial/\partial z = 0$, $\partial/\partial\theta = 0$. Then (10.23)–(10.25) are reduced to:

$$\frac{1}{r}\frac{\partial}{\partial r}(rB_{s\theta}) = \mu_0 J_z \qquad -\frac{\partial B_{sz}}{\partial r} = \mu_0 j_\theta.$$

Expressing j_θ and j_z in terms of space-charge density and corresponding components of velocity one obtains:

$$B_{sz} = -\mu_0 \int_r^b \rho v_\theta dr \tag{10.26}$$

$$B_{s\theta} = \frac{\mu_0}{2\pi r} \int_0^r \rho v_z 2\pi r dr. \tag{10.27}$$

The longitudinal component of self-field B_{sz} caused by rotation of electrons has a direction opposite to that of the external magnetic field, which produces electron rotation. So, it is possible to consider an intense relativistic Brillouin beam as a diamagnetic medium.

Approximate Analysis of a Brillouin Beam

Assuming that $v_\theta \ll v_z$ one can neglect the longitudinal component of the self-magnetic field ($B_{sz} \approx 0$) and calculate the relativistic factor by $\gamma \approx \left(1 - v_z^2/c^2\right)^{-1/2}$. It is possible to show that under this assumption an equilibrium Brillouin-type beam will be realized as the following conditions are fulfilled $\rho = $ const, $v_z = $ const, $j = $ const, that is, the same conditions as in the case of a nonrelativistic beam (see Sect. 7.1).

The balance equation of radial forces is written as:

$$eE_r + ev_\theta B - ev_z B_{s\theta} + \frac{m_0 \gamma v_\theta^2}{r} = 0, \tag{10.28}$$

where $E_{\rho r} = \frac{\rho}{2\varepsilon_0} r$ – radial component of space charge field, B – induction of external uniform field, $B_{s\theta}$ – azimuthal component of magnetic self-field, $v_\theta = r\dot\theta = -\frac{1}{2}\frac{e}{m_0\gamma} Br$ – azimuthal electron velocity.

The value of $B_{s\theta}$ is found from (10.27), if we put in it $\rho v_z = j_z = $ const, then $B_\theta = \frac{1}{2}\mu_0 j_z r$. The longitudinal velocity is expressed as:

$$v_z = \sqrt{2\frac{|e|}{m_0} U_0} \left(\frac{\gamma_0+1}{2}\right)^{1/2} \frac{1}{\gamma_0}, \tag{10.29}$$

where U_0 is the potential at the beam axis and $\gamma_0 = 1 + (1.96 \times 10^{-6}) U_0$.

From the balance equation (10.28) one can find the value of the external magnetic field B that provides equilibrium beam motion:

$$B = \frac{830}{b} \frac{I^{1/2}}{U_0^{1/4} \left(\frac{\gamma_0+1}{2}\right)^{1/4}}, \tag{10.30}$$

where B – induction of external field, Gs; b – beam radius, cm; U_0 – axial potential, V.

For low potential U_0, when $\gamma_0 \approx 1$, this equation is reduced to (7.7). The potential at the axis of symmetry U_0 is determined by equation:

$$U_0 = U_a - \frac{I}{4\pi\varepsilon_0 \sqrt{2\frac{|e|}{m_0} U_0} \left(\frac{\gamma_0+1}{2}\right)^{1/2} \frac{1}{\gamma_0}} \left(1 + 2\ln\frac{a}{b}\right).$$

It is presented in the form convenient for practical use:

$$U_0 = U_a \left[1 - 0.0152 P \frac{1}{\left(\frac{U_0}{U_a}\frac{\gamma_0+1}{2}\right)^{1/2}\frac{1}{\gamma_0}} \left(1 + 2\ln\frac{a}{b}\right)\right], \tag{10.31}$$

where $P = I/U_a^{3/2}$ – beam perveance calculated with respect to the accelerating potential U_a and expressed in $\mu A/V^{3/2}$.

As U_0 is included in the right part of this equation, calculation of potential U_0 is produced by the method of successive approximation, $U_0 = U_a$ and $\gamma_0 \approx \gamma_a = 1 + 1.96 \times 10^{-6} U_a$ being taken as a zero approximation.

Results of Rigorous Analysis of Equilibrium Beam

In the rigorous analysis of relativistic equilibrium beam motion both the components of self-magnetic fields $B_{s\theta}$ and B_{z0} are taken into account and the relativistic factor is calculated through the total electron velocity $v^2 = v_\theta^2 + v_z^2$, $\gamma = \left(1 - v^2/c^2\right)^{-1/2}$ (component v_r in the equilibrium beam is equal to zero).

10.2 Intense Relativistic Beams in Vacuum Channels

As is shown in [160] there exist several variants of equilibrium beam motion. For a particular case of a magnetic system with a shielded cathode the following equations have been obtained, which connect total beam current I, external magnetic field B and potentials at the beam axis U_0 and the beam boundary U_b:

$$I = I_n \frac{(\gamma_0^2 - 1)^{1/2}}{2} \left(\frac{\gamma_b^2}{\gamma_0^2} - 1 \right), \qquad (10.32)$$

$$B \approx \frac{m_0 c}{|e| b} \left(\frac{\gamma_b^2}{\gamma_0^2} - 1 \right)^{1/2} \left(\frac{\gamma_b}{\gamma_0} + 1 \right), \qquad (10.33)$$

where I_n – normalizing factor, $I_n = 17 \times 10^3$ A, $\gamma_0 = 1 + (1.96 \times 10^{-6}) U_0$.

From (10.32) and (10.33) one finds the equation for the applied magnetic field providing equilibrium beam motion:

$$B = \frac{415}{b} \left(\frac{\gamma_b}{\gamma_0} + 1 \right) I^{1/2} / U_0^{1/4} \left(\frac{\gamma_0 + 1}{2} \right)^{1/4}. \qquad (10.34)$$

It is seen that as $\gamma_b \approx \gamma_0$ this equation is reduced to (10.30).

The value of current of the equilibrium beam is limited by the space-charge action. The following equation has been obtained for the critical current value for a Brillouin-type relativistic beam [160]:

$$I_{cr} = \frac{1}{2} I_n \left\{ \frac{\gamma_b^2}{2} \left[\left(1 + \frac{8}{\gamma_b^2} \right)^{1/2} - 1 \right] - 1 \right\}^{1/2} \left[\frac{2}{(1 + 8/\gamma_b^2)^{1/2} - 1} - 1 \right],$$

where $I_n = 17 \times 10^3$ A.

Computation of the critical current and value of external (applied) magnetic field is done with the help of the curves presented in Figs. 10.4 and 10.5 [160].

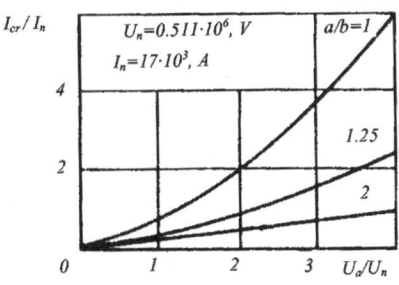
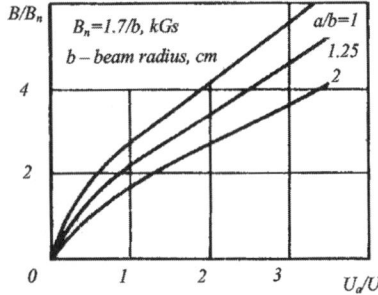

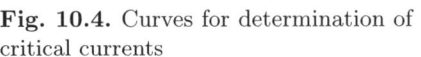

Fig. 10.4. Curves for determination of critical currents

Fig. 10.5. Curves for determination of magnetic induction required for beam focusing

10.3 Neutralized Relativistic Beams

10.3.1 Beams in Space Free from External Fields

The space charge is supposed to be completely neutralized by charges of the opposite sign. Therefore, beam motion is determined only by the self-magnetic field. This effect has been considered in Sect. 4.6.

As the radial force caused by the magnetic field of a beam is proportional to its current and the ratio v^2/c^2, it is obvious that the effect of beam compression (beam pinch) will be especially strongly expressed in high-current relativistic beams. There exists a critical beam current (Alfven's current) at which the beam is collapsed:

$$I_{cr}^A = 17 \cdot 10^3 \gamma_a v_a / c,$$

where I_{cr}^A is the critical current expressed in amperes, γ_a and v_a – relativistic factor and velocity calculated in terms of accelerating voltage U_a.

10.3.2 Neutralized Magnetically Confined Beams

It is supposed that an electron beam moves in an infinitely large uniform magnetic field, with its space charge being neutralized by the ion background. It seems that in this case there are no causes for limitation of the beam-current value. However, more sophisticated analysis shows that there exists a limitation of beam current, which is caused by growth of an instability. Two types of instability are significant for the confined neutralized beams: Pierce type and Buneman–Budker type. The limiting current, at which the beam ceases to propagate in the forward direction owing to Pierce instability, is given by [13]:

$$I_{cr}^P = 24 \cdot 10^3 \left(\gamma_a^2 - 1\right)^{3/2}.$$

The value of the limiting current of Buneman–Budker instability is the same order as I_{cr}^P.

10.4 Beam-Motion Computation

In accelerating techniques and in installations for physical experiments intense relativistic electron beams are frequently transported for a long distance. Magnetic-lens focusing (or transport) systems are used for this purpose. The systems consist of a set of axially symmetric or quadruple magnetic lenses.

For analysis of beam motion in such systems different computer codes are used. The simplest of them are based on the laminar-beam models and the concept of boundary trajectory. The action of the transverse component of the self-magnetic field as a rule is taken into account by decreasing by

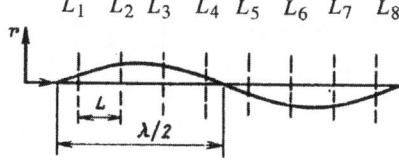

Fig. 10.6. For calculation of advance of phase

γ^2 times (γ is relativistic factor) value of the transverse component of the electrostatic self-field (see 1.18).

The results obtained are considered to be the first-approximation ones. They are usually made more precise by using computer codes based on more complicated discrete beam models.

Computer analysis is also applied for forecasting the stability of particle beams in multiple-lens transport systems. The method of "advance of phase" is used for this. In accordance with this method a single trajectory (in the absence of the beam) is computed. It is a periodically oscillating curve with wavelength λ (Fig. 10.6). The advance of phase, that is, the change of phase trajectory oscillation per period L of the lens system, is calculated, $\mu_0 = 360°(L/\lambda)$. Then the same computation is made for a single trajectory in the presence of the charge particle beam, which gives the phase advance $\mu = 360°(L/\lambda)$. The values of μ_0 and μ are used for forecasting of the potential beam stability. The following criteria of beam stability are recommended in [162, 163]: $\mu_0 \leq 60°$, $\mu/\mu_0 \geq 0.4$.

10.4.1 Axially Symmetric Lens Systems

These systems can be realized with the help of a set of solenoids or ring-shaped permanent magnets, magnetized in axial or radial directions (see Figs. 3.16 and 3.17). Computer-aided trajectory analysis allows us to determine the possibility of their application for transport of beams with accelerating voltage of the order of several MV and currents of the order of kA.

Results of such an analysis for an electron beam with accelerating voltage $U_a = 1$ MV, beam current $I = 1$ kA and initial beam radius $r_0 = 2$ cm are given below. Periodic magnetic systems consist of a set of ring-shaped permanent magnets magnetized in the axial direction located along the axis of the system with interval $L = 25$ cm.

The material of the magnet is barium ferrite, and its geometry is shown in Fig. 10.7. The results of trajectory analysis are presented in Fig. 10.8. The dashed curve of this figure is the envelope of the matched beam injected in the system with zero initial velocities. It is seen that in this case regular periodical beam motion takes place with period L. The envelope oscillates around an average radius, which is approximately equal to the initial one.

The solid curve is the envelope of the beam, which is injected into the system with its axis being displaced from the channel axis by the distance $\Delta r = 0.5$ cm. It is seen that in this case the beam radial dimension remains

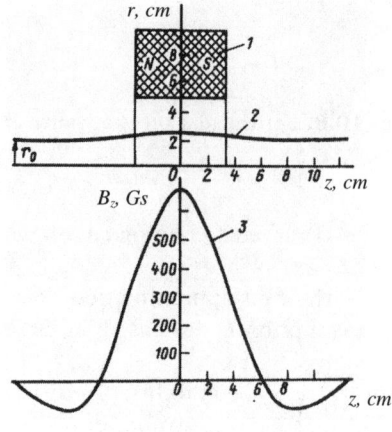

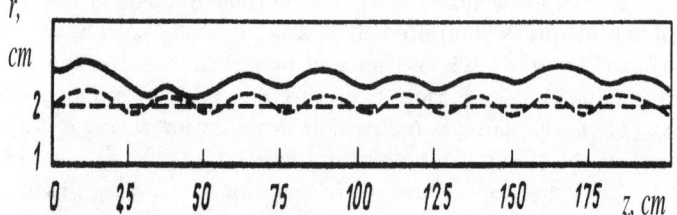

Fig. 10.7. Geometry of the ring-shaped magnet and the curve of axial magnetic induction

Fig. 10.8. Envelopes of the electron beam moving in the periodic magnetic system

a limited one. This shows the stability of the given focusing system. Phase advances for this system are $\mu_0 = 45°$, $\mu/\mu_0 = 0.55$ and, therefore, the criteria of stability mentioned above are fulfilled.

10.4.2 Quadruple Lens Systems

For description of electron beam motion in a quadruple magnetic lens system the following equations are used [161]:

$$\frac{d^2X}{d\tau^2} + \frac{L^2\eta}{v_z\gamma}B_Y - \frac{F_{0X}^2}{X^3} - \frac{4L^2I}{\beta^2\gamma^3 I_?}\frac{1}{X+Y} = 0, \qquad (10.35)$$

$$\frac{d^2Y}{d\tau^2} + \frac{L^2\eta}{v_z\gamma}B_X - \frac{F_{0Y}^2}{Y^3} - \frac{4L^2I}{\beta^2\gamma^3 I_?}\frac{1}{X+Y} = 0, \qquad (10.36)$$

where $\tau = Z/L$ – normalized longitudinal coordinate; L – period of lens system; X, Y – transverse coordinates; F_{0x} and F_{0y} – values of emittance, U_z – longitudinal component of velocity, I_n – normalizing factor, $I_n = 17 \times 10^3$ A, $\beta = \frac{v_z}{c}$ – ratio of electron velocity to velocity of light, γ – relativistic factor.

Components of magnetic induction B_X and B_Y are determined by the following equations:

10.4 Beam-Motion Computation

$$B_X = YG_m \cos 2\pi\tau ; \qquad B_Y = XG_m \cos 2\pi\tau .$$

In the case of zero emittance ($F_{0X} = F_{0Y} = 0$) (10.35) and (10.36) describe the beam-boundary trajectories.

For integration of these equations it is necessary to put the initial coordinates X_0 and Y_0, and initial trajectory slopes $\left(\frac{dX}{d\tau}\right)_0 = X'_0$, $\left(\frac{dY}{d\tau}\right)_0 = Y'_0$ at $\tau = \tau_0$ as well as the value of G_m.

As shown in [164] there exist optimal conditions of injection of a charge particle beam in a quadruple-lens system. For relativistic electron beams they are formulated in the following way:

$$X_0 = r_0 , \qquad Y_0 = r_0 , \qquad X'_0 = 2\pi r_0 \sqrt{2\chi} , \qquad Y'_0 = -2\pi r_0 \sqrt{2\chi}$$

$$\chi = \frac{384 I}{(r_0/L)^2 \, U_a^{3/2} \left(\frac{\gamma+1}{2}\right)^{3/2}} \qquad G_m = \frac{38}{r_0 L} \left[\frac{I}{U_a^{1/2} \left(\frac{\gamma+1}{2}\right)^{1/2}}\right]^{1/2} ,$$

where r_0 – initial beam radius, cm; L – period of the system, cm; I – beam current, A; U_a – accelerating voltage, V; G_m – transverse gradient of magnetic field, Gs/cm.

It follows from these equations for obtaining the minimum beam pulsation that the electron beam should be injected with strictly determined slopes of trajectories.

Results of computation of electron-beam focusing by a quadruple-lens system is given below. Beam and system parameters are as follows: beam current $I = 1$ kA, accelerating voltage $U_a = 1$ MV, initial beam radius $r_0 = 2$ cm, period of lens system $L = 50$ cm. Optimal conditions of beam injection $(dX/d\tau)_0$ and $(dY/d\tau)_0$, as well as the value of magnetic-field gradient G_m were initially used. Boundary envelopes $X(\tau)$ and $Y(\tau)$, and shapes of beam cross section for optimal injection conditions are shown in Fig. 10.9. Regular periodical beam motion takes place with the beam cross section being of elliptical shape.

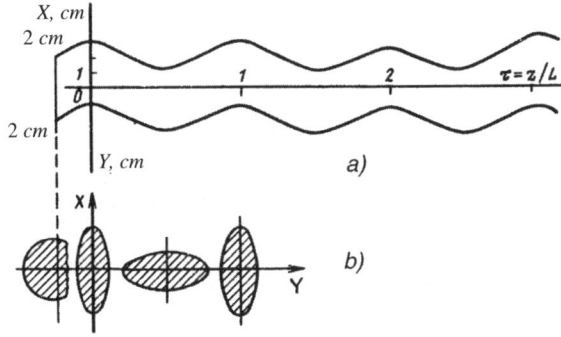

Fig. 10.9. Envelope (a) and cross section (b) of the beam in a quadruple-lens system

10 Intense Relativistic Charged-Particle Beams

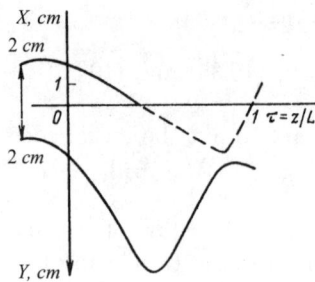

Fig. 10.10. Results of trajectories computation with zero initial conditions

These computations were performed with the optimal value of gradient $G_m = 25$ Gs/cm, which corresponds to the amplitude of magnetic induction $B_m = 50$ Gs at the distance $r = r_0 = 2$ cm from the beam axis.

The results of beam computation with zero initial condition $(dX/d\tau)_0 = (dY/d\tau)_0 = 0$ are presented in Fig. 10.10.

As is seen, the electron beam is practically collapsed at the first period of the system. Analysis of various quadruple-lens systems has confirmed the criticality of these systems for the condition of beam injection. An alternative injection of beam with elliptical cross section is expedient.

11 Multiple-Beam Electron-Optical Systems

11.1 Peculiarities and Application of Multiple Beams

At present, multiple-beam systems are used for forming and focusing (transport) of multiple-beam flows in microwave O-type tubes (klystrons, TWT, etc.) and in particle accelerators. In this case, an electron flow consists of N beams ($N = 2-100$), each of them propagates in a separate metallic channel. Due to this the total perveance of the flow can be increased up to 30 $\mu A/V^{3/2}$ without creation of the virtual-cathode phenomenon. This makes it possible to design power low-voltage wide-band microwave tubes and increase the total power of linear accelerators. Earlier multiple-ion beams were employed in ion jets of cosmic apparatus in order to increase their thrust [36].

The design of a multiple-beam optical system is considered using as an example the optical system of a multiple-beam klystron (Fig. 11.1). The electron flow consists of seven beams, one of them is the central beam, the other six are peripheral ones. Electron beams are formed by the electron gun, which consist of cathode (1) common for all beams, a grid electrode with seven apertures (2) and anode (3) also having seven apertures. The magnetic system includes ring-shaped permanent magnets magnetized in the radial direction (4), magnetic pole pieces (5, 6), magnetic circuit (7) and magnetic

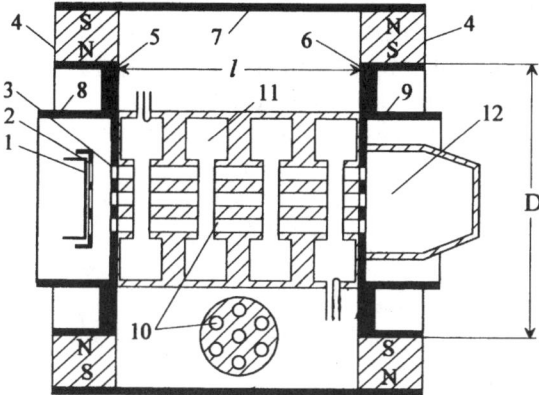

Fig. 11.1. Electron-optical system of a multiple-beam klystron

screens (8, 9). When the diameter of pole pieces D is equal to or greater than the length of the operating gap l (that is, the gap between the pole pieces) the magnetic field produced inside the gap will be almost uniform. Electron beams, after passing through the channels in the klystron cavities (11), enter into the klystron collector (12).

The optical system under consideration has only one row of peripheral beams. There exist tightly packed multiple-row systems. A two-row system contains 12 beams in the second row and the total number of beams is 19. The system with three rows contains 37 beams and so on.

Application of multiple-beam optical systems allows us to decrease considerably the anode voltage that provides the same level of beam power. Let us estimate the saving of voltage when N beams of perveance P_1 are used instead of a single beam of the same perveance to get the same total beam power P. For the single beam of perveance P_1 we have $P = P_1 U_{a1}^{5/2}$, for a multiple-beam system of perveance $P_N = NP_1$ we get $P = P_N U_{aN}^{5/2} = NP_1 U_{aN}^{5/2}$, with U_{a1} and U_{aN} the accelerating potentials of single- and multiple-beam systems, respectively. Then we find $U_{aN} = U_{a1}/N^{2/5}$. As applied to microwave-tube computation an important parameter is the electron-beam resistance, which is determined as $R = U_a/I_a$, with I being beam current, U_a – accelerating voltage. A similar calculation gives the following ratio of resistances of multiple-beam and single-beam systems: $R_N = R_1/N^{4/5}$.

The equations obtained above show that employment of multiple-beam systems gives the possibility to decrease anode voltage with simultaneous increase of the total beam current. From klystron theory it is known that for the efficient interaction of electrons with alternating electric fields of cavities the impedance of cavities should be of the order of the beam resistance. Therefore, in multiple-beam klystrons it is possible to reduce the equivalent impedance of cavities. This can be done by a proper decreasing of quality (Q-factor) of the cavities, or by means of detuning of cavities. This make it possible to increase the bandwidth of an amplifier.

Practical results indicate that application of a multiple-beam system can reduce the anode voltage by 2–3 or more times and increase the bandwidth up to 3% for S-band tubes and up to 10% for Q-band tubes [165].

Multiple-beam systems are also employed in power traveling-wave tubes (TWT) with cavity-type slow-wave structure. Figure 11.2 shows a cell of a

Fig. 11.2. Cell of a power traveling-wave tube

power TWT. It is seen to consist of 19 beam channels and a cavity-coupling slit. Use of the multiple-beam electron flow, of a high total perveance with each beam of low perveance, gives the possibility to improve beam–field interaction and increase efficiency by 1.5 times as compared with that of a single-beam TWT of the same power level [165].

11.2 Multiple-Beam Guns and Magnetic Systems

11.2.1 Electron Guns

Different versions of multiple-beam electron guns are presented in Fig. 11.3a–d [166]. There exist guns with a separate cathode for each beam (Fig. 11.3a). They are used in power tubes with a low number of beams, which are placed over a relatively large area. Initial beam forming and compression is produced due to the concave shape of the cathode surface and action of the cathode focusing electrode. Shapes of focusing electrodes of a partial gun are usually different from classical Pierce-type electrodes because of a limit of space for their location. The same reason limits the value of cathode radius and area and leads to a high cathode current density.

To increase the cathode area partial electron guns with curvilinear beams can be theoretically used (Fig. 11.3b). However, designing such guns is a complicated problem, which requires application of three-dimensional computer codes. Besides this, it is important to note that peripheral beams usually do not possess the necessary axial symmetry.

For electron guns with a high number of electron beams one cathode is used as a common for all beams (Fig. 11.3c). Electron flow emitted from the common cathode is split for separate beams by a special perforated electrode

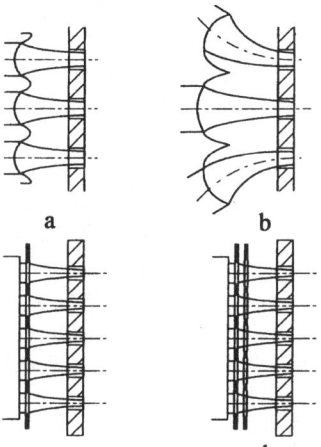

Fig. 11.3. Variants of multiple-beam electron gun

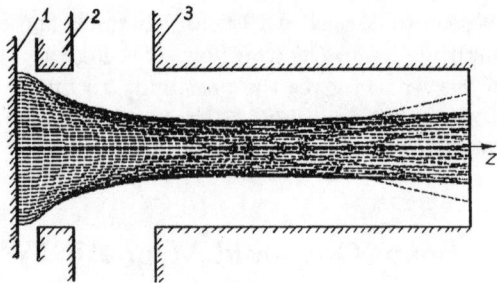

Fig. 11.4. Results of computer simulation of a single partial beam of perveance $P = 0.6 \ \mu\text{A}/\text{V}^{3/2}$: 1 – cathode ($U_k = 0$), 2 – grid (mask) ($U_g = 0$), 3 – anode ($U = U_a$)

placed in the vicinity of the cathode surface and usually operated at the cathode potential. It is called a grid or mask. The electric field produced by this electrode in the vicinity of the cathode surface produces weakly convergent beams. Figure 11.4 illustrates the results of computer simulation of a separate beam of this type of electron gun. It is seen that electron-trajectory crossing takes place, which leads to a nonlaminar structure of the beam. This effect is typical for these guns. Trajectory simulation of many variants of gun shows that the position of the grid electrode and its thickness and potential have a considerable affect on the value of perveance of the partial beam, beam compression and its structure. In some cases, the grid electrode can operate as a control electrode, which changes the value of the beam current.

An electron gun with two grid-type electrodes is presented in Fig. 11.3d. The first grid operates at the cathode potential. The second one can be used as a focusing and control electrode. Its potential can be both negative and positive with respect to the cathode potential. In the case of positive potential this electrode will not intercept the beam current as it is placed "in the shadow" of the first electrode. Variation of the potential of the second electrode allows us to change beam current, compression and structure. All the above refers to guns completely shielded from the magnetic focusing field. In the case of partially shielded guns, beam forming will be essentially affected by the magnetic field.

Designing of multiple-beam guns is performed with the help of computer codes. As a first step, simulation of a single partial beam is produced with the help of two-dimensional codes, the presence of neighboring beams being neglected (Fig. 11.4). In many cases this approach gives acceptable results since distances between gun electrodes are sufficiently small and electrode screening action considerably reduces the interaction of the neighboring beams. As a final stage of gun simulation three-dimensional simulation of the gun as a whole is performed.

11.2.2 Magnetic Focusing Systems

For focusing (transporting) multiple-beams, magnetic systems with uniform or reverse fields are used. Permanent magnetic or solenoids serve as sources of magnetic fields. The design of these systems is essentially the same as that used for a single-beam system. The main difference is the multiple-channel magnetic pole pieces (number of pole channels equal to number of beams) (see Fig. 11.1).

Magnetic systems with a uniform field differ in the rate of magnetic shielding of guns (Fig. 11.5): systems without gun shielding (a), systems with completely (b) and partially (c) shielded guns.

Systems without gun shielding were used in the earlier stage of multiple-beam systems development. They are realized in a uniform magnetic field extending from the tube cathode to its collector. At a high density of magnetic field, electrons emitting from the cathode follow magnetic field lines along spiral trajectories of radius $\rho_c = mv_\perp/eB$, where ρ_c – cyclotron radius, $v_\perp$ – transverse electron velocity of an electron at the cathode surface, B – magnetic induction of uniform field.

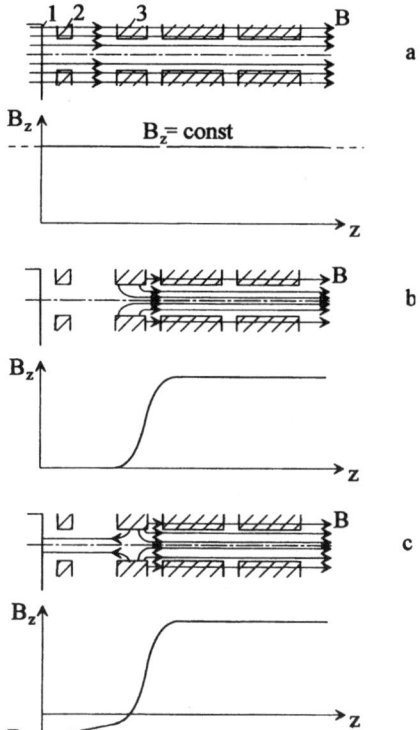

Fig. 11.5. Variants of magnetic focusing systems with different rate of gun shielding

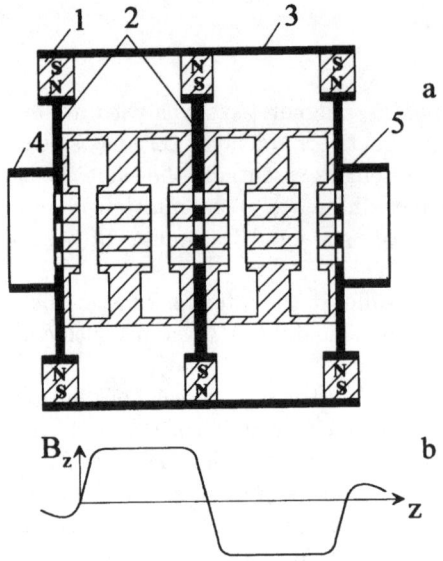

Fig. 11.6. Reversed focusing system of a multiple-beam klystron: 1 – ring magnet, 2 – magnetic pole pieces, 3 – magnetic circuit, 4, 5 – magnetic shields

In a system with a completely shielded gun initial beam forming occurs in a pure electrostatic field produced by the gun electrodes. The radial magnetic force, which balances the space-charge repulsion force, arises as a result of rotation of electrons around the axis of a partial channel. This rotational motion appears when an electron is moving inside a channel of the magnetic pole piece, where magnetic induction has both longitudinal B_z and transverse B_r components. It has the same nature as in the Brillouin focusing system for a single beam (see Chap. 5).

Theoretically it is possible to provide Brillouin focusing of every partial beam, but its practical realization meets some difficulties. In conformity with permanent-magnet focusing systems it is connected with intense stray magnetic fields in the gun region. Application of magnetic shields does not enable us to eliminate completely these fields from the gun region (Fig. 11.5c), and these systems are to be considered as those with partially shielded gun (see Sect. 7.1).

In long electron tubes it is expedient to use a reversed-field focusing system with one or more reversals. Figure 11.6a demonstrates a reversed-field focusing system of a multiple-beam klystron [167]. It includes ring-shaped magnets magnetized in the radial direction (1), pole pieces (2), magnetic circuit (3) and magnetic shields (4) and (5). Distribution of magnetic induction along the exit of the central channel is shown in Fig. 11.6b. Magnetic-field reversal occurs in channels of the intermediate pole piece. A stray magnetic field partially penetrates in the gun region.

Application of reversed system enables us to reduce the distance between the magnetic pole pieces, which in turn allows us to decrease their trans-

11.3 Peculiarities of Multiple Systems

Two main factors determine the distinctive feature of multiple beam-forming and focusing systems: stray fields penetrating in the gun region and transverse components of magnetic field of a different nature.

For analysis of these factors it is expedient to divide the optical system for the three regions: region of initial beam forming, which includes the electron gun and the channels in the cathode pole piece; region of beam transport, i.e. the space between the cathode and collector magnetic pole pieces and collector region, which includes the channels in the collector pole piece and collector.

11.3.1 Analysis of Stray-Fields Action

A typical magnetic-field distribution at the axis of a partial channel is presented in Fig. 11.7. The magnetic field changes direction inside the channel in the magnetic pole piece and in the gun region it is directed opposite to that in the beam-transport region. It is possible to estimate the magnetic-field action on initial beam forming with the help of the equation for radial magnetic force:

$$F_r = -\frac{1}{4}\frac{e^2}{m}B_z^2\left(1 - \frac{B_{zk}^2 r_k^4}{B_z^2 r_k^4}\right),$$

where r – radial coordinate of an electron counted-off from the channel axis, r_k – coordinate of the electron at the cathode surface, $B_z = B_z(z)$ – current value of magnetic induction at the channel axis, B_{zk} – magnetic induction at the cathode surface.

This equation follows from the equation of motion (3.20) within the paraxial approximation, being an analog of (7.1).

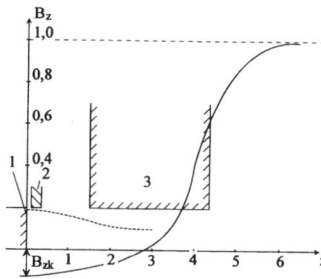

Fig. 11.7. Magnetic-field distribution in gun region: 1 – cathode, 2 – grid, 3 – anode (magnetic pole piece)

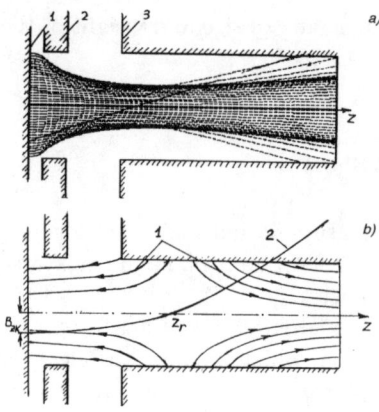

Fig. 11.8. a – results of trajectory computation, **b** – magnetic field distribution in pole-piece channel

As is apparent from the diagram of Fig. 11.7, in the region of initial beam forming $|B_z| < |B_{zk}|$. Therefore, in the region of beam compression $(r < r_k)$ $B_{zk}^2 r_k^4 / B_z^2 r^4 > 1$ and the magnetic force will be defocusing, $F_r > 0$. Figure 10.8 shows the results of trajectory computation with the magnetic field in gun region, the value of B_{zk} being equal to 25% of the B_z used in the transport-beam region.

Comparison of these results with those obtained for a completely shielded gun (Fig. 11.4), indicates the defocusing action of the magnetic field. The physical explanation of this fact is similar to that given in the analysis of the magnetic defocusing lens (Sect. 3.10.3).

As in the case of a single beam focusing (Sect. 7.1.4), the presence of a magnetic field at the cathode involves the increase of the magnetic induction required to balance the space charge repletion. The value of Induction of uniform magnetic field in transport region can be found from:

$$B_z = \frac{B_b}{(1-G)^{1/2}},$$

where B_b – Brillouin magnetic induction, calculated as $B_b = 830 I_1^{1/2} / r U_a^{1/2}$, I_1 – current of a partial beam, r – its radius, U_a – accelerating voltage, $G = B_{zk}^2 r_k^4 / B_z^2 r^4$ – coefficient of cathode shielding.

So, the presence of a stray magnetic field in the gun region involves beam defocusing in the pole-piece channel and increasing of the required induction of the uniform field.

11.3.2 Transverse Fields

Transverse fields are considered to be the components of the magnetic field that are normal to the channel axis, but that do not posses the axial symmetry about it. These fields are of a different nature:

- Local fields arising in pole-piece channels due to finite magnetic permeability of pole-piece materials
- Transverse fields resulting from finite transverse dimensions (diameters) of pole pieces
- Transverse self-fields produced by beam currents

Local Fields

Let us consider the nature and action of local transverse fields in the channels of the second row, which centers are placed at distance R_2 from the axis of a focusing system (Fig. 11.9). Magnetic flux, which enters the beam-transport region (pole pieces gap) through the surface enclosed by the circle of radius R_2, is expressed as $\psi_2 = \pi R_2^2 B_z$, B_z being the induction of the uniform field. In the case of an infinite magnetic permeability of the pole-piece material all this flux passes through the material bridges separating the channels. In systems with permanent magnets a magnetic stray flux $\hat{\psi}_2$ also passes through these bridges.

Assuming approximately that $\hat{\psi}_2 \approx \psi_2$ one can find the value of magnetic induction in the bridge material:

$$B_{br} = \frac{2\psi_2}{S_{br}} = \frac{2\pi R_2^2 B_z}{S_{br}},$$

where S_{br} – area of minimum bridge cross section; $S_{br} = (2\pi R_2 - N_2 d)\Delta$, where N is number of channels of diameter d, Δ – thickness of pole piece.

So, the following equation is obtained for estimation of B_{br}:

$$B_{br} = \frac{R_2 B_z}{\Delta \left(1 - \dfrac{N_2 d}{2\pi R_2}\right)}.$$

For real magnetic materials of a finite value of permeability a part of the bridge magnetic flux is displaced in channels and the transverse magnetic

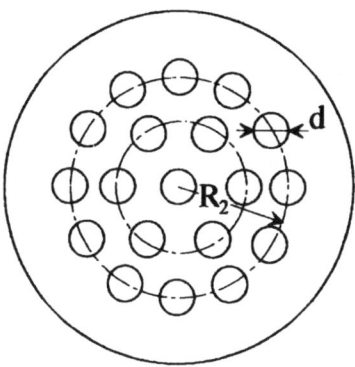

Fig. 11.9. Magnetic pole-piece drawing

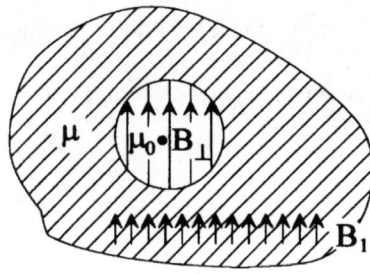

Fig. 11.10. Transverse magnetic field in a channel of a pole piece

field $B_\perp$ is erased in the channels. The value of transverse magnetic induction $B_\perp$ can be found in such a way. Initially, the magnetic field strength in the bridge material is determined:

$$H_{br} = B_{br}/\mu,$$

where μ is the magnetic permeability of the pole-piece material.

Taking into account continuity of the tangential component of magnetic field strength H at the boundary of the two media one can find:

$$B_\perp \approx \mu_0 H_{br} \approx B_{br} \frac{\mu_0}{\mu}. \qquad (11.1)$$

The structure and value of the transverse field in a channel can also be estimated from the following model study. Let us consider a uniform magnetic flux with induction B_1, which passes inside a magnetic pole piece of a finite permeability μ and encounters an aperture of permeability μ_0 (Fig. 11.10).

This problem has the analytical solution from which it follows that the magnetic field inside the aperture is uniform with the value of magnetic induction equal to:

$$B_\perp = \frac{2B_1}{1 + \dfrac{\mu}{\mu_0}}.$$

This equation can be used for estimation of the transverse field induction in a channel of a magnetic pole piece. For this let us take $B_1 = kB_{br}$, with B_{br} being equal to the magnetic induction in the bridges of the pole piece. Coefficient $k < 1$ takes into account increasing bridge magnetic induction B_{br} as compared with the induction B_1 of the falling flux, $k \approx 1 - Nd/2\pi R_2$. Then

$$B_\perp = \frac{2B_1}{1 + \dfrac{\mu}{\mu_0}}.$$

For $\mu/\mu_0 \gg 1$ this is reduced to:

$$B_\perp = 2k\frac{\mu_0}{\mu} B_{br}. \qquad (11.2)$$

For $k = 0.5$ it coincides with (11.1).

11.3 Peculiarities of Multiple Systems

As follows from the above equations for $\mu \to \infty$, $B_\perp \to 0$, so, the existence of the transverse magnetic in pole-piece channels is stipulated by a finite value of magnetic permeability of the pole-piece material.

Transverse Fields in Beam-Transport Region

Results of computer simulation of various magnetic systems using radial and axially magnetized magnets allows us to describe some main peculiarities of the magnetic-field distribution in the space between the pole pieces (working gap).

Figure 11.11 illustrates typical peculiarities of the magnetic field in systems with radial-magnetized permanent magnets. Magnetic field lines bulge in the center of the working gap, as is shown by the line $R = f(z)$. This leads to a decreasing longitudinal component of magnetic field B_z for a value ΔB_z and appearance of a radial component of field B_R.

If a magnetic system possesses total axial symmetry, including axial symmetry of the pole pieces, then at distances from them exceeding one diameter of the pole-piece channels the magnetic field will be of global axial symmetry. But this field is not axially symmetric with respect to the axis of peripheral beams, as well as its radial component B_R. It is considered to be a transverse component $B_R = B_\perp$.

Reducing of the longitudinal field component B_z and the value of amplitude of the radial component B_R depend on the ratio of the gap length l and the diameter of the pole pieces D, with growth of l/D they are increased. At $l/D \approx 1$, $\Delta B_z/B_z = 1 - 2\%$, $B_{R\max}/B_z = 1 - 2\%$.

For magnetic systems with axially magnetized magnets sagging of magnetic lines is typical with B_z and B_R distributions shown in Fig. 11.12. Values of $\Delta B_z/B_z$ and $B_{R\max}/B_z$ for $l/D \approx 1$ are the same order as in the previous case. It should be noted that in a strictly uniform field $B_z = $ const, $B_R = 0$ and field lines are straight ones.

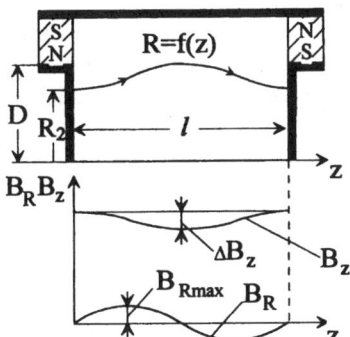

Fig. 11.11. Peculiarities of magnetic field in a system with permanent magnets magnetized in the radial direction

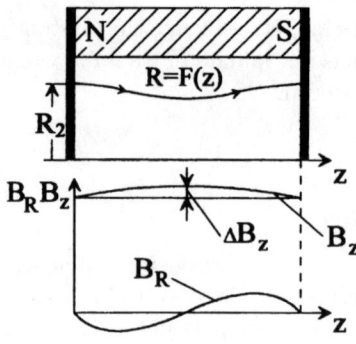

Fig. 11.12. Magnetic field in a system with permanent magnets magnetized in the axial direction

11.4 Estimation of Effects of Transverse Fields

Trajectory computation, performed by using 3D computer codes, shows that in the majority of cases there exists coherent motion of electrons forming a partial beam. This means that under the action of transverse fields the cross section of a partial beam is displaced as a whole. This allows us to estimate its transverse-beam displacement by analyzing the motion of a single central beam electron considering it as the center of gravity of the beam cross section.

Assuming that in a pole-piece channel the magnetic transverse field has a uniform structure and its value is determined by (11.11) and (11.12) one can find the following equations for the central electron displacement $\delta_\perp$ and its transverse velocity $v_\perp$ at the outlet of the channel:

$$\delta_\perp = \frac{1}{2}\frac{|e|}{m}B_\perp\frac{\Delta^2}{v_z}, \quad v_\perp = \frac{|e|}{m}B_\perp\Delta,$$

where Δ – channel length (that is, the thickness of the pole piece), v_z – longitudinal velocity.

These equations can be transformed in the form:

$$\delta_\perp = \frac{1}{2}\omega_c\frac{B_\perp}{B_z}\frac{\Delta^2}{v_z}, \quad v_\perp = \omega_c\frac{B_\perp}{B_z}\Delta,$$

where $\omega_c = |e|B_z/m$ – cyclotron frequency calculated for the longitudinal component of magnetic induction of the working gap.

Let us now estimate the displacement of a partial beam when it is moving in the working gap. Motion of a beam of central electrons occurs in the slightly inhomogeneous field. As analysis shows, this motion can be presented as a superposition of three motions [168]:

- slipping along a field line with velocity $v_\parallel$, which is a component of velocity directed along the field line
- rotation around the magnetic line with transverse velocity $v_\perp$, with $\rho = mv_\perp/|e|B$ being the radius of the orbit (cyclotron or Larmor's radius)
- drift motion with the velocity v_d in the plane normal to the vector of magnetic induction

11.4 Estimation of Effects of Transverse Fields   249

Therefore, the upper limit of displacement of the central electron resulting from this motion can be expressed as:

$$\delta_t \leq \delta_l + \rho + v_{d\,max}(z - z_0)/v_z \,, \tag{11.3}$$

where δ_l – displacement of the central field line, ρ – radius of electron orbit, $v_{d\,max}$ – maximal drift velocity, z_0 – coordinate of pole-piece surface, z – current electron position, $(z - z_0)/v_z$ – transit time.

The displacement δ_l can be found in the following way. For any regular point of the field line it is possible to write:

$$dR/dz = B_R/B_z \,, \tag{11.4}$$

where dR/dz – the slope of tangent of a field line, B_R and B_z – components of induction at the point.

Taking into account a slight inhomogeneity of the field we can put in (11.3) B_z = const and replace B_R at the current point of the field line by the value of B_{R0} at the channel axis. Then, we obtain the following equation for the central field line:

$$\frac{dR}{dz} = cB_{R0}(z) \,, \tag{11.5}$$

where c is constant, $c = 1/B_z$.

Then displacement δ_l is found as:

$$\delta_l = c \int_{z_0}^{z} B_{R0}(z) dz \,.$$

Integration is performed from the pole-piece surface ($z = z_0$) to the current position z.

If the distribution of $B_{R0}(z)$ is found (by measurement or simulation) the last equation allows us to determine δ_l. It should be noted that δ_l is proportional to the area confined by the curve $B_{R0}(z)$.

The velocity of drift motion v_d, which determines the electron-drift displacement, can be found as [168]:

$$v_d = \frac{1}{\omega_c R_c} \left(v_{\parallel}^2 + \frac{v_{\perp}^2}{2} \right) ,$$

where R_c – radius of curvature of a field line.

If the equation of a curve is $R = F(z)$, its radius of curvature is determined by:

$$\frac{1}{R_c} = \frac{\dfrac{d^2 R}{dz^2}}{\left[1 + \left(\dfrac{dR}{dz} \right)^2 \right]^{3/2}} \,.$$

For the case under consideration the field line is a slowly varying function of z, then $(dR/dz)^2 \ll 1$ and

$$\frac{1}{R_c} \approx \frac{d^2R}{dz^2}.$$

Taking account (11.4) one obtains:

$$\frac{1}{R_c} = c\frac{dB_{R0}}{dz},$$

where $c = 1/B_z$.

Then:

$$v_d = \frac{1}{\omega_c B_z}\frac{dB_{R0}}{dz}\left(v_\parallel^2 + v_\perp^2/2\right),$$

and

$$v_{d\,\max} = \frac{1}{\omega_c B_z}\left(\frac{dB_{R0}}{dz}\right)_{\max}\left(v_\parallel^2 + v_\perp^2/2\right).$$

It follows from this equation that the less the field inhomogeneity the less the drift velocity, in the case of a uniform field $dB_{R0}/dz = 0$ and $v_d = 0$.

When $v_d \ll v_z$ it is possible to neglect the drift displacement and estimate the resulting displacement as:

$$\delta_t = \delta_l + \rho. \tag{11.6}$$

Assuming coherent motion of electrons creating a beam it is possible to apply (11.3) and (11.6) for estimation of the displacement of the beam as a whole at the stage of preliminary calculations.

11.5 Analysis of Beam Interaction

An important problem of multiple-beam system design is the interaction of the beams. In the case when the beams propagate in separate metallic channels interaction through self-electric fields is excluded, but interaction by self-magnetic fields still remains.

Let us consider the peculiarities of the interaction using as an example a typical system shown in Fig. 11.3. The self-magnetic field of this beam system can be found as a superposition of the field of separate beams. Assuming that beam length l is considerably greater than its radius it is possible to find the magnetic field of a separate beam from Amperés law:

$$B_\theta = \frac{\mu_0 I}{2\pi r_B^2}r \quad r \leq r_B, \qquad B_\theta = \frac{\mu_0 I}{2\pi r} \quad r \geq r_B,$$

where B_θ – azimuthal component of magnetic induction in the local cylindrical coordinate system r, θ, z, with z – coordinate being coincident with the axis of a channel; I – beam current; r_B – beam radius, r – radial distance from the axis to point of observation.

Let us consider the action of the self-magnetic field on the motion of a separate beam (beam 0 in Fig. 11.13). For the total self-magnetic field of the

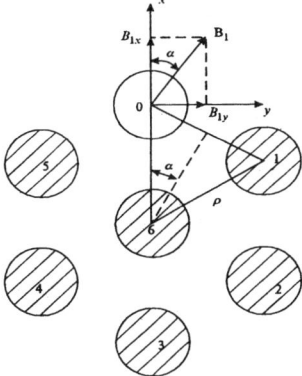

Fig. 11.13. Scheme for computation of beam interaction

beam system it is expedient to separate for the two fields: the self-field of a beam and the field created by the other beams. The magnetic field $B_{s\theta}$ produces the magnetic force $F_{s\mu}$, directed toward the beam axis:

$$F_{s\mu} = -|e|\, v_z B_{s\theta}\,,$$

where v_z – axial velocity.

This force partially compensates the force of space-charge repulsion F_ρ. The resulting radial force is expressed as:

$$F_s = F_{\rho s} + F_{ms} = \frac{|e|\, I}{2\pi\varepsilon_0 r_B v_z}\left(1 - \frac{v_z^2}{c^2}\right)\,.$$

The multiplier $\left(1 - v_z^2/c^2\right)$ in this equation shows the decreasing beam spread due to the action of the self-magnetic field.

To consider the action of the magnetic fields created by the rest of the beams let us introduce a local coordinate system x, y, z, with its center being placed in the center of the beam in question, that is, O-beam, and the z-coordinate directed along the beam axis. The magnetic field in the center of this beam ($x = y = 0$) is the superposition of partial fields created by the rest beams.

The partial fields created in the point of observation ($x = y = 0$) by beams 3 and 6 are determined by

$$B_{3y} = \frac{\mu_0 I}{4\pi R_1} \qquad B_{6y} = \frac{\mu_0 I}{2\pi R_1}\,,$$

where R_1 – distance from the center of the system and the center of the peripheral beams (Fig. 11.13). By means of geometric plotting of this figure it is possible to find the partial fields created by beams 1 and 5, 2 and 4.

$$B_{1y} = \frac{\mu_0 I}{4\pi R_1}, \qquad B_{1x} = \frac{\mu_0 I \cos\alpha}{4\pi R_1 \sin\alpha}$$

$$B_{5y} = \frac{\mu_0 I}{4\pi R_1}, \qquad B_{5x} = \frac{\mu_0 I \cos\alpha}{4\pi R_1 \sin\alpha}$$

$$B_{2y} = \frac{\mu_0 I}{4\pi R_1}, \qquad B_{2x} = \frac{\mu_0 I \cos 2\alpha}{4\pi R_1 \sin 2\alpha}$$

$$B_{4y} = \frac{\mu_0 I}{4\pi R_1}, \qquad B_{4x} = \frac{\mu_0 I \cos 2\alpha}{4\pi R_1 \sin 2\alpha}.$$

So, in the center of the beam under consideration ($x = 0, y = 0$) the magnetic field created by beams (1–6) has only a single y-component equal to:

$$B_y = \frac{7\mu_0 I}{4\pi R_1}. \qquad B_x = 0.$$

Field lines of this field found via a numerical method are presented at Fig. 11.14. It is seen that in the vicinity of the beam 0 the magnetic field is directed along the y-coordinate and close to the uniform one. This field gives rise to beam deflection from the channel axis.

Let us estimate the deflection of the central trajectory of the beam assuming that it moves in a self-magnetic field that is supposed to be the uniform one directed in the y-direction B_y and a uniform longitudinal focusing field B. Additionally, we take into account the evident condition $B \gg B_y$.

Let us introduce a moving coordinate system x', y', z', which is moving along the axis of the channel with velocity v equal to the electron beam

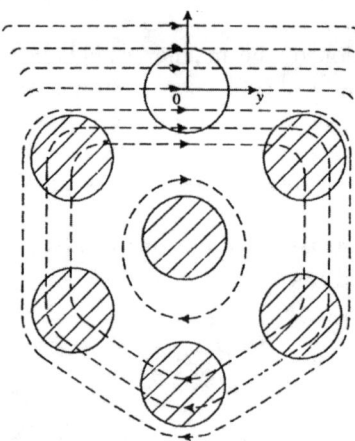

Fig. 11.14. Scheme of field lines of the magnetic field created by 1–5 beams

velocity v_z. Using equations of relativistic electrodynamics it is possible to find the fields that "will act" on the central electron in this new coordinate system [169]. Omitting the details of algebra we obtain:

$$E'_{x'} = -\gamma v B_y, \qquad B'_{y'} = \gamma B_y, \qquad B'_{z'} = B_z.$$

In these equations all primed quantities are related to the moving coordinate system, $\gamma = \left(1 - v^2/c^2\right)^{-1/2}$ – a relativistic factor.

So, motion of the central electron in the moving coordinate system is performed under the action of cross-fields $E'_{x'}, B'_{y'}, B'_{z'}$.

The condition $B \gg B_y$ assumed previously, involves the ratio $B'_{y'} \ll B'_{z'}$ (for γ of order of several units). It allows us to neglect the magnetic induction component $B'_{y'}$ and approximately assume that magnetic induction in the moving coordinate system has only the single component $B' = B'_{z'}$.

The motion of the central electron will obey the following system of equations:

$$\frac{d^2 x'}{dt'^2} = -\frac{|e|}{m'} E'_{x'} - \frac{|e|}{m'} \frac{dy'}{dt'} B'_{z'}$$

$$\frac{d^2 y'}{dt'^2} = \frac{|e|}{m'} \frac{dx'}{dt'} B'_{z'}.$$

A partial solution of these equations with initial conditions $t' = 0$, $x' = y' = z' = 0$, $\frac{dx'}{dt'} = \frac{dy'}{dt'} = \frac{dz'}{dt'} = 0$ is known from the theory of a planar magnetron:

$$x' = \frac{|e|}{m'} \frac{|E'_{x'}|}{\omega'^2_c} \left(1 - \cos \omega'_c t'\right),$$

$$y' = \frac{|e|}{m} \frac{|E'_{x'}|}{\omega'^2_c} \left(\omega'_c t' - \sin \omega'_c t'\right),$$

where $\omega_c = |e| B'_{z'}/m'$.

Thus, in the moving system the trajectory of the central electron is a cycloid.

Using the following relations connecting quantities in moving and laboratory coordinate systems $m = \gamma m'$, $\omega_c = \omega'_c/\gamma$, $t = \gamma t'$, $\omega_c t = \omega'_c t'$, $x = x'$, $y = y'$ we obtain the equation of trajectory in the laboratory coordinate system:

$$x = \frac{v}{\omega_c} \frac{B_y}{B_z} \left(1 - \cos \omega_c t\right)$$

$$y = \frac{v}{\omega_c} \frac{B_y}{B_z} \left(\omega_c t - \sin \omega_c t\right).$$

This equation can be presented in the form:

$$y = z \frac{B_y}{B_z} \left(1 - \frac{\sin \omega_c t}{\omega_c t}\right),$$

where is current z-coordinate of the central electron, $z = vt$. It follows from these equations that the x-displacement of the central electron periodically changes in time, while the y-displacement increases with time. At a high value of time it grows linearly with time as the term $\sin \omega_c t / \omega_c t$ vanishes.

This equation can be used for estimation of the displacement of the beam center from the channel axis.

The same result can be obtained if we consider the drift of the electron in the cross-field. The velocity of the drift is expressed as $v'_d = \frac{|E'_{x'}|}{B'_{z'}} = \frac{\gamma v B_y}{B_z}$.

The y-displacement is expressed as $y' = v'_d t' = \frac{\gamma t' v B_y}{B_z}$, then:

$$y = y' = \frac{\gamma t' v B_y}{B_z} = \frac{tv B_y}{B_z} = z \frac{B_y}{B_z}.$$

12 Methods of Experimental Investigation of Beams

12.1 Classification of Methods

There are many different methods of intense charge particle beam experimental investigation known at present, which reflect different scopes of application and variety of the beam parameters being measured [170–172]. By means of these methods it is possible to determine: geometry and configuration of the electron beam, its current and energy, electron charge in the current pulse, the current density and its distribution in its cross section (the beam profile), electron velocities, high-frequency characteristics of the beam, different kinds of radiation and other effects, which take place when accelerated electrons interact with the media.

With the most common approach to obtaining information about a beam, all investigation methods can be divided into two large groups: direct and indirect ones. Direct methods are based on measurements of characteristics of beams themselves, namely: their current, current density, energy (velocity), distribution of these values in the cross section, etc. Indirect methods suppose registration and analysis of some "mediators", which possess characteristics simply correlated with the characteristics of electron beams, with the relations being well known. The "mediators" can be either the electron beam's own electric and magnetic fields, or different effects taking place when the beams interact with the media or targets.

Direct methods are to be considered as contact or collector ones as they are based on the electron-beam absorption (in part or completely) by the collector of the measuring device, placed in the way of the electron beam. Depending on whether integral or differential data are needed, all direct methods can be divided into those that do not require electron-beam space decomposition, and methods that assume its decomposition into separate elements. The latter, in turn, can be subdivided into methods with simultaneous beam decomposition and with a successive one. The simultaneous beam decomposition can be obtained mechanically, for instance, with the aid of multielement sectional collectors or matrix targets. The successive decomposition can be achieved either by beam electromagnetic scanning relative to a fixed collector or by mechanical scanning of a collector relative to a stationary beam. In the latter case, depending on the collector construction, there are methods of moving the collector with a small aperture; of moving the slit screen; of a rotating (vibrating) probe, etc.

Indirect methods of experimental investigation can be either contact, or noncontact ones. The contact methods are connected with effects occurring when the electron beam interacts with a substance. There can be mentioned the radiation methods, based on measurements of different kinds of radiation (such as gas luminescence, stimulated semiconductor optical radiation, metallic or graphite plate thermal radiation, fluorescing screen luminescence, secondary electron emission luminescence and X-ray radiation). Nonradiation methods suppose measurements of changes in electrical properties and mechanical tensions in a target, the changes taking place when the target is bombarded with the beam being analyzed (the methods of semiconductor stimulated conductivity and acoustic waves in a target).

Indirect noncontact methods can be divided into radiation and field ones. The former include methods, based on analysis of the Vavilov–Cherenkov effect, of synchrotron radiation, bremsstrahlung and transition radiation; the latter are electrostatic, magnetoinductive, resonator methods and methods of probing with the help of electron and laser beams.

To comprehend the variety of experimental investigation methods of charged beams and the connections between them the classification scheme is given, which is based on the discourses mentioned above and presented in Fig. 12.1. It should be noted that the diagram includes only basic investigation methods. Apart from that, in practice their different variations and combinations can be used.

The choice of the specific method of beams experimental investigation depends on the region of beam application, their power and energy, as well as on an investigation problem formulation. In the accelerating techniques the following methods (primary transformers) are the most often used: the field ones (electrostatic and magnetoinductive) and the ionizing one for circulating beams, as well as methods of acoustic waves and of the secondary electron emission for beams being driven out from an accelerator. Besides, to investigate profiles and to measure beam emittance the methods of moving slit screens and diaphragm collector are often applied [170, 171]. To investigate beams in an electron-beam welding apparatus the methods of moving-slit screen and of rotating probe appear to be the most convenient; in vacuum UHF devices the method of a moving collector with a small aperture and its variations are used [171, 172, 174].

To determine the geometry of high-powered high-voltage electron beam cross section and the current-density distribution in the cross section, the method of heated metallic or graphite plate thermal radiation or the method of X-ray radiation is quite often applied [171, 172]. Using these and other radiation methods, the information about the beam is obtained from the analysis of its cross section image fixed on a photograph or X-ray image.

12.1 Classification of Methods 257

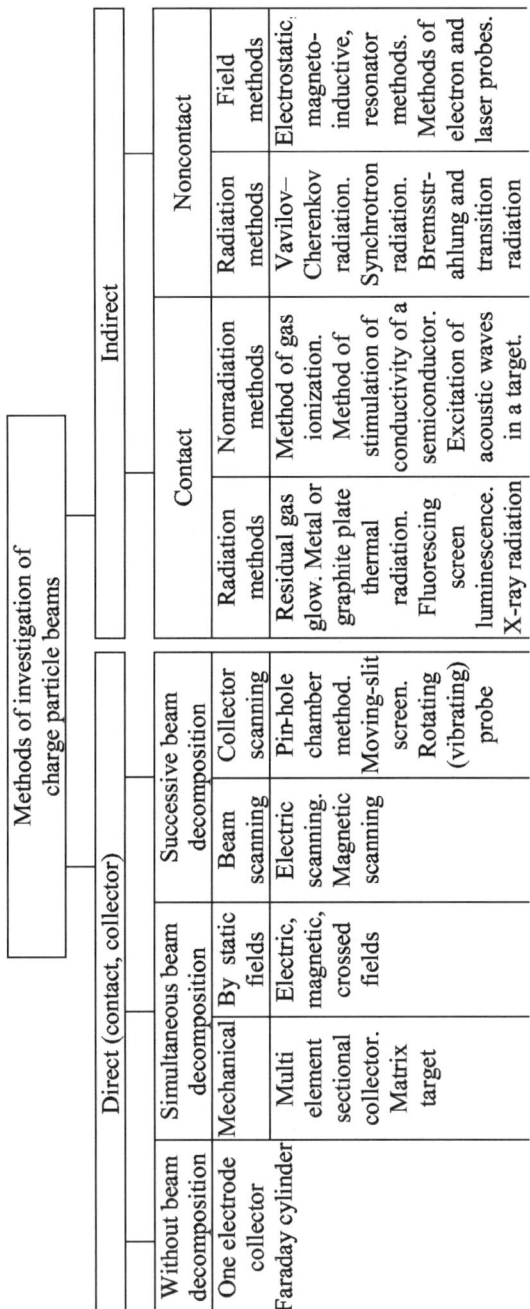

Fig. 12.1. Methods of experimental investigation of charge particle beams

258   12 Methods of Experimental Investigation of Beams

Among all the methods of electron-beam investigation considered, the most informative and precise one is the method of a moving collector with a small aperture, which is often called the pin-hole chamber method. The method allows measurement of different characteristics of the electron beam, including high-frequency ones, when being used in vacuum UHF devices. Measurement results obtained by this method do not require recalculation; the magnitude of error of the method can be less than 10%. These circumstances cause wide application of this method in practice.

12.2 The Pin-Hole Chamber Method

The method of moving a collector with a small aperture consists in successive decomposition (with the aid of a moving aperture) of the electron-beam cross section into small elements and in measurements of these element's current. The dependence of current through the aperture on the aperture location, found in such a way, determines, with a certain error, the function of distribution of the current density along the line of aperture motion. It allows us to find the current-density distribution in different cross sections of the electron beam, and to determine the cross section geometry and its change along the beam length.

To reduce measurement error it is necessary to decrease the radius of the diaphragm aperture in comparison with the electron-beam radius. However, when investigating electron beams with small current density the decrease of the analyzing aperture radius may not lead to increasing measurement precision due to insufficient sensitivity of the measuring devices. To increase the sensitivity, different methods of amplification of the current are used: with the aid of transistor amplifiers, secondary electron multipliers and so on.

Of certain interest among these methods is a method based on the effect of impact ionization of a semiconductor target with accelerated electrons. The main advantage of this method is the opportunity to obtain large coefficients of current amplification $\left(\approx 10^3\right)$ with a very small thickness of target $(10\text{--}10^2\ \mu m)$ [173, 174].

The main point of the method of the electron-beam current amplification can be explained with the aid of a device presented in Fig. 11.2. The basis of the device is a semiconductor diode with a small back bias of a p–n junction and a metallic (usually aluminum) film-electrode, which thickness does not exceed 0.1 µm.

When such a diode is bombarded by accelerated electrons (they are shown as arrows in the diagram) they penetrate deeply into the semiconductor and ionize its atoms on their way. Appearing in the process electron–hole pairs are divided by the electric field of the p–n junction, which leads to an electric current increase in the diode circuit.

If we suppose that the energy of electrons bombarding the semiconductor, E_0, is completely spent for atom ionization, then the number of resultant

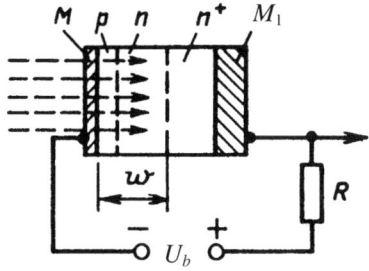

Fig. 12.2. Scheme of semiconductor amplifier of electron-beam current: M – thin metal layer, w – depletion region of p–n junction, M_1 – metal electrode, R – load resistor, U_b – back bias

pairs – charge-carriers – is equal to the relation E_0/ε_i, where ε_i – is mean ionization energy, i.e. the energy of appearance of one electron–hole pair.

To effectively separate charges of different signs, it is very important that their reproduction takes place in the depleted layer, where a strong electric field acts. Therefore the bed depth of the p–n transition is usually chosen to be 1.5–2 times less than the depth of the accelerated electron penetration into the semiconductor. Nonetheless, a certain number of electron–hole pairs are recombining. Therefore, the relation between current in the target circuit I_t and electron-beam current I_b is characterized by:

$$G = \frac{I_t}{I_b} = \frac{E_0 - E_{loss}}{\varepsilon_i},$$

where E_{loss} – is the energy lost by accelerated electrons when they move through the metallic film-electrode and surface p^+-layer.

The parameter E_{loss} can be considered as the threshold energy, which depends on the target construction and technology properties. The bed depth of p–n-transition of contemporary silicon diode targets consists of 0.5–0.7 µm, the threshold value $E_{\text{loss}} = 3\text{--}4$ kV ionization energy is $\varepsilon_i = (4 \pm 0.5)$ kV. Due to this, when the energy of accelerated electrons is $E_0 = 10$ kV, they provide current amplification coefficient values of $G = 1500\text{--}1700$.

The measurement error in this case is determined by the sum of accidental and systematic errors. The accidental error depends on measurement conditions and in the majority of cases can be minimized at the level of 5% [175]. The systematic measurement error, caused by a measuring device construction imperfection, can be decreased by aperture error, i.e. error due to finite sizes of the analyzing aperture in the diaphragm.

To estimate the measurement aperture error let us suppose that the current-density distribution in the beam cross section is expressed by a certain function $J(x, y)$. Then the collector current passing through the diaphragm aperture I_c is determined by the expression:

$$I_c(x, y) = \iint_{S_0} J(x, y)\, dx dy,$$

where $S_0 = \pi r_0^2$ is the area of the collector aperture with radius r_0; then the measured current density in the aperture plane will be

$$J_c(x,y) = J_{av}(x,y) = I_c/\pi r_0^2,$$

and the measurement aperture error

$$\delta = \left(1 - \frac{I_c(x,y)}{\pi r_0^2 J(x,y)}\right) \times 100\%.$$

From the expression obtained it follows that the aperture error depends on the law of current-density distribution in the electron-beam cross section and on the diaphragm aperture radius. To reduce this error it is necessary to decrease the difference between the real and measured current-density values, which requires a decreasing diaphragm aperture radius (in comparison with the electron-beam radius). Under fixed current density, this leads to reducing the value of current through the aperture.

The minimum value of current $I_{c\,\min}$, which can be measured with fixed precision, is to be determined by the sensitivity of the measuring device, level of noise at the device's exit limits the latter.

According to [176], the common level of noise in a device with a diode semi-conductor target consists in general of three components: the shot noise of the electron beam, amplified by the target – $I_{sh.n}^2 = 2I_c G^2 \Delta f$; the target current leakage fluctuation – $I_{l.n}^2 = 2eI_l\Delta f$; and the thermal noise of the load resistance – $I_{R.n}^2 = 4kT\Delta f/R$ Therefore, the square of the total noise current can be presented as follows:

$$I_n^2 = I_{sh.n}^2 + I_{l.n}^2 + I_{R.n}^2 = 2eG^2\Delta f + 2eI_l\Delta f + 4kT\Delta f/R,$$

where e – is electron charge; Δf – is width of frequency band; k – is the Boltzmann constant; I_l – is the target current leak; T – is resistance temperature; G – is coefficient of current amplification by the target.

The relation of amplified signal and noise current is determined by the expression:

$$\beta = \frac{I_c G}{(2e\Delta f(I_c G^2 + I_l + 2kT/eR))^{1/2}}. \qquad (12.1)$$

This gives

$$I_c^2 G^2 - 2e\Delta f\beta^2 G^2 I_c - 2e\Delta f\beta^2 (I_l + 2kT/eR) = 0.$$

Solving this quadratic equation with respect to I_c we obtain the expression for the minimum current passing through the diaphragm aperture, which current can be measured under fixed relation between signal and noise:

$$I_{c\,\min} = e\Delta f\beta^2 \left[1 + \left(1 + \frac{2I_l + 4kT/eR}{e\Delta f\beta^2 G^2}\right)^{1/2}\right]. \qquad (12.2)$$

Expression (12.2) determines the sensitivity of a measuring device with a semi-conductor target. If we assume $G = 1$ and $I_l = 0$ in this expression, then it will determine the sensitivity of a measuring device without a target. In

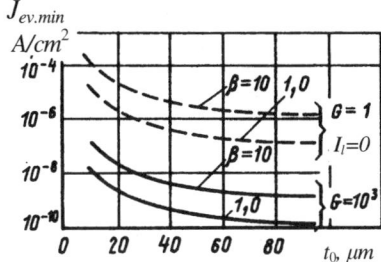

Fig. 12.3. Curves of sensitivity of the measuring device

both cases, the corresponding minimum value of the electron-beam current density in the plane of the diaphragm aperture will be equal to:

$$J_{av\ min} = I_{c\ min}/\pi r_0^2 \ . \tag{12.3}$$

Expressions (12.2) and (12.3) allow us to find the current-density minimum value, which can be determined under a fixed value of the diaphragm aperture radius with fixed precision. In Fig. 12.3 curves are presented that were calculated with formulae (12.2) and (12.3) and that show the connection between $J_{av\ min}$ and r_0. Solid curves correspond to a measuring device with a target, providing current amplification of 1000 times; dashed curves correspond to a device without a target. In the calculations it was assumed that $\Delta f = 1$ Hz, $R = 1$ kOhm, $T = 300$ K, $I_l = 10^{-6}$ A. The curves, calculated with $\beta = 1$, correspond to the threshold sensitivity of the device.

So, the method of a moving collector with small aperture, complemented with the method of current amplification with the aid of a semiconductor structure, allows us to raise significantly the sensitivity of a measuring device and electron-beam parameter measurement precision [177, 178].

12.3 Application of Modified Pin-Hole Chamber

To investigation electron beams by the method of a moving collector with small aperture a vacuum analyzer is used. Its main element is a vacuum chamber, usually made of a wide metallic tube, in one end of which the electron optical system to be analyzed is installed with the aid of a removable flange, and in the other end the moving collector is placed. The required vacuum inside the analyzer (approximately 10^{-4} Pa) is obtained with the aid of vacuum pumps, connected to the tube. The moving collector is made as a sufficiently thin coaxial metallic construction, protruding into the analyzer chamber, with syphon bellows.

The coaxial probe is installed on its end inside the chamber. Its end, being bombarded with the electron beam, is covered with a thin diaphragm with a small aperture, while in the other end it has a special passage allowing connection the probe to the main body of the collector. One such probe, containing

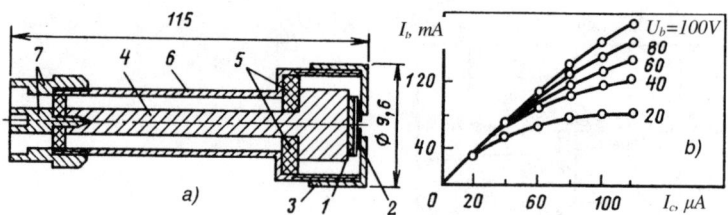

Fig. 12.4. Measuring probe with semiconductor target (**a**), current curves of target (**b**): 1 – semiconductor target, 2 – metal plate with small aperture, 3 – nut, 4 – inner conductor, 5 – insulating bead, 6 – shell, 7 – connecting device

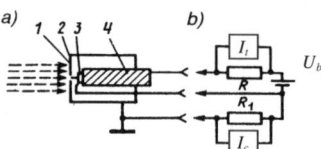

Fig. 12.5. Scheme of measuring device (**a** – coaxial probe, **b** – measuring circuits): 1 – metal plate with small aperture, 2 – shell, 3 – semiconductor target, 4 – inner conductor, U_b – back bias, I_c and I_t – current measuring devices, R and R_1 – resistors

a semiconductor target, is presented in Fig. 12.4a [174]. The semiconductor target 1 is soldered to the inside electron collector 4, which is simultaneously a heat-transfer device. The target presents a passivated planar silicon diode with an operative area equal to 2.4 mm². A changeable diaphragm with an aperture of 10–50 µm diameter is adjacent to the operating side of the target. The characteristic of the target $I_t = f(I_c)$, presented in Fig. 12.4b, shows that by choosing the displacement voltage it is possible to provide linear amplification in a wide range of electron-beam currents.

The probe construction allows connecting it to different electric circuits and measuring electron-beam structures over a wide range of energy and current values, provided, of course, that the power of the probe bombardment does not exceed the permissible dissipation power, which depends on the probe-cooling conditions. Figure 12.5 presents a schematic of the probe and the circuits connected to it; the circuit contains: the source of direct-current voltage for the diode back bias measuring devices, which can be normal gauges, an oscillograph or a plotter, and load resistors, which can be either special measuring resistors, or input resistors of gauges. With the aid of the circuit, it is possible to register simultaneously the current of the electron beam bombarding the target – I_c, and the current through the target – I_t. This allows, in particular, determination of the current-amplification coefficient $G = I_t/I_c$, and its control through all the range of currents being measured.

So, by moving the probe along and across the electron beam, it is possible to measure its variours parameters, including the current distribution in the cross section, which first of all characterizes the beam structure.

To investigate experimentally the sensitivity of the considered method of electron-beam structure analysis, it was used for the measurements of the

Table 12.1. Distribution of beam current density over the beam

N	1	5	9	14	19	22	26	31	35	41
$I_c \cdot 10^6$ A	0	0	0.01	0.20	0.93	0.40	0.05	0.01	0	0
$I_t \cdot 10^6$ A	0.01	1.2	16.3	334	1550	662	82	16.5	8	0.01
$G = I_t/I_c$	–	–	1630	1670	1666	1655	1640	1650	–	–

N – number of measuring probe position

current-density distribution in a beam with total current equal to 10 mA accelerated by potential $U_a = 10$ kV.

In the probe there was a diaphragm installed with an aperture diameter of 50 μm, and as registering devices, galvanometers were used with a scale value of 0.01 μA. During the measurements the probe was moved discretely with a pitch of 50 μm and current values were fixed in 41 points (positions). The measurement results for some positions are presented in Table 12.1.

These results show that the device with a semiconductor target under mean amplification coefficient $G = 1650$ can fix minimum currents $I_{c\ min} = I_{t\ min}/G \approx 6 \times 10^{-12}$ A, which corresponds to the mean current density of 10^{-6} A/cm². It is also seen that the device sensitivity is actually limited by the galvanometer sensitivity.

Figure 12.6a presents the current I_t as a function of radial position r for an electron beam formed by another diode electron gun. The measurements were carried out with the same probe as in the previous case and a recorder as the terminal device. It is seen that the beam has a diameter equal approximately to 2 mm and nonuniform current-density distribution. The followed analysis of the electron gun, which formed this beam, has shown that the "gap" in the center of the curve of the current-density distribution is caused by a layer of oxide covering the center of the cathode.

Investigation of various electron-optical systems has shown that the thin structure of the electron beam can be affected not only by nonuniform cathode emission, but also by construction peculiarities of the systems, for instance, the presence of control grids and their operating conditions, etc.

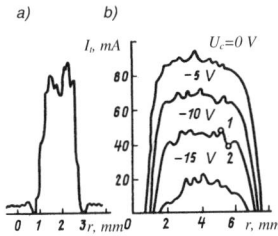

Fig. 12.6. Current-density distribution in the beams formed by a diode gun (*a*) and a triode gun (*b*) for different grid bias ($U_g = 0$ to -15 V)

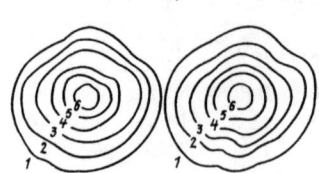

Fig. 12.7. Contours of equal current density, for two times: **a** – immediately after installation switching on, **b** – after 1.5 h: $1 - J/J_{\max} = 0.05$, $2 - 0.1$, $3 - 0.3$, $4 - 0.6$, $5 - 0.8$, $6 - 0.95$

Figure 12.6b shows curves that were obtained when analyzing the structure of the electron beam formed by a gun with a control grid of parquet-type structure.

The gun operates at an accelerating voltage of 10 kV and provides a total beam current of order of 10 mA. The diameter of the analyzing aperture is equal to 200 μm. The curves were drawn from the results of direct-current measurements taken each 100 μm of the probe movement, for different values of the grid potential U_g. They show different levels of the electron-beam structure nonuniformity and particularly, they show that the negative grid potential increase leads to a decrease of the mean value of the beam current. As in current-density oscillation, it can even increase in certain points of the beam cross section. For example, transition from the point 1 to the point 2 in the curve $U_g = -10$ V corresponds to the change of measured current I_t by 9.2 mA, a situation that is not observed in other curves.

The results of measurements of the current-density distribution over the electron-beam diameter gave the reason to suppose that the azimuthal current density distribution will be also nonuniform. Figure 12.7 shows the contours of constant current density, which clearly confirms this assumption.

The disturbance of the electron-beam axial symmetry can be explained by the lens effect of grid meshes, which have a parquet structure.

In conclusion, it should be noted that functional possibilities of a probe with a semiconductor target could be extended, if a target with a grid aluminum film instead of a solid one is used. In this case, the target becomes photosensitive, which allows measurement of not only the electron-beam current-density distribution, but also the distribution of thermal (optical) radiation intensity of the electron gun thermal cathode [178].

Appendix

Emittance and Brightness

The emittance of a charge particle beam is usually determined as phase ellipse area A divided by π:

$$\varepsilon = A/\pi . \tag{A.1}$$

This definition of emittance has evident physical meaning for a beam model known as the Kapchinsky–Vladimirsky (KV) model. In this model, a charge particle beam has sharp boundaries both in coordinate and phase spaces. In the case of axially symmetric beams the phase figures of beams in the phase plane $r - r'$ are ellipses, with phase density being a constant inside the ellipse and vanishing at its boundary.

In real beams, which have nonlaminar structure caused by thermal spread of particle velocities, the phase density is a complicated function of phase coordinates r, r', $f(r, r')$ (Fig. A.1). In this case the boundary of the beam phase figure is determined by a convention. For instance, it can be taken as a contour at which the phase density has a constant value equal to a part of the maximum phase density $f(r, r')_{\max}$. As an alternative the beam-boundary contour in phase plane (r, r') can be taken as the contour enclosing some part of the total beam current.

The emittance resulting from thermal velocity spread can be estimated with the help of the equation for the current density in beam crossover [1]:

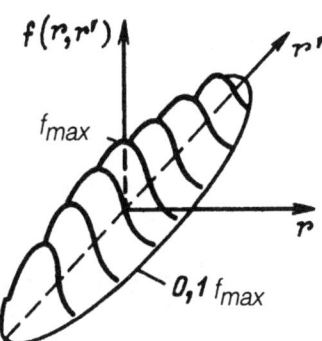

Fig. A.1. For determination of conventional phase contour

$$j = j_k \frac{|e|U_a}{kT_k}\gamma^2 \exp\left(-\frac{|e|U_a}{kT_k}\gamma^2 \frac{r^2}{r_k^2}\right), \tag{A.2}$$

where j – current density in the crossover, j_k – cathode current density, U_a – potential at the crossover, T_c – cathode temperature, γ – angle of beam convergence, k – Boltzmann constant.

If crossover is located behind the anode electrode its, potential is substituted in this equation $U = U_a$. The radial current density distribution obeys the Gauss law. The current density reaches a maximum value at the beam axis, $r = 0$:

$$j_{\max} = j_k \frac{|e|U_a}{kT_k}\gamma^2. \tag{A.3}$$

Let us introduce a conventional radius of the beam r_p as the radius of a circle enclosing a part, ΔI, of the beam current $I = \pi r_k^2 j_k$. Then, from (A.2) we can obtain:

$$r_p = \frac{r_k}{\gamma}\sqrt{\frac{kT_k}{|e|U_a}\ln\frac{1}{1-p}}, \tag{A.4}$$

where $p = \Delta I/I$, $U_T = kT_k/|e|$.
If we take $p = \Delta I/I = 0.63$, then $\ln\frac{1}{1-p} = 1$ and

$$r_p = \frac{r_k}{\gamma}\sqrt{\frac{kT_k}{|e|U_a}}. \tag{A.5}$$

For $p = \Delta I/I = 0.865$, $\ln\frac{1}{1-p} = 2$ and

$$r_p = \sqrt{2}\frac{r_k}{\gamma}\sqrt{\frac{kT_k}{|e|U_a}}. \tag{A.6}$$

Since the area of the phase ellipse with axis γ and r_p is $A = \pi\gamma r_p$ then the equation for the emittance is written as:

$$\varepsilon_p = r_p\gamma = C_p r_k\sqrt{\frac{kT_k}{|e|U_a}}, \tag{A.7}$$

where C_p is a factor depending on the choice of $p = \Delta I/I$, for $p = 0.63$ $C_p = 1$, for $p = 0.865$ $C_p = \sqrt{2}$.

Brightness

In general case brightness (microbrightness) is determined by [12]:

$$B = dI/dSd\Omega, \tag{A.8}$$

where dI is the current of an elementary beamlet separated out from the beam under consideration, dS – area of its cross section, $d\Omega$ – solid angle inside which it propagates.

Since dI/dS is the beam current density the brightness can be written as:

$$B = j/d\Omega \,, \tag{A.9}$$

Brightness defined by these equations is a parameter that is varied inside the beam. For this reason some simplified equations of brightness are used for practical calculation.

In electron microscopy brightness is determined through the current density at beam axis and the solid angle of beam divergence $\pi\gamma^2$:

$$B_0 = j_{\max}/\pi\gamma^2 \,. \tag{A.10}$$

Substitution in this equation value of $j_{\max}$ from (A.3) gives:

$$B_0 = \frac{j_k}{\pi} \frac{|e|U_a}{kT_k} \,. \tag{A.11}$$

If current density in a beam is changed over the beam cross section in accordance with the Gauss law then the value of $j_{\max}$ can be expressed as:

$$j_{\max} = I/\pi r_p^2 \,,$$

where r_p is the conventional beam radius determined by (A.5).

Then, brightness B_0 can be written as:

$$B_0 = \frac{I}{\pi^2 r_p^2 \gamma^2} \,. \tag{A.12}$$

It is possible to introduce an averaged brightness in the following way:

$$\bar{B} = \frac{Ip}{\pi^2 r_p^2 \gamma^2} \,, \tag{A.13}$$

where r_p – conventional beam radius, $p = \Delta I/I$ – relative part of beam current passing through the surface confined by the circle of radius r_p, $Ip/\pi r_p^2$ – averaged current density.

For $p = 0.63$ the values of B_0 and $\bar{B}$ are connected by the relation $\bar{B} = pB_0$.

As $\varepsilon_p = r_p\gamma$, then for the case $p = 0.63$ one finds:

$$\bar{B} = \frac{Ip}{\pi^2 \varepsilon_p^2} \qquad B_0 = \frac{I}{\pi^2 \varepsilon_p^2} \,. \tag{A.14}$$

Root Mean Square Emittance

For a description of the quality of nonlaminar beams the root mean square emittance (rms emittance) is also used [12].

This is determined by:

$$\bar{\varepsilon} = 4\left[\langle r^2\rangle\langle r'^2\rangle - \langle rr'\rangle^2\right]^{1/2} \,. \tag{A.15}$$

In this equation $\langle r^2\rangle, \langle r'^2\rangle, \langle rr'\rangle^2$ are the average values of beam parameters enclosed in the brackets. They are calculated as:

$$\langle r^2 \rangle = \frac{\iint r^2 f(r,r')\,drdr'}{\int f(r,r')\,drdr'}$$

$$\langle r'^2 \rangle = \frac{\iint r'^2 f(r,r')\,drdr'}{\int f(r,r')\,drdr'}$$

$$\langle rr' \rangle = \frac{\iint rr' f(r,r')\,drdr'}{\int f(r,r')\,drdr'}.$$

Integrals are calculated over the area of phase space A filled by phase points. In the case of the Kapchinsky–Vladimirsky beam model, when $f(r,r') = $ const, rms emittance is equal to the emittance determined from (A.1).

In [12] the following equation is obtained for the rms emittance of an electron beam emitted by a thermal cathode:

$$\bar{\varepsilon} = \sqrt{2} r_k \sqrt{\frac{kT_k}{|e|U_a}}, \tag{A.16}$$

where r_k and T_k – cathode radius temperature.

Emittance measurements of an electron beam formed by a Pierce-type gun were fulfilled in [141]. Results of these are in conformity with those obtained with (A.16).

The rms emittance equation (A.15) can be applied for estimation of the beam quality in the case of a discrete beam model [142], with values of $\langle r^2 \rangle, \langle r'^2 \rangle, \langle rr' \rangle$ being calculated by the following equations:

$$\langle r^2 \rangle = \frac{1}{N} \sum_{k=1}^{N} r_k^2,$$

$$\langle r'^2_k \rangle = \frac{1}{N} \sum_{k=1}^{N} r'^2_k,$$

$$\langle rr' \rangle = \frac{1}{N} \sum_{k=1}^{N} r_k r'_k,$$

where r_k and r'_k are radial coordinates and trajectory slopes of the k-th trajectory; summation being performed over all the trajectories creating the beam.

If beam trajectories carry different currents this can be taken into account by introducing the weight factor $\alpha_k = I_k/I$, where I_k is a part of the current carried by the k-th trajectory, I – total beam current. Then the average values included in the equation for rms emittance (A.15) are calculated by the equations:

$$\langle r^2 \rangle = \sum_{k=1}^{N} r_k^2 \alpha_k, \quad \langle r'_2 \rangle = \sum_{k=1}^{N} r'^2_k \alpha_k, \quad \langle rr' \rangle = \sum_{k=1}^{N} r_k r'_k \alpha_k.$$

References

1. Glaser W (1952) Grudlagen der Electronenoptik Fundamentals of Electron Optics. Springer, Wien
2. Kelman VM, Yavor SYa (1969) Electron Optics (in Russian). Nauka, Moscow
3. Howkes PN (1972) Electron Optics and Electron Microscopy. Taylor and Francis, London
4. Wollnik H (1987) Optics of Charge Particles. Academic Press, Orlando
5. Szilagyi (1988) Electron and Ion Optics. Plenum Press, New York, London
6. Hawkes PW, Kasper E (1989) Principles of Electron Optics, Academic Press, New York, London
7. Pierce YR (1954) Theory and Design of Electron Beams. Van Nostrand, Toronto, New York, London
8. Molokovsky SI, Sushkov AD (1965) Electron Optical Systems of Microwave Tubes (in Russian). Energiya, Leningrad
9. Kirstein PT, Kino GS, Water WE (1967) Space-charge Flow. McGraw Hill, New York, Toronto, London, Sydney
10. Molokovsky SI, Sushkov AD (1972) Intense Electron and Ion Beams (in Russian). Energoatomisdat, Moscow
11. Didenko AN, Grigoriev VP, Usov YuP (1977) Power Electron Beams and their Application (in Russian). Energoatomisdat, Moscow
12. Lowson YD (1977) The Physics of Charge Particle Beams. Clarendon Press, Oxford
13. Rukhadse AA, Bogdankevich LO, Rosinsky SE, Rukhlin VG (1980) The Physics of Intense Relativistic Electron Beams (in Russian). Atomisdat, Moscow
14. Miller RB (1982) An Introduction to Physics of Intense Charge Particle Beams. Plenum Press, New York
15. Septier A (ed) (1983) Applied Charge Particle Optics, part C – Very-high-density Optics. Academic Press, New York
16. Abramyan EA, Alterkopf VA, Kuleshov GD (1984) Intense Electron Beams (in Russian). Energoatomisdat, Moscow
17. Gabovich MD (1984) Physics and Techniques of Plasma Ion Sources (in Russian). Energoatomisdat, Moscow
18. Ilyin VP (1985) Numerical Methods of Solution of Electrophysics Problems (in Russian). Nauka, Moscow
19. Humphries S (1990) Charged Particle Beams. Wiley, New York
20. Reizer M (1994) Theory and Design of Charged Particle Beams. Wiley, New York, Chichester, Brisbane, Toronto, Singapore

21. Harris LA (1959) Toroidal electron guns with hollow beams. Journ Appl Phys 30: 826–836
22. Kino GS, Tailor NJ (1962) Design and performance of a magnetron injection gun. IRE Trans ED-9: 1–11
23. Sushkov AD, Molokovsky SI, Zybin GP (1967) Investigation of electron guns with electron beam current control. In: Transactions of 3rd Czechoslovak Conference on Electronics and Vacuum Physics. Academia Publishing House, Prague, pp 577–588
24. Staprans A, McCune EW, Ruetz JA (1973) High power linear beam tubes. In: Clempit LL (ed) Proc IEEE 61; pp 299–330
25. Lavrentiev TuV, Nevsky PV, Kustikov OT (1977) Triode electron gun with shadow grid. Electronnaya technika, serie 1 – UHF elektronika, issue 5: 88–90
26. Iliyushin BD, Galitsky II, Bleivas IM (1980) Forming of multiple beamlet electron flow with beam current control. Electronnaya technika, serie 1 – UHF electronika, issue 2: 19–24
27. Zinchenko NS (1967) Theory, experimental results and application of high perveance electron guns (in Russian) Ukrainien fisichesky journal 12: 1828–1837
28. Brix DL et al. (1986) High brightness test stend. Nuclear Instruments and Methods A250: 49–56
29. Hirshfield JL, Granatstein VL (1977). The electron cyclotron maser – a historical survey. IEEE Trans MMT 25:
30. Harris LA (1952) Axially symmetric beam and magnetic field system. Proc IRE 40: 700–708
31. Chernov ZS (1958) Methods of beam focusing in modern microwave devices (in Russian). Radiotechnika I Electronika 3: 1650–1670
32. Chvertko AI, Nazarenko OK, Sviatsky AM (1973) Equipment for electron beam welding (in Russian). Naukova dumka, Kiev
33. Maurakh MA (ed) (1964) Electron Beam Melting (in Russian). Mir, Moscow
34. Barber GF, Bakish R Design consideration for electron beam systems (1962) In: Bakish R (ed) Introduction in Electron Beam Technology: 96–123. Wiley, New York, London
35. Kreindel YuE (1977) Electron Plasma Sources (in Russian) Atomisdat, Noscow
36. Lengmuir DB, Stuhlinger E, Sellen JM (1961) Electrostatic Propulsion. Academic Press, New York, London
37. Smythe WR (1950) Static and Dynamic Electricity. McGraw Hill, New York
38. Simonyi K (1956) Theoretische Electrotechnik, VEB Deutscher Verlag der Wissenschaften, Berlin
39. Tichonov AN, Samarsky (1966) Equations of Mathematical Physics (in Russian). Nauka, Moscow
40. Volynsky BA, Buchman VE (1960) Models for Solution of Boundary Problems (in Russian). Fizmatgiz, Moscow
41. Vladimirov VS (1967) Equations of Mathematical Physics (in Russian). Nauka, Moscow
42. Milne WE (1953) Numerical Solution of Differential Equations. Wiley, New York

43. Demidovitch BP, Maron IA, Shuvalova EZ (1967). Numerical Methods of Analysis (in Russian). Nauka, Moscow
44. Tozoni OV (1975) Methodes of Secondary Sources in Electrotechnique (in Russian). Energiya, Moscow
45. Tozoni OV (1967) Computation of Electromagnetic Fields. Technika, Kiev
46. Demirchyan KS, Tchechulin VL (1986) Computation of Electromagnetic Fields (in Russian). Vyshaya shcola, Moscow
47. Hockney RW, Eastwood JW (1981) Computer Simulation using Particles. Mc-Graw Hill, New York
48. Tichonov AN, Arsenin Vya (1986) Methods of Solution of Incorrectly Set up Problems (in Russian). Nauka, Moscow
49. Monosov GG (1968) Steady performance of magnetron type tubes with negative emitting electrode (in Russian). Electronnaya Technika, serie 1 – UHF elektronik, issue 10: 78–87
50. Golenizky II, Zakharova AN, Khomitch (1971) Analysis of performance of 0-type tubes at high signal level with account of beam focusing (in Russian). Electronnaya Technika, serie 1 – UHF elektronika, issue 3: 37–48
51. Molokovsky SI, Tregubov VF (1971) Numerical computation of Green function for electrode systems of complicated boundary (in Russian). In: Marchuck GI (ed) Transuctions of All-Union Symposium on Solution of Electron Optics Problems. Novosibirsk.
52. Melnikov YuA (1967) Permanent Magnets Used in Microwave Tubes (in Russian). Sovetskoe Radio, Moscow
53. Koshelev AI (1986) Regularity of Solution of Elliptical-type Equations and Systems (in Russian). Nauka, Moscow
54. Kantarovich LV, Akilov GP (1977) Functional Analysis (in Russian). Nauka, Moscow
55. Merkulov AL, Molokovsky SI (1982) Investigation and application of methods for computation of saturated magnetic systems (in Russian). In: Methods of Calculation of Electron Optical Systems. Publishing House of Computer Center of Academy of Science, Novosibirsk, pp 133–138
56. Blokhin AA, Kosheev BG, Molokovsky SI (1977) Numerical calculation of magnetic focusing systems taking into account real properties of magnetic materials (in Russian). In: Methods of Calculations of Electron Optical Systems. Nauka, Moscow, pp 155–159
57. Blokhin AA, Molokovsky SI (1982) Computer code "Weber" for computation and optimization of magnetic focusing systems taking into account real properties of magnetic materials (in Russian). Electronnaya Technika, serie 1 – UHF electronika, issue 2: 68–74
58. Alamovsky IV, Smirnova NN (1985) Computation of parameters built in type MPFS taking into account saturation of pole pieces (in Russian). Electronnaya Technika, serie 1 – UHF electronika, issue 6: 37–41
59. Landau LD, Lifshitz EM (1988) Theory of Field (in Russian). Nauka, Moscow
60. Buhgolts NN (1966) Fundamentals of Theoretical Mechanics (in Russian). Nauka, Moscow
61. Abramovich M, Stigan N (eds). (1977) Handbook on Special Function (in Russian) Nauka, Moscow

62. Vine J (1960) Numerical investigation of a range of unipotential electron lenses. But J Appl Phys 11: 408–411
63. Durandeau P (1957) Etude sur les lentilles electroniques magnetiques. Ph.D. thesis, Toulouse
64. Zuckermann II (1958) Electron Optics in Television (in Russian). Gosenergoisdat, Moscow
65. Vlasov AA (1950) Theory of Many Particles (in Russian). Gostechisdat, Moscow
66. Lichtenberg AL (1969) Phase-space Dynamics of Particles. Wiley, New York, London
67. Landau LD, Lifshitz EM (1988) Mechanics (in Russian). Nauka, Moscow
68. Kapchinsky IM (1966) Particle Dynamics in Linear Resonant Accelerators (in Russian). Atomisdat, Moscow
69. Devidson RC (1974) Theory of Nonneutral Plasma. Benjamin Inc, Massachusetts, London, Amsterdam, Don Hills, Ontario, Sidney, Tokyo
70. Kusnetsov VS (1967) A method of computation of inner structure of intense charged particle beams. J Technical Physics 37: 835–841
71. Lanczos C (1962) The Variational Principles of Mechanics. University of Toronto Press, Toronto
72. Sturrock PA (1955) Static and Dynamic Electron Optics. Cambridge University Press, Cambridge
73. Spangenberg K (1941) Use of action function to obtain the general equation of space charge motion in more than one dimension. J Frank Inst 232, 365–371
74. Ovcharov VT (1957) Theory of electron beam forming (in Russian). Radiotechnika i Electronika 2: 696–704
75. Meltzer B (1957) Single-component stationary electron flow under space-charge conditions. J Electronics 2: 256–261
76. Syrovoi VA (1964) On single-component beams of charge particles (in Russian). J Applied Mechanics and Technical Physics, issue 3: 24–31
77. Liebman G (1949) An improved method of Numerical ray tracing through electron lenses. Proc Phys Soc Sec B 62: 753–772
78. Ginsburg VE (1977) Optimization of model an electron optical system from point of view of its computation (in Russian). In: Methods of Computation of Electron Optical Systems, Nauka, Moscow
79. Bakhrah LE (1965) Critical current of electron beams with space charge density supposed to be constant (in Russian). Radiotechnika i Electronika 9: 1104–1109
80. Taranenko VP (1965) Effects of positive ions on electron beam forming (in Russian). Izvestia vuzov, Radioelectronika, issue 6: 1104–1109
81. Lowson JD (1958) Perveance and the Bennet pinch relation in partially neutralized electron beams J Electron Control 5: 146–151
82. Bredov MM (1950) Compensation of space charge of electron beams (in Russian). In: Collection of papers devoted to 70th Jubilee of Academician AF Ioffe, pp 155–172
83. Davidov BI, Braqinsky SI (1950) on the theory of gas concentration of electron beams (in Russian). In: Collection of papers devoted to 70th Jubilee of Academician AF Ioffe, pp 72–91

84. Green AES, Sawada T (1972) Ionization cross-section and secondary electron distribution. J Atm and Terr Phys 31: 1719–1728
85. Molokovsky SI, Suchalkin DD (1982) Estimation of effect of secondary electrons on ion focusing of electron beams (in Russian). Bulletin of Leningrad Electrotechnical Institute 315: 20–25
86. Vendik OG, Gorin YuN, Popov VF (1984) Corpuscular-photon Technology (in Russian). Visshaya shkola, Moscow
87. Miller FA, Gerardo JB (1972) Electron propagation in high-pressure gases. J Appl Phys 43: 3008–3013
88. Cuttler CC, Hines ME (1958) Thermal velocity effects in electron guns. Proc IRE 43: 307–314
89. Ovchinnikov AV (1979) Method of analysis of charge particle beams (in Russian). Zarubeznaya electronika, issue 5: 26–41
90. Radley DE (1958) The theory of the Pierce-type electron gun. J Electron and Control 4: 125–132
91. Kino GS (1960) A design method for crossed-field electron gun. Trans IRE ED-7: 420–426
92. Radley DE (1969) Electrodes for convergent Pierce-type electron guns. Electron and Control 45: 730–738
93. Ovcharov VT (1962) Equations of electron optics for plane and axis symmetrical electron beams with a high current density (in Russian). Radiotechnika i Electronika 6: 1367–1378
94. Nevsky PV, Ovcharov VT (1969) Computation of electrostatic forming systems for klystrons (in Russian). Electronnaya technika, serie 1 – UHF electronika issue 7: 62–70
95. Ovcharov VT (1967) External problem for paraxial electron beams (in Russian). Radiotechnika i Electronika 11: 2156–2161
96. Kino GS, Taylor NJ (1962) The design and performance of magnetron-injection gun. IRE Trans ED-9: 1–9
97. Ligin VK, Manuilov VN, Zimring ShE (1983) Numerical analysis of intense electron beams of gyrotrons (in Russian). In: New methods of calculation of electron optical systems. Nauka, Moscow, pp 44–50
98. Goldenberg AL, Petelin MI (1973) Forming of spiral-type electron beams in an adiabatic gun (in Russian). Izvestia vuzov, serie Radiofizika 16: 141–149
99. Daushvily AM, Koshaev BG, Molokovsky SI (1972) Calculation of lens action of grid meshes in triode gun (in Russian). Izvestia vuzov, serie Radioelectronika 15: 1507–1509
100. True R (1984) A theory for coupling gridded gun design with PPM focusing. IEEE Trans ED-31: 353–362
101. True R (1987) Emittance and the design of beam formation, transport and collection systems in periodically focused TWT's. IEEE Trans ED-34: 473–485
102. Morev SP, Pensiakov VV (1984) Methods of calculation of electron beams with a finite phase volume (in Russian). Review on electron technique, serie 1 – UHF electronics
103. Chesnokov VV (1968) Present state of cold cathode technique (in Russian). Obzory po electronnoy technike, series Elektrovakuumnie pribory, issue 9
104. Elinson MI (ed) (1974) Cold Cathode (in Russian). Sovetskoe Radio, Moscow

105. Bondarenko BV, Shishkin EP, Schuka AA (1979) Electron devices and instruments using cold cathodes (in Russian). Zarubeznaya elektronnaya technika, issue 2: 3–43
106. Crewe AV, Eggenberger DN, Wall J Welter LM (1968) Electron gun using a field emission source. Rev Sci Instr 39: 576–583
107. Slaviansky VV (1986) Development of methods of computation and investigation of electron injector systems for some instruments (in Russian). Ph. D. thesis, Leningrad Electrotechnical Institute, Leningrad
108. Spindt CA, Brody J, Humphry L, Westerbery ER (1976) Physical properties of thin-film emission cathode with molybdenum cones. J Appl Phys 47: 5248–5263
109. Slaviansky VV, Slomatov UB, Sushkov AD, Karpov LD (1981) Investigation of field and electron trajectories in microcathode cells (in Russian). Electronnaya technika, series 4 – vacuum and gas discharged devices, issue 4: 3–5
110. Gulyaev YuV, Chernozatonskii LA, Kasokovskaya Zya, Musatov Al, Sinitsyn NI, Torgashov GV (1996) Carbon nanotube structure – a new material of vacuum microelectronics. Technical conference digest of 9th International Vacuum Microelectronics Conference, July 7–12, 1996, St. Petersburg: 5–9
111. Gulyev YuV, Sinitsyn NI, Torgshov GV, Grigiriev YuA, Shesterkin VI, Veselov AG, Shvetsov YuV, Semyonov VS (1996) Emission of low-voltage multi-tip carbon matrices coated by carbon clusters. Ibid: 519–521
112. Yokoo K, Ishizuka H (1996) RF application field emitter arrays. Ibid: 490–498
113. Spindt CA, Holland CE, Schwoebel PR, Brodie I (1996) Field emitter array development for microwave application. Ibid: 638–639
114. Gabovich MD (1983) Liquid metal ion emitters (in Russian). Uspekhy fizicheskih nauk (UFN) 140: 137–151
115. Forbes RG, Djuric Z (1996) Progress in understanding liquid-metal ion source operation. Technical digest of 9th International Vacuum Microelectronics Conference July 7–12, 1996, St. Petersburg: 468–472
116. Valyi L (1978) Atom and Ion Sources. Academia, Budapest
117. Novikov AA (1972) Electron sources on the base of high voltage glow discharge with anode plasma (in Russian). Energoatomizdat, Moscow
118. Kreindel YuK (ed) (1983) Electron Sources with Plasma Emitters (in Russian). Nauka, Novosibirsk
119. Gabovich MD, Pleshivtsev NB, Semashko NN (1986) Ion and Atom Beams (in Russian). Energoatomizdat, Moscow
120. Torii Ya, Shimad M, Watanabe I (1987) Very high current ECR ion source for an oxygen ion implanter. Nucl Instr Meth in Phys Res B 21: 178–181
121. Murphy BT (1958) A method of focusing electron beams. Proc IEE 105 B: 1033–1040
122. Mendel GT (1955) Magnetic focusing of electron beams. Proc IRE 43: 1033–1040
123. Burke PFC (1963) Compensated reversed field focusing of electron beams. Proc IRE 51: 1653–1659
124. Harker KJ (1955) Periodic focusing of beams from partially shielded cathodes. Trans IRE ED – 2: 11–19
125. Chodorow M, Susskind C (1958) Space-charge balanced hollow beam with uniform space charge distribution. Proc IRE 46: 497–498

126. Samual AL (1949) On the theory of axially symmetric electron beams in axial magnetic field. Proc IRE 37: 1252–1258
127. Harris LA (1952) Axially symmetric electron beam and magnetic-field systems. Proc IRE 40: 700–708
128. Alamovsky IV (1960) A strip electron beam in periodic magnetic field with arbitrary shielded cathode (in Russian). Radiotechnika i Electronika 5: 827–833
129. Molokovsky SI (1963) Electrostatic focusing systems for intense electron beam. J Inst Telecommun Engineers 9: 328–335
130. Szilagyi M (1966) Analysis of an axially symmetric beam in a periodic electrostatic field (in Russian). Radiotechnika i Electronika 11: 870–878
131. Tien PK (1954) Focusing of a long cylindrical electron stream by means of periodic electrostatic fields. J Appl Phys 25: 1281–1288
132. Danovich IA (1965) Some problems of periodic electrostatic focusing of intense electron beams (in Russian). Radiotechnika i Electronika 10: 435–442
133. Molokovsky SI (1970) Estimation of stability of electron beam in periodical lens focusing system (in Russian). Bulletin of Leningrad Electrotechnical Institute 90: 85–88
134. Molokovsky SI (1962) Analytical calculation of electrodes for periodical focusing of a strip beam (in Russian). Radiotechnika i Electronika 7: 1048–1050
135. Sirovoi VA (1983) Methods of solution of three dimensional external problem in theory of beam synthesis (in Russian). In: New Methods of Calculation of Electron Beam. Nauka, Moscow, pp 55–60
136. Chang KKN (1957) Confined electron flow in periodic electrostatic fields of very short periods. Proc IRE 45: 66–73
137. Chang KKN (1957) Biperiodic electrostatic focusing for high density electron beam. Proc IRE 45: 1522–1528
138. Johnson CC (1958) Periodic electrostatic focusing of a hollow electron beam. IRE Trans ED – 5: 233–243
139. Sbchevsky SP, Mladenov GM (1988) Criterions and principles of optimization of concentrating electron optical systems (in Russian). Journal of technical physics 58: 2063–2067
140. Nasarenko OK, Kaidalov AA, Kovbasenko SM (1987) Electron Beam Welding. Naukova dumka, Kiev
141. Namkung W, Chojnacki EP (1986) Emittance measurements of space charge dominated electron beams. Rev Sci Instum 57: 341–345
142. Agrittelis C, Chosmen R, Slayters ThJM (1969) Design of low-energy beam transport sytem of Brookhaven 200-MeV injector linac IEEE Trans NS-16: 221–226
143. Molokovsky SI, Tregubov VF (1979) Estimation of the effect nonlinearity of phase curve on the size of focusing spot (in Russian). Journal of Technical Fisiks 49: 202–204
144. Munro E (1980) Electron beam lithography. In: Septier A (ed) Applied Charge Particle Optics, part B. Academic Press, New York
145. Brodie I, Muray JJ (1982) The Physics of Microfabrication. Plenum Press, New York, London
146. Crewe AV, Wall J, Welter LM (1968) A high resolution scanning transmission electron microscope. J Appl Phys 39: 5861–5869

147. Cleaver JRA, Ahmed HA (1981) A 100 kV ion probe microfabrication system with a tetrode gun. Jour Vac Sci and Technol 19: 1145–1148
148. Thode LE, Godfrey BB, Shanahan WR (1979) Vacuum propagation of solid relativistic electron beams. Phys Fluids 22: 747–763
149. Miller RB, Strow DC (1977) Propagation of nonneutralized intense relativistic electron beam in a magnetic field. Journ Appl Phys 48: 1061–1069
150. Birdsall CK, Bridges WB (1967) Electron dynamics of diode regions. Academic Press, New York
151. Twombly JC, Lauer JE (1967) A mathematical model which describes the high perveance instabilities of long drifting electron beams. In: Proc 4-th Congress on Microwave Tubes, Scheveningen, 1967
152. Boxer AS, Ollum JF (1970) A new quasi-static model of space-charge-limiting. IEEE Trans ED – 17
153. Molokovsky SI, Stenukova RE, Tregubov BF (1973) Numerical analysis of virtual cathode creation in a cylindrical channel (in Russian). Bulletin of Leningrad Electrotechnical Institute 136: 17–21
154. Beresin YuA, Breizman BI, Vshivkov VA (1979) Numerical simulation of a power electron beam injection in a vacuum chamber with an intense magnetic field (in Russian). Preprint of the Institute of Theoretical and Applied Mechanies of Siberian Branch of Academy of Science USSR, issue 18
155. Sullivan DJ (1979) Application of the virtual cathode in relativistic electron beams. In: Proc 3-rd Int Conf on High power Electron and Ion Beam Research and Technology, Novosibirsk, 1979
156. Didenko AN, Karasin YaE, Perelman SG, Pisarenko IP (1979) High power microwave generation in a triode electron system (in Russian). Pisma u JTF 5: 321–324
157. Kwan TJT (1984) High power coherent microwave generation from oscillating virtual cathodes. Phys Fluids 27: 228–232
158. Sze H, Benford J, Woo W (1987) High-power microwave emission from virtual cathode. Laser and Particle Beams 5: part 4, 675–618
159. Kalinin VI, Gershtein GM (1957) Introduction in Radiophysics. Gostechisdat, Moscow
160. Reizer M (1977) Laminar-flow equilibra and limiting currents in magnetically focused relativistic beams. Phys Fluids 20: 477–486
161. Kapchinsky IM (1967) Dynamics of charge particles in resonance accelerators (in Russian). Atomisdat, Moscow
162. Very-high-density beams (1983). In: Septier A (ed) Applied Charged Particle Optics, Part C. Academic Press, New York
163. Reiser M. Namkung W, Brennan MA (1979) Electron model experiment to study instabilities in long periodic focusing systems for intense beams. IEEE Trans NS – 26: 3026–3028
164. Bakhrah LE, Kudriavtseva SP, Mursin VV (1979) Magnetic quadrupole periodic focusing of intense electron beams (in Russian). Radiotechnika i Electronika 23: 1187–1193
165. Borisov LM, Gelvich EA, Zaryi EV (1993) Power multiple beam electrovacuum amplifiers (in Russian). Electronnaya technika, series 1 – UHF technika, issue 3: 12–20

166. Kiselev AB, Marchenko NN (1991) Heaters for multiple beam electrovacuum tubes (in Russian). Electronnaya technika, series 1 – UHF technika, issue 9: 3
167. Drozdov SS et al. (1984) Reversible periodic focusing system. US Patent N 443270
168. Artsimovich LA (1976) What Every Physicist should know about Plasma. Atomisdat, Moscow
169. Meerovich EA (1966) Methods of Relativistic Electrodynamics Used in Electrotechnique (in Russian). Energia, Moscow
170. Mockalev VA, Sergeev GI, Shestakov VG (1980) Measurement of Charge Particle Beam Parameters (in Russian). Atomisdat, Moscow
171. Evtifeeva ES, Kibardina XA (1961) Methods of experimental investigation of electron beams (in Russian). Problems of radioelectronique, series 1 – electronique, issue 8: 54–107
172. Aleksandrov AG, Zamorozkov BM, Kalinin YuA (1973) methods of experimental study of electron beam structure in O- and M-type tubes (in Russian). Obzory po electronnoy technike, series 1 – UHF electronika, issue 8
173. Sushkov AD, Vavilova OS, Dmitriev GI (1981) Vacuum devices with electron bombardment of semiconductors (in Russian). Zarubeznaya electronika, issue 1: 60–71
174. Sushkov AD, Dmitriev GI (1981) Probe with semiconductor taget used for electron beam investigation (in Russian). Pribory i technika experimenta, issue 3: 230–231
175. Minkin AM (1963) An instrument for measurements of electron beam parameters (in Russian). Pribory i technika experimenta, issue 5: 177–178
176. Osadochii VI, Osadochii AI (1980) Electron beam storage devices (in Russian). Technika, Kiev
177. Ivanov BV, Sushkov AA, Sushkov AD, Tupitsin AD (1988) Coaxial probe with semiconductor target (in Russian). Pribory, technika experimenta, issue 3: 130–132
178. Sushkov AD (1988) A method of experimental stady of fine structure of electron beams (in Russian). In: Proc 2-nd Int Conf on Electron Beam Technology, 31 May – 4 June 1988, Varna, Bulgaria

Index

Action function, 60, 102
Alfven's current, 232
Ampere low, 250
Angle coefficient of spherical aberration, 84
Angle of trajectory refraction, 82

Beam
  compression, 4
  brightness, 3, 266
Brillouin focusing, 10
Bush theorem, 62
  analog of Bush theorem, 74

Calculation of particle trajectories
  method of Taylor series, 74
  Runge-Cutta method, 75
Cauchy problem
  for Laplace equation, 30
  for Poison equaton, 34
Congruence of trajectories, 99
  normal congruence, 99
  skew congruence, 100
Cyclotron resonance, 163
Cylindrical magnetron, 67

Dynemic virtal cathode model, 228
Duoplasmatrons, 161
  focusing electrodes, 166

Effect of ion neutralization, 110
Effect of thermal (initial) velocities, 117
Electron guns 4, 127
  computer-aided design, 151
  for convergent beam, 4, 134
  for hollow beams 5, 141, 191
  for parallel cylindrical beam, 133
  for parallel strip beam, 126
  for strip beam in crossed fields, 129
  for wedge-shaped beam, 128
  design by Ovcharov's method, 138
  magnetron guns, 5, 142
  method of analysis, 127
  method of synthesis, 127
  Pierce gun design, 137
  with beam current control, 6, 145
  with plasma sourse, 14
Electron probe of welding installation, 215
Electrostatic charge particle lenses, 77
  parameters, 85
  spherical aberration, 84
  thick particle lens, 83
  thin lens, 80
  types, 77
Electrostatic field calculation, 20
  computer calculation, 35
  Dirichlet problem, 20
  method of Green function, 23
  method of finite difference, 27
  method of integral equations, 26
  method of separation of variables, 20
Electrostatic focusing systems, 200
  centrifugal focusing, 213
  for cylindrical beam, 200
  for hollow beams, 211
  for strip beams, 210
  periodic focusing, 201
  with set of unipotential lenses, 203
  stability of beam focusing, 209
  synthesis of electrodes, 202
  synthesis of lenses, 207
  with set of unipotential lenses, 203
  with system of apertured discs, 203
Emittance, 3, 98, 265
Equations of motion

    in curvilinear coordinates, 56
    Hamilton equations, 59
    Hamilton-Jacobi equation, 60
    in axially symmetric fields, 62
    in cylindrical coordinates, 56
    in Newton form, 55
    In rectangular coordinates, 55
    Lagrange equations, 58
    modified equations, 64
    paraxial equations, 65
    trajectory equation, 65
Explosive emission, 160

Field emission cathodes, 157
    matrix cathode, 159
    with graphite nanotubes, 159
Field of ring shaped magnets, 48
Focusing system
    classification, 9
Free electron laser, 8

Gauss theorem, 105
Green function, 23, 25
    discreet Green function, 41
    for cylindrical channel, 24
    matrix form, 42
    of free space, 39
Gyrotron, 5, 10, 144

Hamilton function, 59, 67
Hollow beam focusing, 13
Hydrodynamic approximation, 100
Hydrodynamic beam model, 101

Ion focusing, 112
Ion guns
    with a liquid-metal emitter, 160, 221
    with plasma emitters, 161, 165
    with solid emitters, 16
Ion probes, 221
Ion sources
    cold cathode sources, 162
    douplasmatron, 161
    extracting of charge particles, 165
    high frequency ion sources, 163
    microwave ion sources, 163
    plasma sources, 161

Kaptchisky–Vladimirsky beam, 176
Klystron, 9

Lagrange function, 58
Laplace equation, 19
Laplace operator, 19
Law of Energy Conservation, 56, 224
Liouville theorem, 96
Liquid metal emitters, 160

Magnetic field equation, 43
Magnetic field calculation
    field of a solenoid, 47
    field of ring shaped magnets, 48
    nonlinear magnetic problem, 50
    numerical calculation, 47
Magnetic flux, 45
Magnetic focusing systems
    computer-aided design, 196
    focusing of hollow beams, 188
    focusing of strip beams, 193
    stability of periodic focusing, 187
    systems with uniform field, 169
    with periodic fields, 184
    with reversed fields, 181
Magnetic lenses, 86
    long magnetic lens, 89
    parameters, 88
    thick magnetic lens, 87
    thin or weak lens, 86
    with divergent action, 91
    with permanent magnets, 91
Magnetic periodic focusing, 11
Magnetic scalar potential, 44
Magnetic vector potential, 45
Magnetron guns, 142
    magnetron guns of gyrotrons, 144
Mathieu equation, 187
Method of curvilinear coordinates, 102
Method of pin hole chamber, 258
    modified pin hole chamber, 259
Method of successive approximation, 121
Microwave discharge, 163
Microwave injectors, 158
Model of large particle, 123
Monovelocity models, 98
Motion of electrons in cylindrical magnetron, 67

Multiple beam ion cesium gun, 17
Multiple beam systems, 237
   multiple beam klystron, 237
   analysis of beam interaction, 250
   analysis of stray fields action, 243
   electron guns, 239
   magnetic focusing systems, 241
   reversed field focusing system, 242
   transverse beam displacement, 248
   transverse fields, 244

Nonlinear phase curves, 207, 217
Numerical trajectories computation, 74

Optical systems of technological installations, 215
Paraxial trajectory equation, 105
Particle beams experimental investigation, 256
Periodic electrostatic focusing, 12, 201
Perveance, 1
   microperveance, 1
Phase space, 2, 96
   phase curve(s), 2, 197, 207, 217
   phase ellipse, 2
   phase parallelogram, 217
   phase volume conservation, 97
   transverse phase space, 97, 217

Quasi-hydrodynamic model of charge particle flow, 102

Relativistic equations of motion, 223
   law of energy conservation, 224
Relativistic particle beams, 223
   beam current limitation, 227
   beam motion computation, 232
   Bogdankevich–Rukhadze equation, 227
   Brillowin beam, 229
   dynamic model of virtual cathode, 228

   effective perveance, 226
   neutralized beams, 232
   Pierce's instability, 232
   virtual cathode, 227
Ring-shaped magnets, 92

Scalar magnetic potential, 46
Scattering of electron beam, 114
Self-consistent fields, 95, 98
Single-component flows, 103
Solution of self-consistent problem, 121
   method of "step by step", 123
   method of successive approximation, 121
Spangenberg's equation, 101
Spherical aberration, 84, 158
Strip beam focusing
   periodic magnetic focusing, 195
   uniform field focusing, 193
Synthesis
   electron guns, 126, 137, 138
   electrostatic focusing systems, 202, 207
Systems with cold field emitter, 221

Transverse phase space, 97
Traveling wave tube, 11, 238
Triode electron guns, 7, 145

Uniform magnetic field systems, 180
Universal beam-spread curves, 109

Velocity spreading, 95
Virtual cathode, 109, 227
   dynamic model, 228
Vlasov equation, 96, 177

Welding systems, 217

Zones of stable and unstable focusing, 188